Boris Z. Katsenelenbaum

High-frequency Electrodynamics

Boris Z. Katsenelenbaum

High-frequency Electrodynamics

WILEY-VCH Verlag GmbH & Co. KGaA

The Author
Prof. Dr. Boris Z. Katsenelenbaum
Nahariya, Israel

The Editor/Translator
Prof. Dr. N. N. Voitovich
Institute of Applied Problems of Mechanics and
Mathematics NASU, Lviv, Ukraine

Cover
MaX-1 simulation by Christian Hafner
http://www.wiley.co.uk/max-1/

Library of Congress Card No.:
applied for

**British Library Cataloguing-in-Publication
Data**
A catalogue record for this book is available from
the British Library.

**Bibliographic information published by
Die Deutsche Bibliothek**
Die Deutsche Bibliothek lists this publication in
the Deutsche Nationalbibliografie; detailed
bibliographic data is available in the Internet at
<http://dnb.ddb.de>.

© 2006 WILEY-VCH Verlag GmbH & Co. KGaA,
Weinheim

Printing betz-druck GmbH, Darmstadt
Binding Schäffer GmbH, Grünstadt

Printed in the Federal Republic of Germany
Printed on acid-free paper

ISBN-13: 978-3-527-40529-9
ISBN-10: 3-527-40529-1

Contents

High-frequency Electrodynamics. Boris Z. Katsenelenbaum
Copyright © 2006 WILEY-VCH Verlag GmbH & Co. KGaA, Weinheim
ISBN: 3-527-40529-1

Preface

The subject of this book is the laws of formation and propagation of electromagnetic waves. It is an advanced textbook for college and university students, PhD students, and scientists.

Two main objectives were set by the author when writing this book. The
first one consists in relatively complete description of the modern state of
high-frequency electrodynamics. Studying the book should give a reader an
ability to understand the monographs and journal articles of this field as well
as to take an active part in conferences and seminars. In particular, certain
brevity of the derivation is caused by this objective. Almost throughout the
book all intermediate derivations are omitted and replaced by their descrip-
tions. Carrying out of the omitted derivation by a reader is probably the best
way to have an in-depth study of the subject. In the book the qualitative solution to a problem and its analysis are often given right after formulating the
problem and only then derivation of the solution is carried out. A new idea is
understood easier if it is formulated and analyzed at the same time, whereas
the logic of derivation is more easily grasped if the reader knows to what he or
she is led. Besides, having familiarized himself or herself only with problem
formulation and result, the reader may omit derivations.

The second objective challenging the author is to give a reader who has already studied electrodynamics within a general university course of physics,
the foundation sufficient for enabling him or her to work on some practical
problems of this type. The structure of the book is determined by this objective.

High-frequency electrodynamics is a common area of physics and mathematics (being, on the author's opinion, a branch of physics). The physical approach consists in maximally "nondestructive" simplification permitting us to
replace the real problem by a certain idealized one which preserves all "essential" peculiarities of the real problem, but can be exactly solved. The notion
"essential" cannot be determined formally. Formulation of the idealized problem cannot be algorithmized. It should rely on experience, imagination, and
intuition. The mathematical approach consists in quantitative description of

High-frequency Electrodynamics. Boris Z. Katsenelenbaum
Copyright © 2006 WILEY-VCH Verlag GmbH & Co. KGaA, Weinheim
ISBN: 3-527-40529-1

the idealized model with exactly formulated properties. The question about the degree of consistency between the model and the real object is not a subject of mathematics.

Study of the high-frequency electrodynamics can be carried out in two ways: by following the investigation objects or the solution methods for corresponding problems. The first way, usual in physics textbooks, is chosen in the book.

Methods of diffraction theory are described, as a rule, in connection with solving certain electrodynamic problems arisen when developing different practical devices. Exposition of these methods is always very brief; it only delivers the main idea to a reader and indicates the branch of mathematical physics, and the notions used in the methods. Detailed derivation of the diffraction theory basis would only increase the size of the book almost twice.

The book can be used for getting familiarized with a concrete problem without studying the whole discipline. For instance, it is not obligatory to know the theory of open resonators when developing antireflection coatings. The background material, necessary for reading, is given in several first sections of each chapter.

Of course, the author does not pretend to the complete covering of the subject announced in the book title. Exposition of the whole high-frequency electrodynamics including the numerical methods even in a brief style similar to that in this book would take three to four volumes like this one. Detailed analysis of a number of particular problems requires creation of the whole library such as, for instance, the IEEE series. Selection of problems included into the book is determined not only by their significance but also by the scientific interests of the author.

As an autonomous science, high-frequency electrodynamics was formed in the forties of the last century. Appearance of radiolocation as a way of the target determination in air, on ground, and sea was the main stimulus of works in this field. In these years novel theories were developed such as antenna theory, including multivibrator and parabolic (mirror) antennas, the theory of waveguide elements, theory of radiowave propagation, accounting nonhomogeneity and curvature of the ground surface as well as atmosphere nonhomogeneity, etc. Of course, classical diffraction problems (on cylinder, sphere, half-plane) and those connected with signal transmission by means of megahertz waves have been solved earlier, but this direction has not been developing intensively.

In these years, as well as nowadays, development of science was stimulated by military investments whereas the scientific interests were secondary consequences of works on the "defence" objectives. This situation is not only immoral and economically inexpedient but also illogical. However, probably, it has always existed. Even Archimedes was a military engineer in a modern terminology. Calculating the circumference and determining the carrying power of liquid were secondary, nonessential results of his main activity.

When writing this book, the author used material of two textbooks written on the notes of the lectures taught by him as a professor of the Moscow Institute of Physics and Technology. These books *High-frequency Electrodynamics* and *Backgrounds of Diffraction Theory* (with a co-author) were published by Nauka in Moscow in 1966 and 1982, respectively. The present book, however, is written on the basis of a lot of newly obtained results (see, e. g., the author's overview, "Half-century progress in high-frequency electromagnetics," *Journal of Communications Technology and Electronics*, **49** (2004), 9, 963–972). As a rule, key points are highlighted in different ways. Wherever possible, new trends in the development of this branch of science are accounted in the proposed book.

The reference list containing the main English-written literature in high-frequency electrodynamics was compiled by Professor Manfred Thumm to whom I address my gratitude for this, as well as for other help in publishing this book. The complementary list contains several monographs and papers originally published in Russian and then translated into English. Only few papers are selected from a very long list of those related to the subject. Unfortunately, many extra-class monographs published in Russian are not translated. Such books are not included into the list.

Numbers in the brackets after the book title in the complementary list refer to the book sections which are developed in details there.

The author is deeply grateful to Professor Nikolai N. Voitovich for his careful translation of the manuscript into English and its scientific editing. He has also written Subsections 5.3.6, 6.2.4, and participated in writing Subsection 5.3.3. The author also thanks Oleg Kusyi for his contribution to the translation and technical preparation of the book. Special thanks are due to Ulrike Werner, Melanie Rohn, and Yogesh Kukshal for thorough editing and preparing the manuscript and for the permanent attention to the project.

B. Z. Katsenelenbaum   Israel, January, 2006

1
The Maxwell Equations

1.1
Complex amplitudes

1.1.1
Harmonic (monochromatic) oscillations

High-frequency electrodynamics deals with the electromagnetic field consisting of two vector fields: electric and magnetic. These fields depend on spatial coordinates (i. e., on the vector $\vec{r}(x, y, z)$) and time t. Almost throughout the book only those fields dependent on t by the so-called harmonic law are considered. Each field component $a(\vec{r}, t)$ consists of two summands proportional to $\cos \omega t$ and $\sin \omega t$, respectively:

$$a(\vec{r}, t) = A^c(\vec{r}) \cos \omega t + A^s(\vec{r}) \sin \omega t. \tag{1.1}$$

Fields that have all of their components in the above form are called *harmonic* or *monochromatic* fields.

The parameter ω is called the *circular frequency*. It has the dimensionality s^{-1} and differs from the frequency f, usually used in radioengineering, by the multiplier 2π: $\omega = 2\pi f$, where $f = 1/T$ (T is an oscillation period). In the further formulas a proportional value $k = \omega/c$ is used instead of ω, where c is the light velocity in vacuum, $c = 3 \times 10^8$ m/s. The value k has the dimensionality cm^{-1} and is called the *wave number*. However, we call k itself the frequency, omitting the multiplier $2\pi/c$, by which k differs from f: $k = 2\pi/c \cdot f$. The *wavelength* $\lambda = 2\pi/k$ often participates in formulas. The frequency f is inversely proportional to λ: $f = c/\lambda$.

Monochromatic fields are of interest because the majority of radiating devices create fields with the time dependence close to (1.1). It is explained by resonant nature of radiating devices used in radioengineering. Harmonic oscillations are also of interest because almost every process is described by the time functions which can be expressed as a sum or integral of functions of the type (1.1), that is, as superposition of harmonic oscillations.

High-frequency Electrodynamics. Boris Z. Katsenelenbaum
Copyright © 2006 WILEY-VCH Verlag GmbH & Co. KGaA, Weinheim
ISBN: 3-527-40529-1

1.1.2
Complex amplitudes

There is a mathematical technique which permits us to exclude explicit time occurrence while describing harmonic oscillations. This technique implies that a complex function (*complex amplitude*) introduced as

$$A(\vec{r}) = A^c(\vec{r}) - iA^s(\vec{r}), \quad i = \sqrt{-1}, \tag{1.2}$$

is considered instead of the function $a(\vec{r}, t)$ (1.1), containing two functions of the coordinates $A^c(\vec{r})$ and $A^s(\vec{r})$. Then the function $a(\vec{r}, t)$ is expressed as

$$a(\vec{r}, t) = \text{Re}\left[A(\vec{r}) e^{i\omega t}\right]. \tag{1.3}$$

This equality can be easily verified using the *Euler formula* $\exp(i\omega t) = \cos \omega t + i \sin \omega t$. The complex function $A(\vec{r})$ can be presented in the form

$$A(\vec{r}) = |A(\vec{r})| \exp\left[-i\alpha(\vec{r})\right], \tag{1.4}$$

where

$$|A(\vec{r})| = \sqrt{[A^c(\vec{r})]^2 + [A^s(\vec{r})]^2},$$
$$\cos \alpha(\vec{r}) = \frac{A^c(\vec{r})}{|A(\vec{r})|}, \quad \sin \alpha(\vec{r}) = \frac{A^s(\vec{r})}{|A(\vec{r})|}. \tag{1.5}$$

The function $a(\vec{r}, t)$ can now be expressed as

$$a(\vec{r}, t) = |A(\vec{r})| \cos\left[\omega t - \alpha(\vec{r})\right]. \tag{1.6}$$

The physical meaning of the modulus $|A(\vec{r})|$ of the complex amplitude $A(\vec{r})$ is that it equals the maximal value of $a(\vec{r}, t)$ as a function of time. In fact, the phase $\alpha(\vec{r})$ is not interesting; only the difference of the phases of two harmonic processes is essential. If one of the processes has the complex amplitude A and the other has the amplitude $B = |B| \exp(-i\beta)$, then they are synchronous if $\alpha - \beta = 0$; the first process is ahead of the second one in time if $\alpha - \beta > 0$ and lags behind it if $\alpha - \beta < 0$.

All linear operations on the fields can be performed on the complex amplitudes $A(\vec{r})$ without considering the physical amplitudes $a(\vec{r}, t)$, simply by replacing $a(\vec{r}, t)$ by $A(\vec{r}, t)$. This is possible because of the operation of extracting the real part of a complex number, which connects $a(\vec{r}, t)$ with $A(\vec{r}, t)$ by (1.3), is additive and, therefore, interchangeable with any other additive operation. For instance,

$$\text{Re}\left[(A + B) e^{i\omega t}\right] = \text{Re}\left[A e^{i\omega t}\right] + \text{Re}\left[B e^{i\omega t}\right]. \tag{1.7}$$

Hence, the complex amplitude corresponding to the sum of two functions $a(\vec{r}, t)$ and $b(\vec{r}, t)$ equals the sum of the complex amplitudes corresponding to each of the functions. Similarly, the equality

$$\text{Re}\left[\frac{\partial A(\vec{r})}{\partial x} \cdot e^{i\omega t}\right] = \frac{\partial}{\partial x}\text{Re}\left[A(\vec{r})e^{i\omega t}\right] \tag{1.8}$$

together with (1.1) means that the complex amplitude corresponding to the function $\partial a(\vec{r}, t)/\partial x$ equals the derivative of the complex amplitude of the function $a(\vec{r}, t)$ with respect to x.

It is easy to verify that the complex amplitude of the function $\partial a(\vec{r}, t)/\partial t$ is $i\omega A(\vec{r})$, that is, differentiation of the function $a(\vec{r}, t)$ with respect to t is equivalent to multiplying the complex amplitude with $i\omega$, as follows:

$$\frac{\partial}{\partial t}\text{Re}\left[Ae^{i\omega t}\right] = \text{Re}\left[i\omega Ae^{i\omega t}\right]. \tag{1.9}$$

Introducing the complex amplitudes instead of the harmonic functions simplifies all the calculations. Any operation over the functions of the type (1.1) requires equating the terms at $\cos\omega t$ and $\sin\omega t$, which is more complicated than performing operations on complex functions. There is another advantage in using the complex amplitudes. In many diffraction problems, analytical properties of functions of a complex variable are used. Applying this technique to the functions which are not complex themselves would require very cumbersome derivations. This difficulty does not arise when operating with the complex amplitudes.

Introducing the complex amplitudes by (1.3) is usually described by the expression "Time dependence is taken in the form of $\exp(i\omega t)$." In nearly 50 per cent of monographs and papers the time dependence is taken as $\exp(-i\omega t)$. While reading a paper or a book, you must find out to which half it belongs. Since the imaginary unit is introduced by choosing the time dependence, the sign of i would be opposite throughout your book if the time dependence is taken as $\exp(-i\omega t)$. Note that the usual definition of the imaginary unit as a square root of -1 is ambiguous; it does not specify which of the two roots i is.

1.1.3
The period-average product of two harmonic functions

Nonlinear operations cannot be performed on the complex amplitudes. The product of two harmonic functions $a(\vec{r}, t)$ and $b(\vec{r}, t)$ equals

$$a(\vec{r}, t) \cdot b(\vec{r}, t) = \frac{1}{2}|A||B|\left\{\cos(\alpha - \beta) + \cos(2\omega t - \alpha - \beta)\right\}, \tag{1.10}$$

that is, it depends on t differently than in (1.1). However, the period-average product, introduced as

$$\overline{a(\vec{r},t) \cdot b(\vec{r},t)} = \frac{1}{T} \int_0^T a(\vec{r},t) \cdot b(\vec{r},t)\, dt, \tag{1.11}$$

equals the first summand in (1.10), which contains moduli and phases of the complex amplitudes, so that

$$\overline{a \cdot b} = \frac{1}{2} \operatorname{Re} [A \cdot B^*], \tag{1.12}$$

where asterisk means the complex conjugation.

Therefore, in almost all calculations related to the monochromatic fields, we can operate with the complex amplitudes $A(\vec{r})$ instead of passing to the physical values $a(\vec{r},t)$.

The complex amplitudes of electric and magnetic fields are denoted by $\vec{E}(\vec{r})$ and $\vec{H}(\vec{r})$, respectively. These two complex vectors are called *electric* and *magnetic fields*.

1.2
The Maxwell equations

1.2.1
The conduction current and the extrinsic current

In this section we give the Maxwell equations describing the laws of the field $\vec{E}(\vec{r})$ and $\vec{H}(\vec{r})$ behavior in space. The equations contain two more fields, namely, the *electric* and *magnetic inductions*. The complex amplitudes of these fields are called the inductions, denoted as $\vec{D}(\vec{r})$ and $\vec{B}(\vec{r})$, respectively. Further, we will consider a connection between the inductions and fields.

Besides, two other vectors, namely, the volume *densities* of the *conduction current* and the *extrinsic* current take part in the Maxwell equations. Their complex amplitudes are denoted by $\vec{j}(\vec{r})$ and $\vec{j}^{\text{ext}}(\vec{r})$, respectively.

The conduction current density is proportional to the electric field $\vec{E}$,

$$\vec{j} = \sigma \vec{E}. \tag{1.13}$$

The proportionality coefficient σ is called the *conductivity*; it is a property of the medium where the field $\vec{E}$ and current density $\vec{j}$ are considered. The conductivity has the dimensionality s^{-1}; $\sigma = 0$ in vacuum and nonconducting media.

The extrinsic current density $\vec{j}^{\,\text{ext}}\,(\vec{r})$ has, in general, a different meaning than $\vec{j}$. In applied electrodynamics problems, the extrinsic current appears as a prescribed value, similar to the outside force in dynamics. It can be specified explicitly; then $\vec{j}^{\,\text{ext}}\,(\vec{r})$ is the known vector function. The current may also be given in an implicit form as a wave of specified magnitude and structure, coming from infinity. In this case there is no term $\vec{j}^{\,\text{ext}}$ in the equation; instead, the solution must contain the oncoming wave.

Separation of the current into the conduction and extrinsic ones depends on the problem formulation. Let, for instance, an antenna in the form of a metallic cylinder be investigated. The current flowing on its surface can be treated as an extrinsic (given) current, and then the field created by it in the surrounding space is studied. Otherwise, the field created by a line feeding the radiating device may be considered as a given field. Then the current flowing on the line and creating the field is found from the requirement that this field together with the prescribed one satisfies specified conditions. The problem formulated in this way is closer to the actual one and the result obtained is more informative. Needless to say that solving the problem in this formulation is more difficult. The current on the cylinder should be considered as an induced one, that is, the current of conductivity created by the field, but not as an exterior one. The density $\vec{j}(\vec{r})$ is found from the solution of the problem.

1.2.2
The Maxwell equations

The *Maxwell equations* are a system of two vector (that is, of six scalar) partial differential equations of the first order. We write them for the complex amplitudes of the fields, inductions, and current densities. The time derivation is replaced by a multiplication by $i\omega$, and k stands for the w/c multiplier. The equations become

$$\text{rot}\,\vec{H} - ik\vec{D} = \frac{4\pi}{c}\vec{j} + \frac{4\pi}{c}\vec{j}^{\,\text{ext}}, \tag{1.14a}$$

$$\text{rot}\,\vec{E} + ik\vec{B} = 0. \tag{1.14b}$$

We give some elementary consequences of these equations. Taking the divergence of (1.14b) and keeping in mind the vector identity

$$\text{div}\,\text{rot}\,\vec{A} = 0, \tag{1.15}$$

we arrive at the scalar equation

$$\text{div}\,\vec{B} = 0. \tag{1.16}$$

Introducing the so-called *charge density* by the formula

$$\rho = \frac{1}{4\pi}\,\text{div}\,\vec{D} \tag{1.17}$$

and replacing (only in this formula) multiplication by $i\omega$ with the time differentiation $\partial/\partial t$, we obtain from (1.14a)

$$\frac{\partial \rho}{\partial t} = -\operatorname{div}(\vec{j} + \vec{j}^{\text{ext}}). \tag{1.18}$$

This is one of the main equations of alternating current theory, the so-called *continuity equation*. There is usually no need for introducing the charge density (1.17) and using equation (1.18) when studying the harmonic fields, that is, in high-frequency electrodynamics. Similar to (1.17), equation (1.16) can be treated as an assertion that the absence of magnetic charges follows from (1.14).

In the domain where neither material objects nor extrinsic currents exist, that is, where $\vec{j} \equiv 0$ and $\vec{j}^{\text{ext}} \equiv 0$, there are also no charges, that is, $\vec{D} \equiv \vec{E}$, and, similar to (1.16),

$$\operatorname{div}\vec{E} = 0. \tag{1.19}$$

Sometimes the set of the Maxwell equations consists not only of equation (1.14) but also of equations (1.16) (1.19), and (1.17). In these cases (1.17) is treated not as a definition of ρ, but as an equation for $\vec{D}$. In the text below, the system of the Maxwell equations is understood as system (1.14) or any of its other notations.

1.2.3
Dielectric permittivity and magnetic permeability of a medium

The inductions are linearly related to the fields. Coefficients of these relations constitute a medium in which the fields $\vec{E}$, $\vec{H}$ and inductions $\vec{D}$, $\vec{B}$ are considered.

In most media, $\vec{D}$ depends only on $\vec{E}$, as well as $\vec{B}$ on $\vec{H}$, so that

$$\vec{D} = \tilde{\varepsilon}\vec{E}, \tag{1.20a}$$

$$\vec{B} = \mu\vec{H}. \tag{1.20b}$$

The difference in notation of $\tilde{\varepsilon}$ and μ will be explained below. The coefficient $\tilde{\varepsilon}$ is called the *dielectric permittivity* (further, permittivity) and μ the *magnetic permeability* (further, permeability) of the medium. We accept that $\tilde{\varepsilon}$ and μ are dimensionless; $\tilde{\varepsilon} = 1$, $\mu = 1$ in vacuum.

In anisotropic media the vector $\vec{D}$ is not parallel to $\vec{E}$, and, generally speaking, $\vec{B}$ and $\vec{H}$ are not parallel, either. In such media, $\tilde{\varepsilon}$ (sometimes also μ) is a tensor, not a scalar, and it is described by all its components $\tilde{\varepsilon}^{\tau\sigma}$, where τ and σ take the values x, y, z. The expression $\tilde{\varepsilon}\vec{E}$ is the product of a tensor and a vector; it is a vector with the components

$$D_x = \sum_{\sigma} \tilde{\varepsilon}^{x\sigma} E_{\sigma}, \quad D_y = \sum_{\sigma} \tilde{\varepsilon}^{y\sigma} E_{\sigma}, \quad D_z = \sum_{\sigma} \tilde{\varepsilon}^{z\sigma} E_{\sigma}, \tag{1.21}$$

where the summation is made over the three values $\sigma = \{x, y, z\}$. The above equalities are basic for crystal optics and interdisciplinary sciences. Further, almost everywhere in the text $\widetilde{\varepsilon}$ and μ are scalar.

In nonhomogeneous media, $\widetilde{\varepsilon}$ and μ are functions of the coordinates. In general, they are complex, that is, for instance, $\vec{D}$ and $\vec{E}$ are not in-phase. In linear electrodynamics, which is only considered in the book, $\widetilde{\varepsilon}$ and μ do not depend on $\vec{E}$ and $\vec{H}$. Such dependence takes place only for large fields.

1.2.4
The polarization current

In order to explain the difference between the induction $\vec{D}$ and field $\vec{E}$, we rewrite (1.14) in the form

$$\mathrm{rot}\,\vec{H} = \frac{4\pi}{c}\left[\vec{j}^{\mathrm{ext}} + \vec{j} + \frac{i\omega}{4\pi}\vec{E} + \frac{i\omega}{4\pi}(\widetilde{\varepsilon} - 1)\vec{E}\right]. \tag{1.22}$$

The magnetic field is created by the currents $\vec{j}^{\mathrm{ext}}$ and $\vec{j}$, the *displacement current* $i\omega/4\pi\vec{E}$, and the current with density $i\omega/4\pi(\vec{D} - \vec{E})$. The assertion that the magnetic field is created by the time-varying electrical field in vacuum, that is, by the displacement current, is the basis of the whole Maxwell theory. The last term in (1.22) is called the *polarization current*. It exists only in material objects.

We explain the process of appearance of the polarization current on a model of a simplest medium. Imagine a medium containing a plenty of small metallic particles. The distance between any two particles is large in comparison with their sizes and small with the distance, where the field $\vec{E}$ varies significantly, for instance, in comparison with the length of the wave propagating in the medium.

A particle is being polarized under the influence of the electric field $\vec{E}$, that is, positive and negative charges are gathering on opposite sides of the particle. If the field $\vec{E}$ varies with time, then the charges displace; for instance, if $\vec{E}$ changes its sign, then the charges interchange. The displacement of the charges is the polarization current. It is proportional to the time derivative of the field, that is, to the vector $i\omega\vec{E}$, and, together with three other summands in (1.22), creates the magnetic field, and, therefore, appears in (1.14).

We give two formulas connecting $\widetilde{\varepsilon}$ with medium parameters. Each particle is characterized by a parameter p (*polarization coefficient*) which is determined from the condition that the field $\vec{E}$ causes the polarization current $i\omega p\vec{E}$. In the general case, p is a tensor and the current is not parallel to $\vec{E}$; however, we assume it to be a scalar in our simplest model. Tensorial nature of p is not essential here because the value $\vec{D}$ is the induction averaged over many particles, and, therefore, the particle orientation disappears. The value of p is obtained from the solution of the electrostatic problem. It has the order of the particle volume (if the particle is not very prolate or flattened). For instance, in the case of a sphere, p is equal to the cube of its radius. A number N of the particles per unit volume of the medium is large; however, the total volume of all the particles in the unit volume (*particle concentration*) is assumed to be small. Under this condition, $\tilde{\varepsilon}$ depends only on the dimensionless quantity Np. All the above considerations as well as the below formulas (1.23) and (1.24) for $\tilde{\varepsilon}$ are valid only at not very high frequencies, more precisely, when the field varies slightly at the distance of the particle size.

The simplest formula for $\tilde{\varepsilon}$ is obtained if Np is so small that the mutual influence of the particles can be neglected. Then the polarization current density is equal to the sum of all the polarization currents caused only by the field $\vec{E}$ influence on the particles of unit volume. In this case

$$\tilde{\varepsilon} = 1 + \alpha, \quad \alpha = 4\pi Np. \tag{1.23}$$

The above formula is known as the *formula of molecular optics*.

We derive a more precise formula by taking into consideration that each particle is influenced not only by the applied field $\vec{E}$ but also by the fields of other polarized particles. Then

$$\tilde{\varepsilon} = \frac{1 + 2\alpha/3}{1 - \alpha/3} \tag{1.24}$$

(the *Lorentz–Lorence formula*). Formula (1.23) coincides with the first two terms on the right-hand side of expansion (1.24) in the series in parameter α. Formula (1.24) is also valid only at $\alpha \ll 1$; however, this restriction is not so strong as for (1.23).

The mutual influence of the particles should be taken into account more properly when considering a larger particle concentration. In this case $\tilde{\varepsilon}$ depends not only on the polarization factor but also on other electrodynamic parameters of the particle. The theory, resulting in formulas (1.23), (1.24) and other more complicated and accurate formulas, was developed for natural materials, in which molecules or atoms play the role of particles in our model. Needless to say that the polarization process in such materials is more complicated than in metallic particles; it cannot be treated as charge separation only. However, the process is also characterized by the quantity (tensor) p, and it causes the appearance of the polarization current $i\omega p\vec{E}$. In crystals the particles are not chaotically positioned and, in principle, the polarization current is not parallel to the field; $\tilde{\varepsilon}$ is a tensor.

The theory based on the model described is also applicable to artificial media, so-called composites. These media contain small particles of macroscopic sizes, incorporated into a material with $\tilde{\varepsilon} \approx 1$, $\mu \approx 1$. If the particle sizes are small, then the composite object behaves like the same object from a natural material having the effective $\tilde{\varepsilon}$ and μ. The above formulas (as well as the similar ones for μ) express the effective parameters of the composite in terms of the particle characteristics. The interest to the composites is caused by the possibility of creating the media with desired electrodynamic parameters by choosing the material, shape, size, and number of particles.

Another model for which the explicit expression for $\widetilde{\varepsilon}$ can be found is a set of free electrons (plasma). Being influenced by an electric field, they are fast oscillating with the field frequency. The amplitude of the oscillations is inversely proportional to the inertia of electrons, that is, to their mass m, and also to the force acting on them, that is, to their charge e. The amplitude decreases when the frequency increases. The field of the unit volume is proportional to the amplitude multiplied by the charge and the number N of electrons per unit volume. The permittivity of such medium is

$$\widetilde{\varepsilon}(\omega) = 1 - 4\pi \frac{e^2}{m\omega^2} N. \tag{1.25}$$

If the collisions and effects like "friction" are taken into account, then $\widetilde{\varepsilon}$ also acquires the imaginary part. According to (1.25), $\widetilde{\varepsilon} = 0$ at $\omega = \omega_0$, where

$$\omega_0^2 = 4\pi \frac{e^2}{m} N. \tag{1.26}$$

If $\omega < \omega_0$, then $\widetilde{\varepsilon} < 0$. In the medium of free electrons, the electric field and induction are oppositely directed at low frequencies. If $\omega > \omega_0$, then $0 < \widetilde{\varepsilon} < 1$. At $\omega \to \infty$, $\widetilde{\varepsilon} \to 1$, the sign of the field acting on the electrons changes fast and they have no time to displace; the polarization current is small at the high frequencies. This remark does not relate to this model only. The permittivity tends to unity as $\omega \to \infty$ in any media.

The permeability μ behaves in a similar way; $\mu \to 1$ as $\omega \to \infty$. The frequency value at which $|\widetilde{\varepsilon} - 1|$ and $|\mu - 1|$ become small depends on the physical mechanism of appearance of the polarization. For plasma, such frequencies are noticeably larger than ω_0.

The mechanism causing the distinction between the induction $\vec{B}$ and field $\vec{H}$ can be explained similarly to that causing the distinction between $\vec{D}$ and $\vec{E}$, which has been described for our simple model. Rewrite equation (1.14b) in the form

$$\text{rot}\,\vec{E} = -ik\vec{H} - ik(\vec{B} - \vec{H}). \tag{1.27}$$

The electric field $\vec{E}$ is created not only by the time-varying magnetic field in vacuum (the first term in (1.27)) but also by the time-varying field $(\vec{B} - \vec{H})$, which exists only in the material medium. A set of permanent magnets can serve as a model of the medium for describing the appearance of the additional magnetic field. If there is no external magnetic field, then the magnet

axes are oriented chaotically and magnetic fields created by the magnets mutually cancel. Under the influence of the external magnetic field, the axes become mostly oriented along it and the magnetic fields created by the magnets
are not fully compensated. An additional magnetic field is created. When
the applied magnetic field changes, then the extra magnetic field also changes
and, therefore, it creates the electric field. This effect is described by the second term in (1.27). The additional field $\vec{B} - \vec{H}$ is proportional to $\vec{H}$, so that if
$\vec{B} - \vec{H} \neq 0$, then the factor μ in (1.20b) differs from unity. It is clear that the
model describes only one (probably, the most demonstrative) process leading
to $\vec{B} \neq \vec{H}$.

1.2.5
The frequency dispersion and the spatial dispersion (chirality)

Formula (1.20a) states that at a given point and time moment, the electric induction is determined only by the value of the electric field at the same point
and time moment. This statement needs two specifications.

The first specification implies that the induction depends also on the time
derivative of the field. For the fields varying by the harmonic law, the specification does not require more complicated formula (1.20a), but leads to $\tilde{\varepsilon}$ becoming the function of frequency, $\tilde{\varepsilon} = \tilde{\varepsilon}(\omega)$. This effect is called the *frequency
dispersion*.

The second specification is connected with the phenomenon of *spatial dispersion*. The induction $\vec{D}$ depends not only on the field $\vec{E}(\vec{r})$ but also on its
spatial derivatives. At the arbitrary dependence $\vec{E}(\vec{r})$, formula (1.20a) is not
valid in media where the above effect is essential. It remains valid only if the
field varies in space just like in the plane wave. However, in this case $\tilde{\varepsilon}$ should
be treated as a function of the normal direction to the wave front; in the same
manner as for the frequency dispersion it is a function of the frequency. If we
do not restrict ourselves to the fields of plane waves (as we did to the harmonic fields above), then it is not possible to take into account the existence of
the spatial dispersion by assuming $\tilde{\varepsilon}$ to be dependent on the normal.

It can be shown that the first derivatives of $\vec{E}(\vec{r})$ appear in $\vec{D}(\vec{r})$ only as a
combination of rot $\vec{E}$ if the fields depend on $\vec{r}$ arbitrarily. Since at the points
where no extrinsic currents exist, the fields $\vec{E}(\vec{r})$ and $\vec{H}(\vec{r})$ satisfy equations
(1.14) with right-hand side zero. The relation between induction and fields
can be written in the symmetric form, without derivatives, as follows:

$$\vec{D} = \tilde{\varepsilon}\vec{E} - i\varkappa\vec{H}, \quad \vec{B} = \mu\vec{H} + i\varkappa\vec{E}. \tag{1.28}$$

It is a generalization of formula (1.20), valid only for the media where the
spatial dispersion may be neglected. The "cross" terms $-i\varkappa\vec{H}$ and $i\varkappa\vec{E}$ are
introduced into (1.28) in a special form: with the same factor $\varkappa$ (*chirality factor*),

opposite signs in front of them, and separated multiplier *i*. The expediency of this form of equations connecting inductions with fields will be shown later.

The appearance of the cross terms in the constitutive equations (1.27) for chiral media can be explained by physical reasons without referring to a formally nonlocal dependence between $\vec{D}$ and $\vec{E}$ (and $\vec{B}$ and $\vec{H}$, respectively). Return to the model of dielectric as a set of metallic particles. The presence of the term proportional to $\vec{E}$ in the expression for $\vec{B}$ means that the current induced by the field $\vec{E}$ creates not only the electric field (described by the difference $\tilde{\varepsilon} - 1$) but also the magnetic field described by the term $-i\varkappa\vec{E}$ in the expression for $\vec{B}$.

This magnetic field is created only by the particles not having the plane of symmetry. For instance, if the particles are ball- or ellipsoid-shaped, then the field $\vec{E}$ creates a very low magnetic field proportional to $(ka)^2$, where *a* is the particle size. The induction-to-field connection in such "nonchiral" dielectric, natural or artificial (that is, composite), is given by (1.20). The simplest example of the element possessing the pronounced chiral property is a planar open ring with free ends connected to the two oppositely directed linear wires lying in the plane perpendicular to the ring plane. The electric field parallel to the wires excites the current in them, flowing also along the ring. The current in the wires creates the electric field (resulting in $\tilde{\varepsilon} \neq 1$), and the current in the ring creates the magnetic field, and, therefore, $\varkappa \neq 0$. In a similar way, one can explain the appearance of the term $-i\varkappa\vec{H}$ in the expression for $\vec{D}$ when the particles have no plane symmetry.

In the chiral media, the electrodynamic processes are going on in a different way than in the nonchiral ones. The most known chirality phenomenon is the rotation of the polarization plane. It will be considered in Chapter 2.

1.2.6
Complex permittivity

Equation (1.14) is simplified after substituting the conduction current density and induction with use of (1.13) and (1.20), respectively. Then (1.14) becomes

$$\text{rot}\,\vec{H} - ik\varepsilon\vec{E} = \frac{4\pi}{c}\vec{j}^{\text{ext}},$$
$$\text{rot}\,\vec{E} + ik\mu\vec{H} = 0. \tag{1.29}$$

In such a form the Maxwell equations are applicable only to the nonchiral media.

The factor ε in (1.29) equals

$$\varepsilon = \tilde{\varepsilon} - i\frac{4\pi\sigma}{\omega}. \tag{1.30}$$

We call it the *complex permittivity,* sometimes, however, omitting the word "complex." The Maxwell equations are understood as system (1.29). Rewrite it in the extended form in the Cartesian coordinates x, y, z:

$$
\begin{aligned}
\frac{\partial H_z}{\partial y} - \frac{\partial H_y}{\partial z} - ik\varepsilon E_x &= \frac{4\pi}{c} j_x^{\text{ext}}, & \frac{\partial E_z}{\partial y} - \frac{\partial E_y}{\partial z} + ik\mu H_x &= 0, \\
\frac{\partial H_x}{\partial z} - \frac{\partial H_z}{\partial x} - ik\varepsilon E_y &= \frac{4\pi}{c} j_y^{\text{ext}}, & \frac{\partial E_x}{\partial z} - \frac{\partial E_z}{\partial x} + ik\mu H_y &= 0, \quad (1.31) \\
\frac{\partial H_y}{\partial x} - \frac{\partial H_x}{\partial y} - ik\varepsilon E_z &= \frac{4\pi}{c} j_z^{\text{ext}}, & \frac{\partial E_y}{\partial x} - \frac{\partial E_x}{\partial y} + ik\mu H_z &= 0,
\end{aligned}
$$

and in the cylindrical coordinates r, ϑ, φ:

$$
\begin{aligned}
\frac{1}{r}\frac{\partial H_z}{\partial \varphi} - \frac{\partial H_\varphi}{\partial z} - ik\varepsilon E_r &= \frac{4\pi}{c} j_r^{\text{ext}}, & \frac{1}{r}\frac{\partial E_z}{\partial \varphi} - \frac{\partial E_\varphi}{\partial z} + ik\mu H_r &= 0, \\
\frac{\partial H_r}{\partial z} - \frac{\partial H_z}{\partial r} - ik\varepsilon E_\varphi &= \frac{4\pi}{c} j_\varphi^{\text{ext}}, & \frac{\partial E_r}{\partial z} - \frac{\partial E_z}{\partial r} + ik\mu H_\varphi &= 0, \quad (1.32) \\
\frac{1}{r}\frac{\partial (rH_\varphi)}{\partial r} - \frac{1}{r}\frac{\partial H_r}{\partial \varphi} - ik\varepsilon E_z &= \frac{4\pi}{c} j_z^{\text{ext}}, & \frac{1}{r}\frac{\partial (rE_\varphi)}{\partial r} - \frac{1}{r}\frac{\partial E_r}{\partial \varphi} + ik\mu H_z &= 0.
\end{aligned}
$$

1.2.7
The radiation condition

System (1.29) should be complemented by the condition that the fields are created only by the current $\vec{j}^{\text{ext}}$, that is, there are no waves incoming from infinity. This condition can be written in the form of an asymptotic equality, which implies that the field has the structure of an outgoing wave in any domain outside the sphere containing all the objects and currents. This condition is called the *radiation* (or *Sommerfeld*) *condition.* In the spherical coordinates R, ϑ, φ it is written for the meridional E_ϑ, H_ϑ and azimuthal E_φ, H_φ components of the vectors $\vec{E}$ and $\vec{H}$, and has the form

$$
\begin{aligned}
E_\vartheta \equiv -H_\varphi &= F_1(\vartheta, \varphi) \exp(-ikR)/kR \cdot \left[1 + O\left(\frac{1}{kR}\right)\right], \\
E_\varphi \equiv H_\vartheta &= F_2(\vartheta, \varphi) \exp(-ikR)/kR \cdot \left[1 + O\left(\frac{1}{kR}\right)\right].
\end{aligned}
\quad (1.33)
$$

This dependence of the complex amplitudes on R means that the physical fields are proportional to the functions

$$
\frac{\cos(\omega t - kR)}{kR}, \quad \frac{\sin(\omega t - kR)}{kR} \quad (1.34)
$$

in the higher order with respect to the parameter $1/kR$. The spheres $R = ct + const$ are (asymptotically) equiphase ones. The surfaces expand with the velocity c as t increases, that is, (1.33) represents a divergent spherical wave. The complex amplitudes of the convergent wave incoming from the infinity would be proportional to $\exp(ikR)/kR$. Condition (1.33) means that such waves are not present in the field. The more simpler requirement that the field decreases not slower than $1/kR$ as $kR \to \infty$ would not be enough to exclude the incoming waves. For doing this it is also necessary to specify how the wave phase depends on R.

The functions $F_1(\vartheta, \varphi)$ and $F_2(\vartheta, \varphi)$ are not given in conditions (1.33). They can be found only after solving the problem, that is, after finding the fields $\vec{E}(\vec{r})$ and $\vec{H}(\vec{r})$ at the given $\vec{j}^{\text{ext}}$.

There is another way to exclude the waves incoming from infinity. It can be required that the field tends to zero as $R \to \infty$, if the real k is replaced by $k = k' + ik''$, $k'' < 0$. Obviously, at such k the field decreases only for waves which depend on R as $\exp(-ikR)/kR$, that is, for the outgoing waves and infinitely increases for the incoming waves as $R \to \infty$. One can prove that at $k'' \to 0$ the solution of the problem with the above requirement passes to the solution of the problem with $k'' = 0$ satisfying the radiation condition (1.33). The last statement is called the *limiting absorption principle*. This way of excluding the incoming waves is also applicable to the problems with object expanding to infinity, for which the radiation condition should be modified to exclude not only the incoming spherical waves but also the plane waves.

There are no waves with amplitudes of all components decreasing at $kR \to \infty$ faster than in (1.33), for instance, as $1/(kR)^2$. It follows from (1.29) that the fields with all components decreasing at infinity faster than $1/kR$ are identical zeros. Further, we will specify this assertion because under certain idealized conditions the fields are possible which exist only in the finite domain (closed resonators).

From (1.33), (1.16), (1.19) it follows that the asymptotic dependence of the radial field components on R also has a universal form. In spherical coordinates the equation div $\vec{E} = 0$ is written as

$$\frac{1}{R^2}\frac{\partial}{\partial R}(R^2 E_R) + \frac{1}{R\sin\vartheta}\frac{\partial}{\partial\vartheta}(\sin\vartheta E_\vartheta) + \frac{1}{R\sin\vartheta}\frac{\partial E_\varphi}{\partial\varphi} = 0. \tag{1.35}$$

Substituting the asymptotic expressions (1.33) for E_ϑ and E_φ, we obtain that $\partial(R^2 E_R)/\partial R$ is asymptotically proportional to $\exp(-ikR)$; hence,

$$E_R = \Phi(\vartheta, \varphi)\exp(-ikR)/(kR)^2\left[1 + O\left(\frac{1}{kR}\right)\right]. \tag{1.36}$$

A similar dependence of H_R on R follows from the equality div $\vec{H} = 0$, except that the angular dependence is given by the different function than for E_R. These angular functions are expressed by the angular functions from (1.33) and they can be found only after the problem is solved.

1.2.8

The wave equations

Equations (1.29) contain two vector functions, that is, six coordinate functions. Sometimes it is expedient to eliminate one of the fields from (1.29) and pass to the second-order equations in only either $\vec{E}$ or $\vec{H}$. The equations of such type are called the *wave (Helmholtz) equations*.

Let no extrinsic currents exist in a domain and the medium be homogeneous and nonchiral, that is, $\varepsilon = const, \mu = const, \varkappa = 0$. Acting on both equations (1.29) by the operation "rot" and replacing the obtained terms rot $\vec{E}$ and rot $\vec{H}$ by the corresponding expressions from the second equation, we obtain the two independent equations

$$\begin{aligned} \operatorname{rot} \operatorname{rot} \vec{H} - k^2 \varepsilon \mu \vec{H} &= 0, \\ \operatorname{rot} \operatorname{rot} \vec{E} - k^2 \varepsilon \mu \vec{E} &= 0, \end{aligned} \tag{1.37}$$

each containing only one field.

Transform the first terms in these equations. For any vector $\vec{A}$, we introduce a vector $\Delta \vec{A}$ defined as

$$\Delta \vec{A} = \operatorname{grad} \operatorname{div} \vec{A} - \operatorname{rot} \operatorname{rot} \vec{A}. \tag{1.38}$$

One can verify that its Cartesian components equal ΔA_x, ΔA_y, ΔA_z, that is, they are the result of the Laplace operator

$$\Delta = \frac{\partial^2}{\partial x^2} + \frac{\partial^2}{\partial y^2} + \frac{\partial^2}{\partial z^2} \tag{1.39}$$

acting on the Cartesian components of the vector $\vec{A}$. This fact is not valid for the non-Cartesian components; for instance, $(\Delta \vec{A})_R \neq \Delta A_R$. In domains where ε and μ are constant, (1.16) and (1.19) follow from (1.29), that is, rot rot $\vec{E} = -\Delta \vec{E}$, rot rot $\vec{H} = -\Delta \vec{H}$, and equations (1.37) become

$$\begin{aligned} \Delta \vec{E} + k^2 \varepsilon \mu \vec{E} &= 0, \\ \Delta \vec{H} + k^2 \varepsilon \mu \vec{H} &= 0. \end{aligned} \tag{1.40}$$

The first equation is separated into three independent scalar equations in the Cartesian components E_x, E_y, E_z, as follows:

$$\begin{aligned} \Delta E_x + k^2 \varepsilon \mu E_x &= 0, \\ \Delta E_y + k^2 \varepsilon \mu E_y &= 0, \\ \Delta E_z + k^2 \varepsilon \mu E_z &= 0. \end{aligned} \tag{1.41}$$

The second equation (1.40) can be separated in a similar way.

Equations (1.41) follow from system (1.29), that is, any solution to (1.29) with right-hand side zero satisfies (1.41) and similar equations in H_x, H_y, H_z. However, the inverse statement is not valid; not all solutions to (1.40) satisfy system (1.29) because (1.29) is not a consequence of (1.40). System (1.40) does not contain the connection between the Cartesian components of the fields at all. For instance, $\vec{H} \equiv 0$, $\vec{E} \neq 0$ may be its solution, but the system of Maxwell equations cannot have such a solution. After finding some field components from (1.40) we have to substitute them into system (1.29). The solutions for the found components are valid if the system for the remaining components is consistent. The solutions for the remaining components can be obtained from this consistent system.

In the domains where $\vec{j}^{\text{ext}} \neq 0$, the wave equations would involve the term $\text{rot}\, \vec{j}^{\text{ext}}$. In such domains it is simpler to introduce the auxiliary functions, so-called *potentials*. The fields are expressed in terms of the derivatives of these functions. Potentials may be introduced in a way that they satisfy the wave equations with extrinsic currents standing on the right-hand side themselves, instead of their derivatives.

The *Hertz electric vector* $\vec{\Pi}\,(\vec{r})$ is such a potential. It is proportional to the vector potential, usually introduced in electrodynamics. The Hertz vector satisfies the nonhomogeneous wave equation

$$\Delta \vec{\Pi} + k^2 \varepsilon \mu \vec{\Pi} = \frac{4\pi i}{\omega} \vec{j}^{\text{ext}}. \tag{1.42}$$

The fields are expressed in terms of the Hertz vector by the formulas

$$\vec{E} = k^2 \varepsilon \mu \vec{\Pi} + \text{grad div}\, \vec{\Pi}, \qquad \vec{H} = ik\varepsilon\, \text{rot}\, \vec{\Pi}. \tag{1.43}$$

It is easy to check that the fields (1.43) satisfy system (1.29) if $\vec{\Pi}$ satisfies equation (1.42); we suppose that $\varepsilon = const$, $\mu = const$.

In Cartesian coordinates, equation (1.42) is equivalent to three scalar equations in Π_x, Π_y, Π_z having the form

$$\Delta u + k^2 \varepsilon \mu u = f. \tag{1.44}$$

This equation follows from the Maxwell system when the medium is homogeneous. However, (1.44) is also of interest in investigation of the field in nonhomogeneous media; in particular, it is valid for certain two-dimensional nonhomogeneous media. The equation itself and the properties of the function $u(x, y, z)$ satisfying it are interesting mainly because they describe the qualitative peculiarities of propagation and diffraction of the electromagnetic wave, which are not determined by the vector nature of the wave. If the wave polarization (that is, directions of $\vec{E}$ and $\vec{H}$) and the mutual connection between different components of the fields are not essential, then many qualitative properties of the field, as well as many peculiarities of various methods of their finding, may be investigated on the instance of the scalar wave equation (1.44).

1.2.9
The reciprocity conditions

Consider the connection between the fields created by the two different extrinsic currents $\vec{j}^{\text{ext}(1)}$ and $\vec{j}^{\text{ext}(2)}$ in some medium. The fields created by these two currents are denoted as $\vec{E}^{(1)}$, $\vec{H}^{(1)}$ and $\vec{E}^{(2)}$, $\vec{H}^{(2)}$, respectively. The functions $\varepsilon(\vec{r})$, $\mu(\vec{r})$ (and $\varkappa(\vec{r})$, if the medium is chiral) in the constitutive equations are the same for both cases, that is, the currents $\vec{j}^{\text{ext}(1)}$ and $\vec{j}^{\text{ext}(2)}$ are placed into the same medium.

Consider the expression $\text{div}(\vec{E}^{(1)} \times \vec{H}^{(2)} - \vec{E}^{(2)} \times \vec{H}^{(1)})$. According to the vector identity

$$\text{div}\left(\vec{A} \times \vec{B}\right) = \vec{B}\,\text{rot}\,\vec{A} - \vec{A}\,\text{rot}\,\vec{B}, \tag{1.45}$$

the above expression equals

$$\vec{H}^{(2)}\,\text{rot}\,\vec{E}^{(1)} - \vec{E}^{(1)}\,\text{rot}\,\vec{H}^{(2)} - \vec{H}^{(1)}\,\text{rot}\,\vec{E}^{(2)} + \vec{E}^{(2)}\,\text{rot}\,\vec{H}^{(1)}. \tag{1.46}$$

Substituting the value of the rotor of the fields from the Maxwell equations (1.29) into (1.46), we obtain

$$\frac{4\pi}{c}\left(\vec{E}^{(2)}\vec{j}^{\text{ext}(1)} - \vec{E}^{(1)}\vec{j}^{\text{ext}(2)}\right)$$
$$+\, ik\left(\vec{H}^{(1)}\vec{B}^{(2)} - \vec{H}^{(2)}\vec{B}^{(1)} + \vec{E}^{(2)}\vec{D}^{(1)} - \vec{E}^{(1)}\vec{D}^{(2)}\right). \tag{1.47}$$

We find the conditions under which the second bracket in (1.47) equals zero. If the constitutive equations are $\vec{D} = \varepsilon\vec{E}$, $\vec{B} = \mu\vec{H}$, then the bracket equals

$$\left(\vec{H}^{(1)} \cdot \mu\vec{H}^{(2)} - \vec{H}^{(2)} \cdot \mu\vec{H}^{(1)}\right) + \left(\vec{E}^{(2)} \cdot \varepsilon\vec{E}^{(1)} - \vec{E}^{(1)} \cdot \varepsilon\vec{E}^{(2)}\right). \tag{1.48}$$

If ε and μ are scalar, then the above two brackets equal zero. However, if ε or μ is a tensor, then the brackets are zero only if the tensor is symmetric, that is, if $\varepsilon^{xy} = \varepsilon^{yx}$, and so on. The media with ε or μ being a nonsymmetrical tensor do not possess the property of "reciprocity" which other media do.

If the medium is chiral, that is, the constitutive equations have a form (1.28), then the second bracket in (1.47) remains zero. If there were different factors $\varkappa$ in the coefficients at the cross terms of (1.28), then this bracket would not be zero, and the chiral medium also would not possess the reciprocity property. Such situation is possible if, for instance, the particles forming a composite consist of the material for which ε or μ is a nonsymmetrical tensor.

Hence, if ε and μ are scalars or symmetrical tensors, then

$$\text{div} \left(\vec{E}^{(1)} \times \vec{H}^{(2)} - \vec{E}^{(2)} \times \vec{H}^{(1)} \right) = \frac{4\pi}{c} \left(\vec{E}^{(2)} \vec{j}^{\text{ext}(1)} - \vec{E}^{(1)} \vec{j}^{\text{ext}(2)} \right) \qquad (1.49)$$

for any $\vec{j}^{\text{ext}(1)}$ and $\vec{j}^{\text{ext}(2)}$.

Integrate (1.49) over a sphere of radius a so large that there are no currents $\vec{j}^{\text{ext}(1)}$, $\vec{j}^{\text{ext}(2)}$ and material objects outside the sphere. The asymptotic radiation conditions (1.33) hold on the surface of such a sphere. On the left-hand side of the obtained integral form of equation (1.49) we have the flux of the vector standing under the "div" operation in (1.49), through the sphere surface. The vector flux is equal to the integral of the radial component of this vector, that is, of the quantity

$$E_\vartheta^{(1)} H_\varphi^{(2)} - E_\varphi^{(1)} H_\vartheta^{(2)} - E_\vartheta^{(2)} H_\varphi^{(1)} + E_\varphi^{(2)} H_\vartheta^{(1)}, \qquad (1.50)$$

taken over the angles. The terms of the order $1/a^2$ in this sum mutually cancel whatever the functions $F_1(\vartheta, \varphi)$, $F_2(\vartheta, \varphi)$ are in (1.33), so that the above quantity becomes of the order not lower than $1/a^3$. Since the surface element is proportional to a^2, the integral over the sphere surface decreases at $a \to \infty$ not slower than $1/a$. It equals the volume integral of the right-hand side of equality (1.49). However, this value differs from zero only inside the sphere of finite radius; it does not depend on the radius a if a is sufficiently large. Hence, the integral of (1.50) is zero at large a, and it follows from (1.49) that if the integrals are taken over the volume V containing all the extrinsic currents, then

$$\int_V \vec{E}^{(1)} \vec{j}^{\text{ext}(2)} \, dV = \int_V \vec{E}^{(2)} \vec{j}^{\text{ext}(1)} \, dV. \qquad (1.51)$$

This formula is called the *reciprocity condition*.

The physical sense of (1.51) is most clearly seen if both the currents are the point sources, that is, they are described by the δ-functions. Then (1.51) becomes

$$\vec{E}^{(1)}(\vec{r}^{(2)}) \cdot \vec{a}_2 = \vec{E}^{(2)}(\vec{r}^{(1)}) \cdot \vec{a}_1. \qquad (1.52)$$

Here $\vec{r}^{(1)}$ and $\vec{r}^{(2)}$ are the points where the currents $\vec{j}^{\text{ext}(1)}$ and $\vec{j}^{\text{ext}(2)}$ are located, respectively; $\vec{a}_1$ and $\vec{a}_2$ are the unit vectors along which the sources (elementary dipoles, see Subsection 6.1.1) are oriented; $\vec{E}^{(1)}(\vec{r})$ and $\vec{E}^{(2)}(\vec{r})$ are the fields created by the currents. The field at the point $\vec{r}^{(2)}$, created by the elementary dipole located at the point $\vec{r}^{(1)}$, equals the value of the field at the point $\vec{r}^{(1)}$, created by the elementary dipole located at $\vec{r}^{(2)}$. In this statement the specification about the field directions at the points $\vec{r}^{(1)}$ and $\vec{r}^{(2)}$ is omitted for brevity. This is the simplest formulation of the reciprocity condition.

Condition (1.51) holds for any set of bodies if the medium possesses the properties mentioned above. If all the parts of the medium have these properties, then the whole medium has them, either. A set of reciprocal elements is reciprocal itself. The nonreciprocity of an element is a property caused by the existence of a particular direction in its material, for instance, an internal magnetostatic field.

It follows from the reciprocity condition that for transmitting energy in a channel only in one direction it is necessary to insert a nonreciprocal element into the channel.

1.2.10
Average energy losses

We derive the expression for average energy losses, that is, the quantity of the electromagnetic field energy transforming into other types of energy in the unit volume per unit time. We begin with the expression for *thermal (joule) losses*. In the unit volume, they equal the conduction current density multiplied by the electric field. Denote the period-average losses by P_E. According to (1.13) and (1.11), if expressed by the amplitudes of the current $\vec{j}$ and the field $\vec{E}$, the losses are $P_E = \sigma|\vec{E}|^2/2$. Following (1.30), the conductivity factor σ can be expressed by the imaginary part of permittivity as $\sigma = -\varepsilon''\omega/4\pi$, so that

$$P_E = -\frac{\omega}{8\pi}\varepsilon''|\vec{E}|^2. \tag{1.53}$$

We show that the above definition of P_E is, in general, valid for any physical sense of ε''; it is not necessarily related to the joule losses only. The energy flux through a closed surface is the integral of the Poynting vector, equal (with accuracy to the multiplier $c/4\pi$) to the vector product of the electric and magnetic fields. The average value of the flux expressed by the complex amplitudes $\vec{E}$ and $\vec{H}$ is

$$\vec{S} = \frac{c}{8\pi}\,\mathrm{Re}\left(\vec{E} \times \vec{H}^*\right), \tag{1.54}$$

where the average mark over $\vec{S}$ is omitted. Integral of div $\vec{S}$ over any volume is the energy decrease in this volume. For infinitely small volume the decrease equals the density of the average losses, that is

$$\mathrm{div}\,\vec{S} = -(P_E + P_H), \tag{1.55}$$

where P_H is the average losses describing the decrease of the magnetic field.

We express div $\vec{S}$ in terms of $\vec{E}$ and $\vec{H}$ with use of equations (1.29). Applying identity (1.45), substituting rot $\vec{E}$ and rot $\vec{H}$ into (1.45) and taking the real part, we obtain

$$\mathrm{div}\,\vec{S} = \frac{\omega}{8\pi}\left(\varepsilon''\left|\vec{E}\right|^2 + \mu''\left|\vec{H}\right|^2\right) - \frac{1}{2}\,\mathrm{Re}\left(\vec{j}^{\,\mathrm{ext}*} \cdot \vec{E}\right). \tag{1.56}$$

According to (1.55), it follows from this equality that formula (1.53) and the similar expression

$$P_H = -\frac{\omega}{8\pi}\mu''|\vec{H}|^2 \tag{1.57}$$

are valid for any losses proportional to the squared electric and magnetic fields.

Media exist for which $\varepsilon'' > 0$, that is, "losses" are negative. In such media not the absorption, but generation of the electromagnetic energy occurs, which is proportional to the squared electromagnetic field. Formula (1.53) describes not the process of the field energy decreasing, but the inverse process of its increasing. The physical sense of the above phenomenon is that the population of quantum levels is inverse in such media: the higher levels are more populated than the lower ones. Sometimes such a situation is formally called as "negative absolute temperature." It is clear that its creation requires energy consumption, so-called pumping. The processes occurring in coherent sources of the electromagnetic oscillations are based on the energy generation caused by the field influence.

In chiral media, the Maxwell equations must be written not in the form (1.29), but in another one with $\vec{D}$ and $\vec{B}$ expressed in terms of $\vec{E}$ and $\vec{H}$ by (1.28). However, it is easy to prove that formula (1.56) remains valid, that is, the losses are expressed in terms of fields by the same formulas (1.53), (1.57) as for nonchiral media. For proving this statement the fact is used that the coefficients of the cross terms in (1.28) coincide (with accuracy to the sign). As was mentioned, this follows from the fact that the chiral media are, in general, reciprocal. In order that the expressions for P_E and P_H were the same in the chiral media as those in the nonchiral ones, it is sufficient that the real parts of the coefficients coincide.

1.2.11
The dispersion relations

The real and imaginary parts of the permittivity $\varepsilon = \varepsilon' + i\varepsilon''$ are functions of the frequency ω. At any fixed frequency they can be independent of each other. However, in the whole frequency interval $0 < \omega < \infty$ the functions $\varepsilon'(\omega)$ and $\varepsilon''(\omega)$ are dependent. Setting one function, we thereby set the other.

We show in an example that the inconsistent setting of both the functions may lead to a paradox. It is clear that the time-dependent impulse, containing the oscillations of the wide frequency band, should be considered instead of the harmonic oscillations.

Assume that a material exists for which $\varepsilon'(\omega) = const$ at any frequency, and $\varepsilon''(\omega)$ is zero at almost all frequencies except for a narrow band near a frequency ω_0. An impulse (the field equal to zero at $t < 0$) falls onto a thin plate of such material at $t = 0$. This field can be expanded in the Fourier integral, that is, written as an integral of the harmonic fields or, more precisely, an integral of the factor $\exp(i\omega t)$ multiplied by a function of ω, and taken over ω. The form of the function is not essential here. The fact that the integral equals zero at $t < 0$ means that its components corresponding to different frequencies compensate each other at $t < 0$. After passing the plate, the components with frequencies not close to ω_0 obtain the same phase factor because they have the same values of ε' and their amplitudes do not change because $\varepsilon'' = 0$ for them. The amplitudes of the other components (with frequencies close to ω_0) for which $\varepsilon'' \neq 0$ decrease. The conditions of the mutual compensation of all harmonics at $t < 0$ are violated. This fact means the same that, besides the initial impulse equal to zero at $t < 0$, a narrow group of harmonics appears after passing the plate, equal (with opposite sign) to the values by which the harmonics of these frequencies decreased. These harmonics exist at any t, and they do not compensate each other at $t < 0$. If the material with the above-mentioned properties of the functions $\varepsilon'(\omega)$ and $\varepsilon''(\omega)$ existed, then the field outside the plate would appear before the impulse incidence onto the plate. The consequence would precede the cause; the *causality principle* would be violated. The paradox also takes place if assuming that $\varepsilon'(\omega) = const, \varepsilon''(\omega) > 0$ in a narrow frequency band and $\varepsilon''(\omega) = 0$ outside.

We outline the mathematical technique allowing us to deduce the connection between the two functions $\varepsilon'(\omega), \varepsilon''(\omega)$ of the frequency using the causality principle. This connection is called the *dispersion relation*. It holds for any material medium and it should be considered while creating composites. The requirements imposed on $\varepsilon'(\omega)$ and $\varepsilon''(\omega)$ should be subjected to the relation.

The induction $\vec{D}(t)$ is the medium response to the field $\vec{E}(t)$. It may depend on the field value at the given time moment and before it, but it cannot depend on the field at the succeeding moments. This requirement can be expressed in the form

$$\vec{D}(t) = \vec{E}(t) + \int_0^\infty \vec{E}(t - \tau)f(\tau)\,d\tau. \tag{1.58}$$

Separation of the first term on the right-hand side is not important. It only simplifies the formulas given below. The essential is that only positive values of the integration variable τ participate in the integral, so that only the field values at the time moments less than t are considered. The function $f(\tau)$ may be arbitrary but such that the integral converges.

For nonharmonic oscillations $\vec{D}(t)$ can be expressed as

$$\vec{D}(t) = \int_{-\infty}^\infty \vec{D}(\omega)e^{i\omega t}\,d\omega \tag{1.59}$$

(for simplicity, we denote the function and its Fourier spectrum by the same symbol). The field $\vec{E}(t)$ can be written in a similar form. Inverting (1.59) gives

$$\vec{D}(\omega) = \frac{1}{2\pi} \int\limits_{-\infty}^{\infty} \vec{D}(t) e^{-i\omega t} \, dt. \tag{1.60}$$

A similar formula exists also for $\vec{E}(\omega)$. Multiply both sides of equality (1.59) by $\exp(-i\widetilde{\omega}t)$ and integrate from $t = -\infty$ to $t = +\infty$. Here $\widetilde{\omega}$ is any frequency. Interchanging the integration order over τ and t on the right-hand side, introducing a new integration variable $t' = t - \tau, dt' = dt$, instead of t in the inner integral, and using (1.60) and similar expression for $\vec{E}(\widetilde{\omega})$, we obtain

$$\vec{D}(\widetilde{\omega}) = \vec{E}(\widetilde{\omega}) + \vec{E}(\widetilde{\omega}) \int\limits_{\tau=0}^{\infty} f(\tau) e^{-i\widetilde{\omega}\tau} \, d\tau. \tag{1.61}$$

From the function $\varepsilon(\omega)$ definition it follows that

$$\vec{D}(\widetilde{\omega}) = \varepsilon(\widetilde{\omega}) \vec{E}(\widetilde{\omega}). \tag{1.62}$$

Hence, there exists the integral representation of $\varepsilon(\omega)$:

$$\varepsilon(\omega) = 1 + \int\limits_{0}^{\infty} f(\tau) e^{-i\omega\tau} \, d\tau. \tag{1.63}$$

Due to the causality principle, the integration in this formula is made over only $\tau \geq 0$. We determine the mathematical properties of the functions represented in the above form. If ω is replaced by the complex variable $\Omega = \omega' + i\omega''$ in (1.63), then the multiplier $\exp(i\omega''\tau)$ with $\tau \geq 0$ appears under the integral. At $\omega'' < 0$ this multiplier is smaller than unity and it tends to zero as $\omega'' \to -\infty$. Hence, for every value of the complex variable Ω lying in its lower half-plane, the integral in (1.63) converges and tends to zero on the infinitely remote half-circle. Thus, $\varepsilon(\omega) - 1$ is a function analytical in the lower half-plane ($\omega'' < 0$) of the complex frequency Ω including its real axis, and vanishing at $\omega'' \to -\infty$.

The value of the function having such properties at any point of the real axis (at $\omega'' = 0$) can be expressed in the form of the Cauchy integral (as the principal value) of its values on the whole real axis. Separating the real and imaginary parts in the obtained equality, we get the sought formulas, that is, the function $\varepsilon'(\omega) - 1$ as the integral of $\varepsilon''(\omega)$ and the function $\varepsilon''(\omega)$ as the integral of $\varepsilon'(\omega) - 1$. For instance, at any fixed frequency ω_0,

$$\varepsilon'(\widehat{\omega}_0) - 1 = \frac{2}{\pi} \lim_{\alpha \to +0} \left\{ \int\limits_{0}^{\omega_0-\alpha} + \int\limits_{\omega_0+\alpha}^{\infty} \right\} \frac{\omega \varepsilon''(\omega)}{\omega^2 - \omega_0^2} \, d\omega. \tag{1.64}$$

In the above example leading to a paradox, it is assumed that $\varepsilon''(\omega)$ equals zero not for all frequencies, and $\varepsilon'(\omega)$ is the same for all frequencies, which contradicts (1.64).

Equation (1.64) and the similar expression for $\varepsilon''(\omega)$ as an integral of $\varepsilon'(\omega)^{-1}$ are known as the principal *Kramers–Kronig equations*. It must be kept in mind that in the theoretical physics books where the time dependence is taken in the form of $\exp(-i\omega t)$, the function $\varepsilon(\Omega)$ is analytic and vanishes in the upper half-plane of the complex frequency Ω.

1.3
Idealized objects

In the high-frequency electrodynamics just like in any other physical theory, a number of idealizations is accepted, that is, some mathematical models are introduced having properties close to those of the real objects. It is easier to solve problems for such models than those for the real objects. Solutions of the idealized problems are assumed to be close to those of the real problems taking into account the details dropped at the idealization. Use of the idealized objects is possible if these details are not essential. However, this condition cannot always be formalized. As in any other physical theory, the introduction of idealized objects is highly based on the intuition.

1.3.1
Interface of two media

It follows from the Maxwell equations and the inductions finiteness that the fields $\vec{E}$ and $\vec{H}$ are continuous (even differentiable) at points where ε, μ, and $\varkappa$ are continuous. A surface on which ε and μ are discontinuous, that is, an interface of the two different media, can be considered as a limit image of the thin layer with very large gradients of ε and μ. On such a surface, some components of the fields $\vec{E}$ and $\vec{H}$ are discontinuous, that is, they have different values on the surface sides. Conditions for tangential and normal components are different. Conditions for tangential components follow from relations on both sides of the thin layer. As is known, these relations are obtained from the Maxwell equations written in the integral form. According to the *Stokes theorem*, the integral of the tangential component H_{t_1} taken over a closed contour passing through both the sides of the thin layer is equal to the integral of $(\mathrm{rot}\,\vec{H})_{t_2}$ over the area inside the contour (shaded area in Fig. 1.1). Equation (1.14a) implies that it is equal to the integral of $ikD_{t_2} + (4\pi/c)\,j_{t_2}$. Here t_1 and t_2 are two tangential unit vectors perpendicular to each other. Since D_{t_2} and j_{t_2} are finite (we will return to this statement), the surface integral is proportional to the thickness of the layer and has the same order. Hence, the difference of the components H_{t_1} has also the same order. If the thickness of the transition layer tends to zero, then this difference also does. This conclusion does not require for the gradients of ε and μ to be finite. It also holds for models with interface, in which the gradients are infinite and the result does not depend on the way of passing to the discontinuous functions ε and μ. The tangential components of the field $\vec{H}$ (as well as of the field $\vec{E}$) are equal on both the sides of the interface

$$H_t^+ - H_t^- = 0, \tag{1.65a}$$
$$E_t^+ - E_t^- = 0. \tag{1.65b}$$

Here t is any tangential unit vector, that is, each equation contains two conditions. The signs $+$ and $-$ in the superscript refer to the different sides of the interface.

The normal components of the fields $\vec{E}$ and $\vec{H}$ may be discontinuous. The normal components of the inductions $\vec{D}$ and $\vec{B}$ are continuous

$$D_N^+ - D_N^- = 0, \tag{1.66a}$$
$$B_N^+ - B_N^- = 0. \tag{1.66b}$$

For instance, if μ is the same and only permittivity ε is different on the interface sides, then five of six components of the fields $\vec{E}$ and $\vec{H}$ are continuous and only the component E_N is discontinuous.

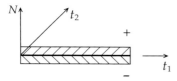

Fig. 1.1 Illustration to the Stokes theorem

Formulas (1.66) do not impose any additional conditions; they are a consequence of formulas (1.65): if conditions (1.65) are valid, then (1.66) are valid, too. For instance, D_N^+ can be expressed by the integral of rot $\vec{D}^+$ taken over a closed contour lying on the interface, that is, by the integral of H_t^+. Since $H_t^+ = H_t^-$, then $D_N^+ = D_N^-$.

Note that in electrostatics, where no magnetic field exists, condition (1.66a) does not depend on (1.65b). It is seen after accurate transferring to the electrostatic equation. In electrostatics both conditions (1.65b) and (1.66a) must be fulfilled.

1.3.2
The impedance (one-side) boundary conditions

An often used idealization is the conditions connecting the tangential components of $\vec{E}$ and $\vec{H}$ on one side of the surface. The fields on the other side are
not involved in such one-side conditions in contrast to the two-side conditions
(1.65). The one-side conditions are the boundary conditions for the fields in a
domain bounded by the surface.

In the simplest case, the *impedance conditions* have the form

$$E_{t_1} = -wH_{t_2}, \qquad E_{t_2} = wH_{t_1}. \tag{1.67}$$

The factor w is called the *surface impedance*. It does not depend on the fields. It
characterizes the medium lying on the other side on the surface. The unit vectors $\vec{t}_1$, $\vec{t}_2$ together with the normal $\vec{N}$ directed outward the domain bounded
by the surface on which (1.67) holds, make up a right-hand triple. In the local
Cartesian coordinate system (x, y, z), in which the body occupies the domain
$z < 0$, we get $t_1 = x$, $t_2 = y$, $N = z$.

In general, the impedance w is complex $w = w' + iw''$. According to (1.54)
and (1.67), the period-average flux of the energy outgoing from the domain
through the unit area of the boundary equals

$$S_N = -\frac{c}{8\pi} w' \left| \vec{H}_t \right|^2 . \tag{1.68}$$

If $w' > 0$, then $S_N < 0$, that is, the energy goes into the domain.

If the surface is anisotropic, that is, its properties are different in different directions, then w is not a scalar, but a two-dimensional tensor. The impedance
boundary conditions (1.67) keep the simple form only if $\vec{t}_1$ and $\vec{t}_2$ are directed
along the principal axes of the tensor. In this case, the coefficients in both
equalities of (1.67) are equal to the principal values of the tensor and, in general, they are not equal to each other.

1.3.3
Skin layer

Conditions (1.67) are fully valid only in the case when the fields in the given
domain do not depend on the fields on the opposite side of the boundary. The
more precisely this requirement holds, the larger any parameter describing
the medium on the opposite side is. An example of such a medium is the
homogeneous material with

$$|\varepsilon\mu| \gg 1. \tag{1.69}$$

The stronger this inequality, the more accurately conditions (1.67) hold on the
boundary of such a medium.

Introduce the Cartesian coordinate system originating on the surface, in which the axes x and y are directed along the unit vectors $\vec{t}_1$ and $\vec{t}_2$, respectively, and z-axis is normal to the surface and directed toward the medium with property (1.69) (Fig. 1.2). From the existence of boundary conditions (1.65) it follows that in this coordinate system, the fields in the tangential direction, that is, along the axes x and y, vary just as in the main domain. The velocity of varying in the z-direction is significantly larger. This follows from the wave equations (1.41), since under condition (1.69), the action of the operator Δ (see 1.39) is equivalent to multiplication by $-k^2\varepsilon\mu$ only if at least one of the derivatives is large. Hence, $\partial^2/\partial z^2 \approx -k^2\varepsilon\mu$, or

$$\frac{\partial}{\partial z} = \pm ik\sqrt{\varepsilon\mu}; \tag{1.70a}$$

$$\frac{\partial}{\partial z} \gg \frac{\partial}{\partial x}, \quad \frac{\partial}{\partial z} \gg \frac{\partial}{\partial y}. \tag{1.70b}$$

In this approximation the wave equation becomes an ordinary differential equation

$$\frac{d^2u}{dz^2} + k^2\varepsilon\mu u = 0, \tag{1.71}$$

where u is any Cartesian component of the field. Its solution is a linear combination of the two functions $\exp(\pm ik\sqrt{\varepsilon\mu}z)$. Only the second function describes the wave going deep into the medium and decreasing due to $\varepsilon'' < 0$ in this direction $z \to -\infty$. Therefore, all the components of $\vec{E}$ and $\vec{H}$ depend on z as

$$\exp(-ik\sqrt{\varepsilon\mu}z). \tag{1.72}$$

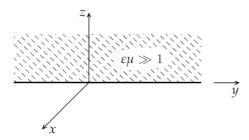

Fig. 1.2 Interface of two media

We derive the relations between different field components, which exist under conditions (1.72), (1.69), (1.70b). At $\vec{j}^{\text{ext}} = 0$, from (1.31) it follows that the ratio of E_z to E_x has the same order as that of $\partial/\partial x$ to ∂/dz, that is

$$\frac{E_z}{E_x} = O\left(\frac{1}{|\sqrt{\varepsilon\mu}|}\right), \quad \frac{H_z}{H_x} = O\left(\frac{1}{|\sqrt{\varepsilon\mu}|}\right), \tag{1.73}$$

where the second equality is written by analogy. Dropping in (1.31) the terms $\partial E_z/\partial x$, $\partial E_z/\partial y$, and so on, and considering (1.72), we obtain the following relation between the tangential components of the fields:

$$E_x = -\sqrt{\frac{\mu}{\varepsilon}}H_y, \qquad E_y = \sqrt{\frac{\mu}{\varepsilon}}H_x. \tag{1.74}$$

The tangential components of the fields coincide on both the sides of the surface; therefore, the above relations are valid on the other side, as well. Thus, if (1.69) holds in some medium, then the impedance conditions (1.67) are valid on its boundary, and the impedance is equal to

$$w = \sqrt{\frac{\mu}{\varepsilon}}. \tag{1.75}$$

As usual, $|\varepsilon|$ is large because such is the conductivity σ, and $|\varepsilon''| \gg \varepsilon'$. It means that the conduction currents $\sigma\vec{E}$ are much larger than the displacement currents $i\omega\vec{E}$. For simplicity, we ignore ε' in comparison with ε'' in (1.30), that is, we assume that $\varepsilon = -4\pi i\sigma/\omega$. Then the exponential multiplier (1.72) can be written as

$$\exp\left[-(1+i)\frac{z}{d}\right], \tag{1.76}$$

where the quantity d having the dimension of length equals

$$d = \sqrt{\frac{c}{2\pi k\mu\sigma}}. \tag{1.77}$$

The quantity d describes the velocity of the amplitude decreasing in the medium in the direction normal to the interface. It is called the *thickness of the skin layer* and is the main characteristics of the well-conducting material placed into the high-frequency field. The thickness d is zero for a conductor with infinite conductivity ($\sigma = \infty$). With increasing frequency, d decreases in inverse proportion to its square root ($d \sim \omega^{-1/2}$). The thickness of the skin layer is not large. For instance, $d = 0.4 \times 10^{-6}$ m for copper ($\sigma = 5 \times 10^{17}$ s^{-1}) at the frequency of 30 GHz ($\lambda = 0.01$ m). The skin layer is a thin layer on the good conductor surface, under which the high-frequency field almost does not penetrate. For a good conductor (with $\varepsilon \approx -4\pi i\sigma/\omega$) the impedance is complex and proportional to the thickness of the skin layer:

$$w = \frac{1+i}{2}\mu kd. \tag{1.78}$$

1.3.4
Ideal conductor

The medium with $\sigma = \infty$ is called the *ideal conductor*. The impedance w is zero on its surface, and the boundary conditions (1.67) become

$$E_{t_1} = 0, \qquad E_{t_2} = 0 \tag{1.79}$$

The field does not penetrate inside the ideal conductor, but from this it does not follow that the tangential magnetic field equals zero on its surface. At any depth and $\sigma = \infty$, $d = 0$, the field is detached from the exterior surface by the skin layer on which the tangential magnetic field has a jump. The current flowing in the skin layer remains finite if passing $\sigma \to \infty$, $d \to 0$. According to (1.74), the electric field tends to zero in the skin layer if $\sigma \to \infty$ as $\sigma^{-1/2}$. The conduction current density σE tends to infinity as $\sigma^{1/2}$. The thickness of the skin layer has the order $\sigma^{-1/2}$ (1.77), and the current flowing in the skin layer at $\sigma \to \infty$ does not depend on σ and remains finite.

The current with infinite volume density but with finite surface one flows on the ideal conductor surface. Denote the two-dimensional vector of the current volume density by $\vec{I}$, so that $I_t = \lim_{\sigma \to \infty}(j_t \cdot d)$. It can be expressed by the tangential magnetic field. From the integral form of the Maxwell equations it follows that

$$I_x = -\frac{c}{4\pi}H_y, \qquad I_y = \frac{c}{4\pi}H_x. \tag{1.80}$$

The surface current density is equal (with accuracy to the multiplier $c/4\pi$) to the magnetic field, and perpendicular to it on the surface. The field $\vec{H}$ can be found after the problem with boundary conditions (1.79) is solved in the domain adjacent to the ideal conductor. The surface density of the current (1.80) can be determined by H_t.

The normal component B_N of the magnetic induction, as well as the tangential component of the electric field, equals zero on the surface of the ideal conductor. The normal component D_N of the electric induction, together with the tangential component of the magnetic field, differs from zero. According to (1.14a) and (1.80), D_N is proportional to the two-dimensional divergence of the surface current:

$$D_N = \frac{4\pi i}{\omega}\left(\frac{\partial I_x}{\partial x} + \frac{\partial I_y}{\partial y}\right). \tag{1.81}$$

Since the quantity in the brackets equals $-i\omega\rho_{surf}$, where ρ_{surf} is the surface density of the charge (1.18), equality (1.81) means that the surface charge density appears on the surface of the ideal conductor.

The above results are also valid for material with large displacement currents instead of the conduction currents, that is, for material in which $\varepsilon' \gg 1$.

For simplicity, we assume that $\varepsilon'' = 0$, that is, the conductivity of the material is zero, and there are no other losses in it. In this case the field is not concentrated in the thin layer on the surface; it exists in the entire volume, but it changes the sign fast with the z-coordinate increasing. According to (1.72), the thickness of the layer, in which the field is of the same sign, has the order $1/(k\sqrt{\varepsilon\mu})$, being small. The tangential components of the induction $\vec{D}$ are large, $D_t = \varepsilon E_t$, that is, they have the order $\sqrt{\varepsilon\mu}H_t$. The product of D_t and the thickness is finite; it does not depend on ε if ε is large. Therefore, the variation of the magnetic field H_t along the thickness of the layer is finite. This fact is analogous to the finiteness of the magnetic field variation over the thickness of the skin layer at $\sigma \gg 1$. The assertion "the impedance conditions (1.67) taking the form (1.79) at the limit $|\varepsilon| \to \infty$, are valid on the surface of the object with $|\varepsilon| \gg 1$," does not depend on the relation between ε' and ε''. It is true either for the metallic objects or for the objects with large real permittivity.

1.3.5
Singularities of fields near the edge or vertex

One of the idealized objects is also an edge which is a fracture line of the two media interface (Fig. 1.3(a)). One of the surface curvatures is infinite on this line. As will be shown below, some of the field components become infinite at the edge, that is, they have a singularity there. Behavior of the fields when becoming infinite at some point is a local property, that is, it only depends on the medium structure near the point. When approaching the edge, the field increases with the velocity depending only on the angle between planes, material of the media separated by these planes, and polarization of the field near the edge. Therefore, the field structure at the edge can be investigated with use of a two-dimensional model, when considering the edge to be a straight line, and the fields to be independent of the coordinate parallel to this line.

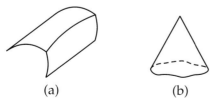

(a)   (b)

Fig. 1.3 Geometry of the edge and vertex

The fields become infinite not only on the edge but also on the linear source. The distinction between the fields structure near the edge and that in the source neighborhood consists in the fact that the density of the electromagnetic field energy is integrable near the edge, that is

$$\int \left| \vec{E} \right|^2 dV < \infty, \tag{1.82a}$$

$$\int \left| \vec{H} \right|^2 dV < \infty, \tag{1.82b}$$

whereas in the source neighborhood these conditions are violated. The integrals are taken over an arbitrary volume containing the edge segment, or the source.

Conditions (1.82) are necessary and sufficient for the line on which some components of the fields become infinite, not to be a linear source, but an edge. Below we show that if these conditions hold, then the energy flux out of the line is zero. Note that in contrast to (1.82) the equality to zero of the energy flux is only necessary (but not sufficient) for the line to be an edge. In certain cases it is valid for a linear source, too. Although the linear source does not radiate in such cases, the field is so large near it that condition (1.82) violates.

Assuming that (1.82) holds, we find the field structure near the edge. We restrict ourselves to the case of a metallic wedge, that is, a wedge on the edges of which condition (1.79) is valid (Fig.1.4). Align the z-axis of the cylindrical coordinate system along the wedge edge. The angle φ is measured from a wedge face. Denote the angle between the wedge faces by α, and its complement to 2π by $\tau\pi$, that is, $\tau = (2\pi - \alpha)/\pi$. Then the equations of the wedge faces are $\varphi = 0$ and $\varphi = \tau\pi$, respectively. For the half-plane, $\alpha = 0$, $\tau = 2$; for the wedge, $\alpha > 0$, $\tau < 2$.

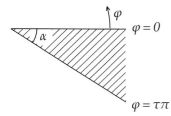

Fig. 1.4 Cross-section of the wedge

For the fields independent of z, the equation system (1.32) with $\vec{j}^{\text{ext}} = 0$ split into two independent systems. One of them contains the components E_z, H_r, H_φ, and the other – H_z, E_r, E_φ. If $\partial/\partial z \equiv 0$ and $H_z \equiv 0$ for the field in which the wedge is placed, then only the components E_z, H_r, H_φ appear (the *electric polarization*). The second component triple appears for the *magnetic polarization*.

For the electric polarization,

$$H_r = -\frac{1}{ik\mu}\frac{\partial E_z}{r\partial\varphi}, \tag{1.83a}$$

$$H_\varphi = \frac{1}{ik\mu}\frac{\partial E_z}{\partial r}, \tag{1.83b}$$

and for the magnetic one,

$$E_r = \frac{1}{ik\varepsilon}\frac{\partial H_z}{r\partial\varphi}, \tag{1.84a}$$

$$E_\varphi = -\frac{1}{ik\varepsilon}\frac{\partial H_z}{\partial r}. \tag{1.84b}$$

The components E_z and H_z satisfy the same wave equation (1.41), which at $\partial/\partial z \equiv 0$ obtains the form

$$\frac{\partial^2 u}{\partial r^2} + \frac{1}{r}\frac{\partial u}{\partial z} + \frac{1}{r^2}\frac{\partial^2 u}{\partial\varphi^2} + k^2\varepsilon\mu u = 0, \tag{1.85}$$

where u is one of the components E_z, H_z. In the edge neighborhood (i.e., at $kr \ll 1$), the last summand in (1.85) can be neglected at $kr \ll 1$. Then the functions $\Phi(r)\cos(n\varphi)$, $\Phi(r)\sin(n\varphi)$ satisfy equation (1.85) with the conditions $E_z = 0$ (electric polarization) and $E_r = 0$ (magnetic polarization) at $\varphi = 0$, $\varphi = \tau\pi,$.

Begin from the case of the electric polarization. In general, E_z can be a superposition of the terms

$$r^n\sin(n\varphi), \qquad r^{-n}\sin(n\varphi), \tag{1.86}$$

where $n = 1/\tau, 2/\tau, \ldots$ However, condition (1.82b) implies that near the edge, E_z contains only the positive powers of r, with $r^{1/\tau}$ to be the lowest one. Indeed, if E_z contained terms proportional to r^{-n}, then, according to (1.83), H_r and H_φ would contain terms proportional to r^{-n-1}, and the integral (1.82b) would have a form $\int_0 r^{-(2n+1)}\,dr$ and diverge. The magnetic field has a singularity $r^{1/\tau-1}$ at $\tau > 1$, that is, for acute angles ($\alpha < \pi$). The singularity order is higher for the half-plane ($\tau = 2$) than for the wedge with $\alpha > 0$. The magnetic field has a singularity $1/\sqrt{r}$ near the half-plane edge. The current I_z proportional to H_z has a singularity as well. On both planes near the edge, the current flows in the same direction parallel to the edge; its surface density becomes infinite when approaching the edge.

The energy flux out of the edge is zero at any α. Since $H_\varphi^* \sim \partial E_z^*/\partial r$, the energy flux (1.54) is proportional to the imaginary part of the integral $\int_0^{2\pi} E_z \cdot \partial E_z^*/\partial r \cdot r\,d\varphi$. The integrand involves only the positive powers of r;

therefore, the integral tends to zero as $r \to 0$. The energy flux equals the integral of $\mathrm{Re}(\mathrm{div}\,\vec{S})$ over the volume inside the surface. According to (1.56), $\mathrm{div}\,\vec{S} = 0$ and therefore the integral cannot depend on r; hence it equals zero.

Thus, if the field does not have a singularity in the domain or its singularity is so weak that condition (1.82) holds, then the energy flux out of the domain equals zero. We show that this assertion is also valid in the magnetic polarization case, that is, at $E_z = 0$. Since the component H_z consists of the terms $r^{\pm n} \cos n\varphi$, the boundary condition $E_z = 0$ at $\varphi = 0$ and $\varphi = \tau\pi$ holds for the same values of n ($n = 1/\tau, 2/\tau, \ldots$) as for the electric polarization. For condition (1.82a) to hold, H_z must not contain the terms r^{-n}. When approaching the edge, a singularity appears in the electromagnetic field (1.84), which depends on r as $r^{1/\tau-1}$ at small r. For the half-plane $\tau = 2$, this dependence is the strongest when the field is proportional to $1/\sqrt{r}$. Then the integral in (1.82a) converges, and the energy flux outgoing from the edge is equal to zero.

The surface current flows perpendicularly to the edge, $I_r \neq 0$, for this polarization. The current has no singularities; the function I_r is finite. The surface density ρ_{surf} of the charge proportional to E_φ at $\varphi = 0$ and $\varphi = \tau\pi$ (i.e., to the normal component of the electric field) has a singularity. According to (1.81), ρ_{surf} is proportional to the two-dimensional divergence of the surface current density. Since I_r is proportional to $\sqrt{r}$ for the half-plane , ρ_{surf} is proportional to $1/\sqrt{r}$.

The simple formulas for the field near the edge exist only in the case of the edges with the ideal-conducting faces. If the faces are impedance or they separate two dielectrics, then the explicit expressions for the field exist only in the electrostatic approximation, analogous to (1.86). Just as for the metallic wedge, from these formulas it follows that if the conditions (1.82) hold, then the field can have only such singularity, at which there is no energy flux out of the edge. In this case the singularity can be even a little stronger than for metallic faces. However, for the edge with arbitrary boundary conditions on its faces, conditions (1.82) are necessary and sufficient for the field singularities in the edge neighborhood to be weaker than those near the linear source.

Similar result is valid for another idealized model, namely, for a pyramid or cone vertex (Fig. 1.3(b)), that is for a point in which both the surface curvatures are infinite. At this point the fields are also singular; some of their components become infinite. At a singular point of another kind, namely, at the point source, the fields are infinite, as well. Similarly as for the linear source or edge, the same condition (1.82) holds for the vertex and does not hold for the point source.

We give another, more formal, proof of the fact that if in the neighborhood of a point or line the fields are finite or have singularities weak enough for satisfying condition (1.82), then the energy flux out of this point or line is zero. This proof is shorter, but it does not allow us to determine nature of the surface currents and charges.

It follows from the Maxwell equations and identity (1.45) that

$$\text{div}\left(\vec{E} \times \vec{H}^*\right) = ik\left(\left|\vec{E}\right|^2 - \left|\vec{H}\right|^2\right) \tag{1.87}$$

at $j^{\text{ext}} = 0$, $\varepsilon = 1$, $\mu = 1$. If the fields are finite everywhere, then this equality is valid at any point and it can be integrated over any volume. Transforming the integral on the left-hand side and taking the real part of the obtained equality, we get $\int S_N \, dS = 0$, where $\vec{S}$ is the Poynting vector and the integral is taken over a closed surface having no singularities inside. The surface interior surrounds a singular point (vertex) or singular line (edge). Conditions (1.82) imply that the integral has a limit when the surface interior is contracting to a point or line. The integral over the exterior part of the surface remains zero, which means that the energy flux out of the vertex or edge equals zero.

1.3.6
The line current and the point current

Above we have considered the additional conditions which are required in the formulation of electromagnetic problems in the case when the idealized objects are introduced into the theory. Since the Maxwell equations are not satisfied on such objects, the fields can have singularities or discontinuities. On the surfaces, where the medium properties are discontinuous, the normal components can also be discontinuous, and only the tangential components remain to be continuous. For instance, on the impedance surfaces the fields are subjected to the one-side boundary conditions. At points and on lines, where the media interface has infinite curvature, the fields tend to infinity but not so fast for the energy density to be nonintegrable. The radiation condition (1.33) implies that if the field exists at the infinite point, then it must have a concrete structure in the infinity "neighborhood."

In each of the above cases, the local conditions imposed on the fields near the idealized objects exclude the existence of the extrinsic currents on them. The continuity of H_t on the interface implies that there is no extrinsic surface current on it. Recall that the current flowing on the surface of the ideal conductor is an induced one. The radiation conditions mean that there is also no extrinsic current at the infinite point. The conditions near the edge or vertex imply that there are neither line nor point extrinsic currents there.

In this subsection we consider the field structure near the true extrinsic currents. If the volume density $\vec{j}^{\text{ext}}$ does not become infinite, then the fields have no singularities. In the neighborhood of both the *line extrinsic current* and the *point extrinsic* one, the fields tend to infinity so fast that the conditions (1.82) do not hold. Nature of the infinity does not depend on the existence of the other objects; it is local for the edges and vertices.

In the case of the line extrinsic current, the fields can be found from the problem about a straight-line current of a constant value. Such a current is an idealized model. The volume density j_z^{ext} is infinite; it exists in an infinitely thin straight-line cylinder. The product of j_z^{ext} and the cross-section area of the cylinder is finite. It is called the *line current density* and has the only component I_z, independent of the cross-section shape.

The fields are infinite near the line current. They are found from the Maxwell equations in which $\vec{j_z}$ is proportional to $\delta(r)$, so that the integral of $\vec{j_z}$ over any part of the surface $z = const$ containing the point $r = 0$, is finite and equal to I_z^{ext}. The fields depend neither on z nor on the azimuthal coordinate φ, and have only the components H_φ and E_z. In the domain of interest, where $kr \ll 1$, the field E_z is much smaller than H_φ; it has a singularity so weak that the electric energy density is integrable. The Maxwell equations are reduced to the equation for the direct currents $\mathrm{rot}\,\vec{H} = (4\pi/c)\vec{j}^{ext}$. Integrating it over a circumference of a small radius a, and using the definition of I^{ext}, we get

$$H_\varphi = \frac{2}{c} \cdot \frac{1}{a} I_z^{ext}. \tag{1.88}$$

The magnetic energy density is proportional to a^{-2}, the integrand has the order a^{-1}, and the integral diverges at $a \to 0$.

The question about the energy flux from the line current is not as simple as in the case of the edge. In general, this flux is not zero. It cannot be determined by the near field and, therefore, it cannot be calculated using values of the fields at $kr \ll 1$. The flux depends on the fields created by other sources at the place where the line current is located. The integral of the vector $\vec{E} \times \vec{H}^*$ flux through the surface $z = a$ is proportional to $\ln ka$; it tends to infinity as $ka \to 0$. However, at $kr \ll 1$ the vectors $\vec{E}$ and $\vec{H}$ are in-quadrature, that is, the real part of their product equals zero. The energy flux is determined by the next terms of the expansion of these fields in powers of kr, that is, by the terms having no singularities.

If the line current is located in vacuum and there are no other objects in its neighborhood, then the energy flux out of it can be calculated by expressing $H_\varphi(r)$ and $E_z(r)$ by exact formulas valid for the direct straight-line current at all r. These formulas contain the Hankel functions. Omitting the derivations, we only point out that the flux per unit length equals $\pi k/(2c) \cdot I^2$.

If there are any other objects in the field of the extrinsic current, then the energy flux created by this current changes, so that the factor at I^2 differs from $\pi k/(2c)$. The fields of currents, induced on those objects, interfere with the field of the line current. For instance, if the current is located near a mirror, then the induced field is the field of the current reflected in this mirror. Since in the neighborhood of the line current the field of other currents is finite, the type of the field singularity does not change. However, in this case, at the same value of J^{ext} the radiated energy can be larger or smaller than in the case when the other fields are absent, or even equal to zero.

To determine the energy flux, given up by a given extrinsic current located in the field of other sources, it is sufficient to know the value of this field in the place where the current is located. This statement follows from formula (1.56), according to which the energy flux from the domain V through its surface S equals

$$\int_S S_N \, dS = -\frac{1}{2} \int_V \text{Re} \left(\vec{j}^{\text{ext}*} \cdot \vec{E} \right) dV \tag{1.89}$$

if there are no losses in the medium. The immediate use of this formula for the line current is practically impossible, since $\vec{E}$ has singularities at the points where $\vec{j}^{\text{ext}}$ is infinite, but the volume of this domain is infinitesimal. Divide the field $\vec{E}$ into the field $\vec{E}_0$ of the current $\vec{j}^{\text{ext}}$ in vacuum and field $\vec{e}$ of all other currents, $\vec{E} = \vec{E}_0 + \vec{e}$. Since the field $\vec{E}$ participates in (1.85) linearly, the expression for the flux is divided into two terms. The first of them is the energy flux created by the field of the current J^{ext} itself in vacuum. The second one is easily calculated in any concrete problem, since the field $\vec{e}$ has no singularities near the line current. The first term was given above for the direct straight-line current $\vec{j}^{\text{ext}}$, so that

$$\int_S S_N \, dS = \frac{\pi k}{2c} \left| \vec{I}^{\text{ext}} \right|^2 - \frac{1}{2} \text{Re} \int_V \vec{j}^{\text{ext}*} \cdot \vec{e} \, dV. \tag{1.90}$$

In the second term the field $\vec{e}$ of the induced currents can be taken out of the integral, so that

$$\int_S S_N \, dS = \frac{\pi k}{2c} \left| \vec{I}^{\text{ext}} \right|^2 - \frac{1}{2} \text{Re} \left[\vec{j}^{\text{ext}*} \cdot e_z(0) \right]. \tag{1.91}$$

For a nonstraight-line current the first term in (1.91) is different, the second one is an integral over the line length, and the flux should be calculated per whole line, not per its unit length as in the last two formulas. However, the structure of the formula for the flux of the energy radiated by the line current in the presence of other fields is kept.

Finally, we consider the last idealized model introduced into the theory, namely, the *point current* (elementary electric dipole). Formally, we have already used this notion when considering the reciprocity conditions.

Let the volume density $\vec{j}^{\text{ext}}$ differ from zero only in some domain, and $\vec{j}^{\text{ext}}$ do not change its direction there. Introduce a notion of the *elementary dipole* as a source whose largest linear size tends to zero, and the amplitude of its volume current increases infinitely in such a way that the product of the source volume and its density remains finite. This product is called the *dipole momentum* $\vec{P}$. The field of the elementary dipole does not depend on the intermediate shapes which the domain obtains while passing to limit. The same image can be obtained if considering an infinitely short segment l of the line current with infinite density I, such that the product $I \cdot l = P$ is finite.

The fields have strong singularities near the elementary dipole. Approaching the dipole, the electric field increases as an inverse cube of the distance to it, whereas the magnetic field as an inverse square. The densities of both electric and magnetic energies are nonintegrable, that is, the two conditions (1.82) are violated. This violation distinguishes singularities of the fields near the elementary dipole from those near the vertex.

Similarly as for the line current, the energy flux from the dipole cannot be calculated using its near field. If the dipole is not located in the field of other sources so that its field does not interfere with another field, then the flux equals $k^2/(3c)|\vec{P}|^2$. This formula will be derived later in the chapter devoted to the spherical waves. In the presence of other objects and fields created by the currents induced on them, the factor at $|\vec{P}|^2$ can be smaller or larger than $k^2/3c$, or even zero. Similarly as for the line current, if the field $\vec{e}(0)$ existing (in the absence of the dipole) at the point of the dipole location is known, then the radiated energy can be found by the formula

$$\int_S S_N \, dS = \frac{k^2}{3c} \left| \vec{P} \right|^2 - \frac{1}{2} \, \mathrm{Re} \left[\vec{P}^* \cdot \vec{e}(0) \right] \tag{1.92}$$

following from (1.56).

1.4
Uniqueness and existence of solution

1.4.1
Uniqueness of solution

There are two types of problems in high-frequency electrodynamics – *direct* and *inverse* problems. In direct problems the fields created by the prescribed extrinsic currents in the given set of objects are found. In inverse problems the currents and objects should be found such that the arisen fields possess prescribed properties. For instance, some functionals of fields have to reach the maximal possible values under the given conditions.

Before solving problems of both types, we should clarify if the solution exists under formulated conditions, and if it is unique. The existence of a solution means that there are not too many conditions imposed and not too many

idealizations accepted, that is, they do not contradict each other. The uniqueness implies that there are enough conditions imposed in order that the only
one solution exists, that is, there are not too few conditions. Below we show
that the conditions formulated earlier and idealizations accepted provide the
uniqueness, but still, only *almost* always. The case when this is not true, that
is, when there exist more than one solution, is considered in the next subsection. Now we note that in this case it may turn out that there is no solution at
all, that is, the accepted idealizations are not consistent.

It is necessary to prove that the Maxwell equations (1.29) supplemented by
the conditions near a set of the idealized objects have the unique solution.
These supplementary conditions are as follows: the radiation condition (1.33)
describing the field near the infinite point, condition (1.65) of continuity of the
tangential components on the interface, conditions (1.67) on the impedance
surface and those (1.79) on the ideal conductor, conditions (1.82) near the
edge or vertex as well as the conditions imposed on the constitutive constants
($\varepsilon'' \leq 0$, $\mu'' \leq 0$) and the impedance ($w' \geq 0$). In fact, the assertion that at any
t ($-\infty < t < \infty$) there exist harmonic fields dependent on t by (1.1) is also an
idealization.

Since the Maxwell equations are linear, the assertion "the only one solution
exists for the given extrinsic currents" is equivalent to "there exist only the
solution $\vec{E} \equiv 0$, $\vec{H} \equiv 0$ if there are no extrinsic currents." If two solutions exist
for the same currents, then their difference is also a solution in the absence of
currents. If there exists a solution for zero extrinsic currents, then its sum with
a solution existing for certain current is also a solution for this current.

The assertion "there are no fields if the extrinsic currents are absent" is
based, in fact, on the reason that if there are no sources increasing the field
energy, and the energy leakage takes place at the same time, then the energy
equals zero. Consequently, the fields are zero, either.

The proof does not use these energetic considerations explicitly. It is based
on formula (1.56), which, at $\vec{j}^{\text{ext}} = 0$, becomes

$$\text{div}\, \vec{S} = \frac{\omega}{8\pi} \left(\varepsilon'' \left| \vec{E} \right|^2 + \mu'' \left| \vec{H} \right|^2 \right). \tag{1.93}$$

As we have noted earlier, this consequence from the Maxwell equations remains to be valid also for the chiral media if their constitutive equations have
the form (1.28).

Integrate (1.93) over the volume V_a of the sphere with the radius a so large
that there are no objects outside the sphere and fields have the form (1.33)
on its surface. The interfaces of different media, as well as the edges and
vertices, can lie inside the sphere. According to (1.82), the volume integrals

over domains where the fields are infinite converge. We use the *Gauss formula*

$$\int_{V_a} \operatorname{div} \vec{A}\, dV = \int_{S_a} A_N\, dS. \tag{1.94}$$

Since it requires that the vector $\vec{A}$ is continuous inside the domain V_a, the interfaces must be separated out during the integration. When integrating over the two domains separated by the interface, the integral of A_N over this interface appears twice. The continuity of S_N follows from the continuity of E_t and H_t. However, the normals to the interface are oppositely directed in these two domains, and, therefore, the surface integrals of S_N cancel each other.

After integrating, formula (1.93) becomes

$$\int_{S_a} S_N\, dS + \int_{S_w} S_N\, dS = \frac{\omega}{8\pi} \int_{V_a} \left(\varepsilon'' \left|\vec{E}\right|^2 + \mu'' \left|\vec{H}\right|^2 \right) dV. \tag{1.95}$$

Here S_w is an impedance surface if presented in the field.

Flux of the vector $\vec{S}$ through the sphere S_a is proportional to the squared modulus of the functions $F_1(\vartheta, \varphi)$, $F_2(\vartheta, \varphi)$ in (1.33) describing the outgoing spherical wave. The flux equals

$$\frac{c}{8\pi k^2} \int_{S_a} \left(|F_1|^2 + |F_2|^2 \right) d\Omega, \qquad d\Omega = \sin \vartheta\, d\vartheta\, d\varphi. \tag{1.96}$$

The flux of the vector $\vec{S}$ into the impedance surface is calculated by (1.68). Hence, equality (1.95) (the *energy conservation law*) has the form

$$\int_{S_a} \left(|F_1|^2 + |F_2|^2 \right) d\Omega + k^2 \int_{S_W} |H_t|^2\, w'\, dS - k^3 \int_{V_a} \left(\varepsilon'' \left|\vec{E}\right|^2 + \mu'' \left|\vec{H}\right|^2 \right) dV = 0. \tag{1.97}$$

The left-hand side of (1.97) contains the sum of three nonnegative values: the radiation losses, the losses through the impedance surface and the volume losses. If at least one of the losses takes place in the system, then the above equality holds only if the field equals zero everywhere. If the system is open, that is, the field exists at infinity, then the radiation losses differ from zero. They would be zero if $F_1 \equiv 0$, $F_2 \equiv 0$. In this case the field would decrease faster than $1/R$. Then, as has already been marked in connection with formula (1.33), the field would be equal to zero in any domain (containing the infinite point), in which the field is analytical.

If the system contains the impedance surface, on which $w' > 0$, then there are no losses on it only if $H_t = 0$, and, consequently, $E_t = 0$ on the surface. Then, as can be shown, the field equals zero in any domain in which it is analytical and which borders on this surface. If there are domains with $\varepsilon'' < 0$ or $\mu'' < 0$, then the losses would be zero if $\vec{E} \equiv 0$ and $\vec{H} \equiv 0$ in these domains. Then the field would be zero also in any larger domain where it is analytical and which contains the domain with absorbent material.

Hence, if the fields $\vec{E}$ and $\vec{H}$ differ from zero, then the presence of any losses makes equality (1.97) impossible, that is, it proves the assertion about the solution uniqueness under the above assumptions.

1.4.2
Violation of the uniqueness theorem: eigenoscillations

The uniqueness can be violated if the field is nonanalytical. If there is an ideal conductor in the domain, then the analyticity on its surface is broken, due to the discontinuity of H_t. If there is a closed surface in the system, on which $E_t = 0$, that is, the closed surface of the ideal conductor, and there are no losses in the material inside this surface ($\varepsilon'' = 0$, $\mu'' = 0$), then equality (1.97) may also hold for nonzero fields there. Then a solution to the Maxwell equations can exist in the domain without the extrinsic currents. Note that this takes place also if conditions (1.67) with $w' = 0$ hold instead of (1.79) on the surface. Further we do not mention about this possibility.

In such a domain (*resonator*) the homogeneous Maxwell equations can have nonzero solutions. In this case the *uniqueness theorem* is violated and, simultaneously, the assertion that the solution exists for any extrinsic currents, that is, the *existence theorem*, may be violated, too. In oscillation theory the absence of a solution is known as the "infinite resonance" in a system without losses. It occurs as a result of introducing certain idealizations turned inconsistent: harmonic time dependence, infinite conductivity, absence of losses, given extrinsic current.

Such a situation occurs in the resonator with fixed parameters only at certain frequencies, or, at a fixed frequency and certain values of some resonator parameter. At other frequencies or values of the parameters, the assertions about the existence and uniqueness of solution remain valid.

These specific combinations of the frequency and parameters are interesting just because they permit the existence of solutions to the homogeneous Maxwell equations, so-called *eigenoscillations*. The corresponding frequency values at the fixed resonator parameters are called the *eigenfrequencies*. A set of fields of eigenoscillations corresponding to different eigenfrequencies makes up a complete set of fields. Any field (with certain reservations) can be expanded in a convergent series by these fields. Introducing in an appropriate way the notion "product of two oscillations," we can show that the product of two eigenoscillations equals zero (this assertion also needs to be specified), which is formulated as the *orthogonality* of eigenoscillations. Here the notion of the orthogonality has more general meaning than the vectorial orthogonality.

At the frequency, being not an eigenfrequency, the solutions of direct problems can be sought in the form of expansions by the fields of eigenoscillations. This method for solving the problems, known as the *spectral method*, will be considered in Section 4.1.

A set of the eigenoscillations may correspond not only to the different values of the frequency k_n ($n = 1, 2, \ldots$) at fixed parameters describing the domain where these oscillations exist. Let, for instance, a dielectric body with permittivity ε be located in a closed resonator with ideal-conducting walls. For any given frequency, the values of the permittivity ε_n ($n = 1, 2, \ldots$) can be found, at which the eigenoscillations exist. The fields of all eigenoscillations corresponding to different values of ε_n at the fixed k also make up a complete set of fields. The field in a resonator with certain permittivity ε excited by an extrinsic current, can be sought in the form of a series by the these fields. In such a method the values ε_n are the eigenvalues and all the eigenoscillations correspond to the same frequency equal to the frequency of excitation current. Other parameters can also play a role of the eigenvalues, for instance, the wall impedance w_n ($n = 1, 2, \ldots$). These generalizations of the spectral method are expedient to use in the theory of *open resonators*, when the field exists at infinity. We return to these questions in Chapter 4.

2
Plane Waves

2.1
Plane waves in an infinite homogeneous medium

2.1.1
The phase velocity

The simplest solutions to the Maxwell equations are plane waves; their fields depend only on one Cartesian coordinate. Denote this coordinate by z, so that

$$\frac{\partial}{\partial x} \equiv 0, \qquad \frac{\partial}{\partial y} \equiv 0. \tag{2.1}$$

Such solutions exist, in particular, in infinite homogeneous media. First, we assume that the medium is nonchiral and isotropic, that is, it is characterized by two scalar quantities ε and μ. It follows from equations (2.1) that the field does not have the longitudinal components, that is,

$$E_z \equiv 0, \qquad H_z \equiv 0. \tag{2.2}$$

For chiral media the system of the equations for the transverse components is divided into two independent subsystems:

$$H_y = -\frac{1}{ik\mu}\frac{dE_x}{dz}, \qquad E_x = -\frac{1}{ik\varepsilon}\frac{dH_y}{dz}, \tag{2.3}$$

for E_x, H_y and

$$E_y = \frac{1}{ik\varepsilon}\frac{dH_x}{dz}, \qquad H_x = \frac{1}{ik\mu}\frac{dE_y}{dz}, \tag{2.4}$$

for H_x, E_y, respectively. The wave equations for each of these components are the same; the wave equation for E_x is written as

$$\frac{d^2 E_x}{dz^2} + k^2 \varepsilon\mu E_x = 0. \tag{2.5}$$

High-frequency Electrodynamics. Boris Z. Katsenelenbaum
Copyright © 2006 WILEY-VCH Verlag GmbH & Co. KGaA, Weinheim
ISBN: 3-527-40529-1

Solutions to these equations are linear superpositions of the two functions

$$\exp(-ihz), \quad \exp(ihz), \tag{2.6}$$

where

$$h = k\sqrt{\varepsilon\mu} \tag{2.7}$$

denotes a *wave number*. If there are no losses in the medium, then ε and μ are real, $\varepsilon\mu > 0$; hence h is real, either.

Passing from the complex amplitudes to physical quantities by (1.3), we obtain the dependence for solutions corresponding to the first function of (2.6) on t and z in the form

$$\cos(\omega t - hz). \tag{2.8}$$

This value is constant if t and z are connected as $\omega t - hz = const$. The wave having such a dependence on t and z propagates in the z-direction. It is called the *direct wave*. Its velocity $v = dz/dt$ equals

$$v = \frac{\omega}{h}; \tag{2.9}$$

it is called the *phase velocity*. The equi-phase plane moves with this velocity. According to (2.7),

$$v = \frac{c}{\sqrt{\varepsilon\mu}}. \tag{2.10}$$

The second solution in (2.6) describes the *back wave*. It propagates in the opposite direction $-z$ with the same velocity.

The phase velocity is a velocity of an observer moving in the same direction as the wave does, so that he is placed in the field having the same value. If the observer moves in the direction making a nonzero angle β with the wave propagation direction, then its velocity is $1/\cos\beta$ times larger. In fact, the phase velocity is not a physical notion, but a geometric one. It can be larger than the light velocity c in vacuum. As will be shown further, the monochromatic oscillation cannot serve for a signal transmission; a more complicated time dependence should be used for this.

If there are losses in the medium, then, according to (2.7), h is complex: $h = h' + ih''$, $h' > 0$, $h'' \leq 0$ and the physical function of the field of the direct wave is

$$e^{h''z}\cos(\omega t - h'z). \tag{2.11}$$

The wave propagates with the phase velocity $v = \omega/h'$ and damps exponentially as $\exp(h''z)$, $h'' < 0$. The back wave propagates with the same velocity and damps as $\exp(-h''z)$ when $-z$ increases.

The connection (1.74) exists between transverse components of the electric and magnetic fields. For back waves the signs in this connection are opposite. The quantity w in (1.75) is called the *wave resistance* of the medium. It is also the impedance of the surface on which (1.67) holds.

The plane wave, that is, the field which does not depend on two Cartesian coordinates, is an idealization. It occupies an infinite domain, and it can be created by extrinsic currents uniformly distributed on the infinite plane $z = const$. Such a wave would transmit an infinite energy flux. This ideal image approximately describes any field varying along two coordinates much slower than along the third one. Such a field can occupy a domain which is although finite, but large in comparison with wavelength, so that properties of this field are close to those of the plane wave in the most part of the domain.

2.1.2
The elliptic polarization

The plane wave may contain either the components E_x, H_y (2.3), or E_y, H_x (2.4). Such waves are *linearly polarized*, that is, the vectors $\vec{E}$ and $\vec{H}$ keep their direction at all t. The waves of both polarizations are *degenerated*: they have the same wave number h (2.7).

The plane waves of both linear polarizations are independent in nonchiral media; each of them can propagate in the absence of another, keeping it linearly polarized. A linear superposition of two plane waves is a plane wave, either. However, if both waves are linearly polarized, then their superposition is also linearly polarized only in the case when the coefficients of this superposition are real. If $E_x = a$, $E_y = b$, and a, b are real, then the components of the vector $\vec{E}$ in the coordinate system, turned by the angle γ with respect to the system (x, y), are

$$E_{\bar{x}} = a \cos \gamma + b \sin \gamma, \qquad E_{\bar{y}} = -a \sin \gamma + b \cos \gamma, \tag{2.12}$$

where $\bar{x}, \bar{y}$ are new coordinates. If $\tan \gamma = b/a$, then $E_{\bar{y}} = 0$, that is, the field keeps the linear polarization, only its direction is changed becoming intermediate between the polarization directions of the initial waves.

However, if

$$E_x = a \exp(-i\alpha), \qquad E_y = b \exp(-i\beta), \qquad \alpha - \beta \neq 0, \tag{2.13}$$

then the polarization is nonlinear, that is, the vector $\vec{E}$ does not keep its direction over the period. Omitting the common factor $\exp(-ihz)$, pass to the physical quantities.

According to (2.13), we obtain

$$E_x = a \cos(\omega t - \alpha), \qquad E_y = b \cos(\omega t - \beta). \tag{2.14}$$

Eliminating the parameter t gives the equation connecting the components E_x and E_y:

$$\left(\frac{E_x}{a}\right)^2 + \left(\frac{E_y}{b}\right)^2 - 2\frac{E_x}{a}\frac{E_y}{b}\cos(\alpha - \beta) = \sin^2(\alpha - \beta). \tag{2.15}$$

If $\alpha - \beta = 0$ (which is the same that the coefficients of two waves, in which $E_y \equiv 0$ or $E_x \equiv 0$, are real), then it follows from (2.15) that $E_y = b/a \cdot E_x$. The wave remains linearly polarized. However, at $\alpha - \beta \neq 0$ and $\alpha - \beta \neq \pi$ the end of the vector $\vec{E}$ does not draw a straight-line segment, but an ellipse over the period. The *elliptically polarized* wave is the most general form of the plane wave. It can be obtained as a superposition of two waves, linearly polarized in different directions, with complex coefficients.

In order to find the directions in which the vector $\vec{E}$ reaches its maximal and minimal values, and these values themselves, that is, the directions of the polarization ellipse axes and lengths of its half-axes, we put $E_x = |\vec{E}|\cos\gamma$, $E_y = |\vec{E}|\sin\gamma$. Substituting these expressions into (2.15), we obtain the dependence of $|\vec{E}|$ on γ. Omitting elementary but cumbersome derivations, we give the final expressions for the angles $\gamma_{\min}, \gamma_{\max}$ made by the ellipse axes with the x-axis, and for the maximal and minimal values of $|\vec{E}|$, that is, for the lengths of its half-axes. The angles are calculated by the formula

$$\tan(2\gamma) = \pm\frac{2ab}{a^2 - b^2}\cos(\alpha - \beta), \tag{2.16}$$

and the lengths of the half-axes are

$$|\vec{E}|_{\substack{\max \\ \min}} = \left[\frac{abp\sin^2(\alpha - \beta)}{1 \mp \sqrt{1 - p^2\sin^2(\alpha - \beta)}}\right]^{1/2}, \tag{2.17}$$

where $p = 2ab/(a^2 + b^2)$. If $\alpha = \beta$, then, according to (2.17), $|\vec{E}|_{\min} = 0$, $|\vec{E}|_{\max} = \sqrt{a^2 + b^2}$. If a device receiving the wave measures only one linear component of electromagnetic field, then it should be oriented in the $\gamma_{\max}$-direction.

Consider two particular cases of (2.16), (2.17). If $\alpha - \beta = \pm\pi/2$, that is, two linearly polarized waves, making up an elliptically polarized one, are *in-quadrature*, then the ellipse axes are directed along the axes x and y, and equal to a and b, respectively. If $a = b$ ($p = 1$), then, at any value of the difference $\alpha - \beta$, the ellipse axes equal $a\sqrt{2}|\sin[(\alpha - \beta)/2]|$ and $a\sqrt{2}|\cos[(\alpha - \beta)/2]|$ and make the angles $\gamma = \pm\pi/4$ with the x- and y-axes. Summing two linearly polarized waves which have the same amplitudes and the phase shift $\pi/2$, we obtain a *circularly polarized* wave.

The elliptically polarized waves are characterized not only by the directions of the ellipse axes (see (2.15)) but also by the direction of the vector $\vec{E}$ turning (clock- or anticlockwise). The latter depends on the sign of $\alpha - \beta$. This characteristic is lost in (2.15); it is seen in (2.14). For instance, if $\alpha - \beta = \pm \pi/2$, then the vector $\vec{E}$ is x-directed ($E_y = 0$) at $t = \alpha/\omega$ and y-directed ($E_x = 0$) at $t = \beta/\omega$. If $\alpha - \beta > 0$, then the turning proceeds from the x-axis to the y-axis, that is, anticlockwise, if we look into the direction of the wave propagation. This turning is called the right turning. If $\alpha - \beta < 0$, then the turning proceeds from the axis x to y, that is, clockwise (left turning). This difference is essential for devices receiving not the linearly polarized, but the circularly polarized waves.

The field of the circularly polarized wave has the components

$$E_x = 1, \quad E_y = i, \quad H_y = 1, \quad H_x = -i \tag{2.18}$$

at the right turning (we put $w = 1$ in (1.74)) and

$$E_x = 1, \quad E_y = -i, \quad H_y = 1, \quad H_x = i \tag{2.19}$$

at the left turning; the normalization is taken at which $E_x = 1$.

When reflecting from any body the right-turning wave generates a left-turning wave, and vice versa. This follows only from the fact that the propagating direction is changed after reflection without changing the turning direction in the space. The same turning, which is right for the incident wave, is left for the reflected wave.

2.1.3
Eigenwaves in chiral media

Plane waves of any linear polarization are the *eigenwaves* in a nonchiral infinite medium, that is, they propagate retaining their shape. All eigenwaves are degenerated, that is, they have the same wave number h.

In the chiral medium, the linearly polarized plane wave is not an eigenwave. The eigenwaves have another polarization in the chiral medium: they are circularly polarized. The circularly polarized waves, which differ from each other by the turning direction, are not degenerated; they have different values of h.

Consider a solution to the Maxwell equations, satisfying condition (2.1), in the chiral medium, that is, in a medium with the constitutive equations (1.28). Two of these equations give $D_z = 0$, $B_z = 0$, which, at $\varepsilon \mu - \varkappa^2 \neq 0$, yield $E_z = 0$, $H_z = 0$, that is, the conditions (2.2) are satisfied. The remaining equations involve the transverse components of the fields together with their derivatives with respect to z. All components in the plane wave depend on z by means of the factor $\exp(-ihz)$, so that $d/dz = -ih$. The Maxwell equations give the following homogeneous system of four linear algebraic equations for the transverse components:

$$ihH_y = ik(\varepsilon E_x - i\varkappa H_x),$$
$$-ihH_x = ik(\varepsilon E_y - i\varkappa H_y),$$
$$ihE_y = -ik(\mu H_x + i\varkappa E_x),$$
$$-ihE_x = -ik(\mu H_y + i\varkappa E_y)$$

(2.20)

This system has nontrivial solutions for the fields only if its determinant equals zero. This leads to a fourth order algebraic equation for h. Substituting one of its roots into (2.20), we obtain relations between four components of the eigenwave corresponding to this eigenvalue h.

Applying this general procedure to the simple equation system (2.3), (2.4) for nonchiral media would give equations having solutions $h = k\sqrt{\varepsilon\mu}$ and $h = -k\sqrt{\varepsilon\mu}$; each of them is a double root. It is convenient to use a simpler artificial approach to system (2.20), which allows us to obtain for chiral media both the eigenvalues h and the structures of the eigenwave fields. This approach is suggested by formulas (2.18), (2.19) for the circularly polarized waves.

Introduce the quantities p_{rg}, g_{rg} and p_{lf}, g_{lf}:

$$p_{rg} = E_x - iE_y, \qquad g_{rg} = H_x - iH_y, \tag{2.21a}$$
$$p_{lf} = E_x + iE_y, \qquad g_{lf} = H_x + iH_y. \tag{2.21b}$$

Expressing the field components by these four functions and substituting them into (2.20), we divide it into two independent subsystems:

$$(h + k\varkappa)g_{rg} + ik\varepsilon p_{rg} = 0,$$
$$-ik\mu g_{rg} + (h + k\varkappa)p_{rg} = 0$$

(2.22a)

for p_{rg}, g_{rg}, and

$$(h - k\varkappa)g_{lf} - ik\varepsilon p_{lf} = 0,$$
$$ik\mu g_{lf} + (h - k\varkappa)p_{lf} = 0$$

(2.22b)

for p_{lf}, g_{lf}. According to these formulas, eigenwaves in the chiral medium are waves of circular polarization with opposite turning directions. As follows from (2.18), for the right-polarized wave described by system (2.22a), $p_{lf} = 0$, $g_{lf} = 0$, that is, there exists the connection $E_x = -iE_y$, $H_x = -iH_y$ between the field components. The wave number of this wave is found from the existence condition for solutions to (2.22a), that is, from the equation $(h + k\varkappa)^2 = k^2\varepsilon\mu$. For direct waves,

$$h = k\sqrt{\varepsilon\mu} - k\varkappa. \tag{2.23}$$

Subsystem (2.22b) has only zero solution at this value of h. For back waves, h has the opposite sign.

According to (2.21), it follows from $g_{lf} = 0$, $p_{lf} = 0$ that $p_{rg} = 2E_x$, $g_{rg} = 2H_x$. Substituting these expressions into (2.22a) and using (2.23), we obtain

$$H_x = -\frac{i}{w}E_x, \qquad H_y = -\frac{i}{w}E_y \tag{2.24}$$

for the right polarization. Note that the parameter $\varkappa$ does not present in these relations.

The existence condition for solutions to subsystem (2.22b) gives the expression

$$h = k\sqrt{\varepsilon\mu} + k\varkappa \tag{2.25}$$

for the wave number of the eigenwave of the left circular polarization. The relations between the field components differ from (2.24) in sign.

2.1.4
Turning of the polarization plane in the chiral media

Assume that a linearly polarized field with $E_x = 1$, $E_y = 0$ is created in a chiral medium at $z = 0$. In order to find the field at $z > 0$, we should express the field at $z = 0$ as a superposition of eigenwaves, that is, the waves of two circular polarizations. The amplitudes of these waves are found from the demand that the sum of their fields equals the given field. According to (2.18), (2.19), both these amplitudes are equal to $1/2$. The fields of the eigenwaves at $z > 0$ are obtained from the fields at $z = 0$ by multiplying by $\exp(-ih_{rg}z)$ and $\exp(-ih_{lf}z)$, respectively, so that the resulting field has the components

$$\begin{aligned}
E_x(z) &= \frac{1}{2}[\exp(-ih_{pg}z) + \exp(-ih_{lf}z)], \\
E_y(z) &= \frac{i}{2}[\exp(-ih_{pg}z) - \exp(-ih_{lf}z)].
\end{aligned} \tag{2.26}$$

Since $h_{rg} + h_{lf} = 2k\sqrt{\varepsilon\mu}$, $h_{rg} - h_{lf} = -2k\varkappa$, we have

$$\begin{aligned}
E_x(z) &= \exp(-ik\sqrt{\varepsilon\mu}z)\cos(\varkappa z), \\
E_y(z) &= \exp(-ik\sqrt{\varepsilon\mu}z)\sin(\varkappa z).
\end{aligned} \tag{2.27}$$

The components of the vector $\vec{E}$ remain in-phase. The field (2.27) is linearly polarized in the direction making the angle $\varkappa z$ with the x-axis. The linearly polarized wave is not the eigenwave of the chiral medium, but it remains linearly polarized when propagating only with changing polarization direction. The polarization plane is turned by the angle, proportional to the covered distance z. The chirality parameter $\varkappa$ is a coefficient in this relation. The vector $\vec{E}$ turns anticlockwise (right direction) at $\varkappa > 0$, and clockwise (left direction) at $\varkappa < 0$. Of course, here the term "turning" has the different meaning than in definitions of the right and left polarizations, where we imply the turning over the period.

2.1.5
The group velocity

The field of a monochromatic plane wave propagating along the z-axis is proportional to $\exp(i(\omega t - hz))$. Such a wave cannot transmit any signal. The signal can be transmitted only by the impulse, that is, by a field having more complicated dependence on t. The impulse can be considered as a result of the interference of a group of monochromatic waves. The dependence of the field of this group on z and t is determined not by the number h, but by the function $h(\omega)$, defined in the frequency band to which the group belongs.

There exists a general rule, according to which (under certain condition) the impulse propagates along a line. By this rule, it is possible to find the field at $z > 0$ as a function of time if we know this function at $z = 0$. The rule describes propagation of not only the plane electromagnetic waves but also the waves of more complicated structure or another nature, such as acoustic, seismic, etc. Only the following property of a transmitting line or medium is used: if some monochromatic oscillation of the frequency ω with the unit amplitude is created at $z = 0$, then the oscillation amplitude at $z > 0$ is $\exp(-ihz)$, where $h = h(\omega)$ is a known function of the frequency. If $F(\omega, z)$ is an amplitude of the oscillation of frequency ω at point z, then

$$F(\omega, z) = F(\omega, 0) \exp[-ih(\omega)z]. \tag{2.28}$$

This condition is satisfied for all homogeneous lines, in particular, for the homogeneous media.

In order to use the property (2.28) for the determination of the dependence of the impulse shape, given at $z = 0$, on the distance z covered by the impulse, we should expand the impulses $f(t, 0)$ at $z = 0$ and $f(t, z)$ at $z > 0$ in the Fourier integrals

$$f(t, 0) = \int F(\omega, 0) \exp(i\omega t) \, d\omega, \tag{2.29a}$$

$$f(t, z) = \int F(\omega, z) \exp(i\omega t) \, d\omega. \tag{2.29b}$$

Formally, the integrals are taken here over an infinite frequency interval $-\infty < \omega < \infty$ (this also relates to similar integrals below). However, in fact, the integration is made over the interval of the order $1/T$, where T is an impulse width. The center of the interval is a frequency ω_0 in the neighborhood of which the frequencies of oscillations, interference of which makes up the impulse, are placed. The function

$$F(\omega, 0) = \frac{1}{2\pi} \int f(\tau, 0) \exp(-i\omega\tau) \, d\tau, \tag{2.30}$$

obtained from (2.29a) as the Fourier integral inversion, has the same order as $F(\omega_0, 0)$ for this frequency.

Substituting (2.30), (2.28) into (2.29b), and interchanging the integration order with respect to ω and τ, we obtain the sought connection between $f(t, z)$ and $f(t, 0)$:

$$f(t, z) = \int f(\tau, 0) I(t - \tau, z) \, d\tau, \tag{2.31}$$

where the kernel

$$I(t - \tau, z) = \frac{1}{2\pi} \int \exp[i\omega(t - \tau) - ihz] \, d\omega \tag{2.32}$$

does not depend on the function $f(t, 0)$; it characterizes only the line. As was expected, $I(t - \tau, 0) = \delta(t - \tau)$.

If the function $h(\omega)$ is proportional to ω in the frequency interval over which the integration in (2.32) is actually made, that is, if the phase velocity $v = \omega/h$ does not depend on the frequency, then, substituting $h = \omega/v$ into (2.32), we have $I(t - \tau, z) = \delta[\tau - (t - z/v)]$, so that

$$f(t, z) = f\left(t - \frac{z}{v}, 0\right). \tag{2.33}$$

Just as was expected, any impulse propagates without deformations with the velocity $dz/dt = v$ (i.e., with the phase velocity) along the line in which the phase velocity does not depend on the frequency.

It is impossible to find a nonintegral expression of the function $f(t, z)$ by $f(t, 0)$ at an arbitrary $v(\omega)$. Since the impulse components corresponding to different frequencies obtain the phase not proportional to ω, the impulse distorts when propagating. The result of the interference of these components at $z > 0$ differs from that at $z = 0$; hence the impulse shape at $z > 0$ is not the same as that at $z = 0$.

However, if in the frequency range $\omega_0 - 1/T < \omega < \omega_0 + 1/T$ the function $h(\omega)$ can be substituted by the first two terms of its Taylor series

$$h(\omega) = h(\omega_0) + \left.\frac{dh}{d\omega}\right|_{\omega=\omega_0} \cdot (\omega - \omega_0), \tag{2.34}$$

then the integral in (2.32) is reduced to the δ-function again. With accuracy to a nonessential phase factor (independent of t), it follows from (2.31) that

$$f(t, z) = f\left(t - z\left.\frac{dh}{d\omega}\right|_{\omega=\omega_0}, 0\right). \tag{2.35}$$

Therefore, when condition (2.34) holds, then the impulse propagates without distortion, however, with the phase velocity $dz/dt = (dh/d\omega|_{\omega=\omega_0})^{-1}$ instead of (2.9). If the phase velocity depends on the frequency, that is, h is not proportional to ω, then this velocity differs from v. It is called the group velocity. Substituting $\omega = kc$, we obtain

$$v_{\mathrm{gr}} = c\frac{1}{dh/dk}. \tag{2.36}$$

This result is valid if the remainder dropped in (2.34) is small. The product of the remainder and the length z of the line should be small in comparison with unity. When increasing z, the impulse distortion becomes more essential, the notion of the impulse velocity loses the sense, and formula (2.35) ceases to be valid.

2.1.6
The relativity restrictions on the propagation factor

The known relativity restriction that the velocity of the signal transmitting is smaller than the light velocity in vacuum leads to the demand

$$\frac{dh}{dk} > 1. \tag{2.37}$$

The similar restriction on the phase velocity, namely, $h/k > 1$, does not exist for any homogeneous medium. For instance, it does not follow from (2.10) that $\varepsilon\mu > 1$. For plasma (see (1.25)), as well as for many other media, $\varepsilon < 1$, however, the dependence of ε on the frequency is such that condition (2.37) holds.

However, there exists another (not local) relativity demand on the function $h(\omega)$. The condition which will be found below has a global nature; it imposes a restriction on the function $h(\omega)$ in the whole infinite frequency range.

Let the impulse have the δ-function shape at $z = 0$:

$$f(t,0) = \delta(t). \tag{2.38}$$

It follows from (2.31) that $f(t,z) = I(t,z)$ at $z > 0$, that is, according to (2.32),

$$f(t,z) = \frac{1}{2\pi}\int \exp(i\omega t - ihz)\,d\omega. \tag{2.39}$$

At $z = 0$, the impulse did not exist for $t < 0$. The relativity restriction means that the impulse cannot propagate with a velocity larger than c; hence, at the point with coordinate $z > 0$ it should be absent at $t < z/c$, that is,

$$\int \exp(i\omega t - ihz)\,d\omega = 0 \quad \text{at} \quad t < z/c. \tag{2.40}$$

This condition is necessary for the function $h(\omega)$ to fulfil the relativity demand, because otherwise there would exist the impulse (2.38) propagating with a velocity larger than c.

Condition (2.40) is also sufficient, that is, if (2.40) holds, then any impulse not existing at $z = 0$ for $t < 0$ is absent at $z > 0$ for $t < z/c$. This means that for any impulse, for which the equality

$$f(t, 0) = 0 \quad \text{at} \quad t < 0 \tag{2.41a}$$

is valid, the equality

$$f(t, z) = 0 \quad \text{at} \quad t < z/c \tag{2.41b}$$

is valid, either. If the function $h(\omega)$ is such that (2.40) holds, then (2.41b) follows from (2.41a) for any impulse.

The proof of this statement is elementary. If $f(t, 0) = 0$ at $t < 0$, then the integrand in (2.31) equals zero at $\tau < 0$, and the integral is taken over $\tau > 0$ only. Therefore, the first argument $t - \tau$ in the function $I(t - \tau, z)$ in this integral is smaller than t. At $t < z/c$ this argument is smaller than z/c, and, according to (2.40), $I(t - \tau, z) = 0$. Hence, $f(t, z) < 0$.

Condition (2.40) can be rewritten in another form. At the high frequencies, any medium and any transmitting line behave as free space, and any signal propagates with the velocity c, that is,

$$h(\omega)|_{\omega \to \infty} \simeq \omega/c. \tag{2.42}$$

This condition introduces the velocity c into the theory of homogeneous media (and of general homogeneous lines). Denote

$$\rho(\omega) = h(\omega) - \frac{\omega}{c}. \tag{2.43}$$

According to (2.42),

$$\lim_{\omega \to \infty} \frac{\rho(\omega)}{\omega} = 0. \tag{2.44}$$

Expressing the function $h(\omega)$ from equation (2.43) and substituting it into (2.40), we obtain the condition

$$\int_{-\infty}^{\infty} \exp(i(\omega\alpha - z\rho)) \, d\omega = 0, \quad \alpha < 0, \quad z > 0, \tag{2.45}$$

equivalent to (2.40), where $\alpha = t - z/c$.

This condition can be transformed once more in the same way as condition (1.64) for the dependence of the dielectric permittivity on the frequency was obtained. For this purpose we introduce a plane of the complex variable Ω, $\mathrm{Re}\,\Omega = \omega$. If the function $h(\omega)$ can be analytically continued from the real axis onto the whole lower half-plane in such a way that $|\rho(\Omega)/\Omega|$ tends to zero as $|\Omega| \to \infty$, then condition (2.45) holds because $i\Omega\alpha = -\Omega''\alpha + i\alpha\Omega'$, and $\exp(i\Omega\alpha)$ tends (exponentially) to zero as $\Omega'' \to -\infty$, at $\alpha < 0$.

An infinite dielectric medium is a transmitting line with the wave number $h = k/\sqrt{\varepsilon}$. For the relativity demand "(2.41b) follows from (2.41a)" to be fulfilled, it is sufficient that $\sqrt{\varepsilon}$, as a function of frequency, can be analytically continued onto the whole lower half-plane of the complex frequency Ω, and that $\rho(\Omega)/\Omega = (\sqrt{\varepsilon(\Omega)} - 1)/c$ tends to zero in this half-plane as $|\Omega| \to \infty$. The second demand is fulfilled for all the functions $\varepsilon(\omega)$; the first one coincides, in fact, with the demand imposed on $\varepsilon(\omega)$ by the causality principle. This principle and the relativity restriction lead to the same demands.

For arbitrary lines, $h(\omega)$ is not expressed by $\varepsilon(\omega)$, and the causality principle, in contrast to the relativity restriction, does not lead to any restrictions on the function $h(\omega)$.

Condition (2.40) is less informative than (2.37). The first condition is equivalent to the existence condition for the analytical continuation of the function $h(\omega)$. If this condition does not hold for some function $h(\omega)$, then there exists a function "close" to $h(\omega)$, for which this condition holds. A small perturbation of almost any function $h(\omega)$ can lead to the function satisfying the necessary and sufficient conditions following from the relativity demand (2.41). The demand (2.37) is stronger than (2.40); small perturbations of the function $h(\omega)$ are not sufficient for obtaining the function which fulfills this condition at all frequencies.

2.2
Plane waves in a plano-layered medium

2.2.1
Medium with constant wave resistance

The waves, whose fields satisfy condition (2.1) exist not only in the infinite homogeneous medium but also in the nonhomogeneous one if its ε, μ (and $\varkappa$ for a chiral medium) depend only on one coordinate. We begin with a nonchiral medium. If (2.1) holds, then it follows from the Maxwell equations at $\varepsilon = \varepsilon(z)$, $\mu = \mu(z)$ (similarly as at $\varepsilon = const$, $\mu = const$) that the longitudinal components are not present in the field (i. e., (2.2) holds), and the fields can be "polarizationally divided" onto two independent groups, for one of which equations (2.3) hold and for the other (2.4) do. However, the wave equations for the components E_x, E_y and H_x, H_y are different; they have the form

$$\frac{d}{dz}\left(\frac{1}{\mu}\frac{dE}{dz}\right) + k^2\varepsilon E = 0 \qquad (2.46)$$

and

$$\frac{d}{dz}\left(\frac{1}{\varepsilon}\frac{dH}{dz}\right) + k^2\mu H = 0 \qquad (2.47)$$

for the components of the fields $\vec{E}$ and $\vec{H}$, respectively. If μ does not depend on z, then (2.46) becomes (2.5), and if ε does not depend on z, then (2.47) becomes (2.5). As in the general case $\varepsilon = \varepsilon(z)$, $\mu = \mu(z)$, in these two particular cases the dependences of $\vec{E}$ and $\vec{H}$ on z are different.

According to (2.3), (2.4), the functions

$$\frac{1}{\mu}\frac{dE}{dz}, \quad \frac{1}{\varepsilon}\frac{dH}{dz} \qquad (2.48)$$

are continuous on the interface of the media with different values of ε and μ.

The wave equations have no explicit solutions for arbitrary functions $\varepsilon(z)$, $\mu(z)$. However, such solutions exist when $\varepsilon(z)$, $\mu(z)$ are proportional to each other, that is, if the wave resistance $w = \sqrt{\mu/\varepsilon}$ does not depend on z. Then system (2.3) and the wave equations (2.46), (2.47) are satisfied by the functions

$$E_x = \exp(-i\gamma(z)), \qquad H_y = \frac{1}{w}\exp(-i\gamma(z)), \qquad (2.49a)$$

$$E_x = \exp(i\gamma(z)), \qquad H_y = -\frac{1}{w}\exp(i\gamma(z)), \qquad (2.49b)$$

where

$$\gamma(z) = k\int_0^z \sqrt{\varepsilon(\varsigma)\mu(\varsigma)}\,d\varsigma. \qquad (2.50)$$

The corresponding formulas for the fields satisfying (2.4) instead of (2.3) are obtained from (2.49) by replacing E_x with E_y and H_y with $-H_x$.

Formulas (2.49a) describe the direct waves propagating in the z-direction, whereas (2.49b) do the back ones propagating in the opposite direction. The lower limit of integral (2.50) can be arbitrary, it only gives a constant factor in the fields (2.49). At $\varepsilon = const$, $\mu = const$, we have $\gamma = k\sqrt{\varepsilon\mu}z$. Not only any wave having E_x, H_y and E_y, H_x is a possible solution to the Maxwell equations in the medium with $w = const$, but also any superposition of such waves, that is, the wave with the elliptic polarization is a solution to these equations.

When a wave falls onto the interface of the media with different constant wave resistances, then it partially transmits into the second medium and partially reflects by the interface. Let the medium with $w = w_-$ be located at $z < 0$ and the medium with $w = w_+$ be located at $z > 0$. A wave with $E_x = 1$ falls onto the boundary $z = 0$ from the medium with $w = w_-$. Denote the *reflection coefficient* by R and the *transmission* one by D. Then the fields on both sides of the interface have the form

$$\left.\begin{array}{l} E_x = \exp(-i\gamma_-) + R\exp(i\gamma_-), \\ H_y = \frac{1}{w_-}[\exp(-i\gamma_-) - R\exp(i\gamma_-)] \end{array}\right\} \quad \text{at } z < 0, \qquad (2.51a)$$

$$E_x = D\exp(-i\gamma_+), \qquad H_y = \frac{1}{w_+}D\exp(-i\gamma_+) \quad \text{at } z > 0, \qquad (2.51b)$$

where γ_+, γ_- are defined by (2.50). Equating the fields at $z = \pm 0$, we obtain two equations for R, D, from which

$$R = \frac{w_+/w_- - 1}{w_+/w_- + 1}, \qquad (2.52a)$$

$$D = \frac{2w_+/w_-}{w_+/w_- + 1}. \qquad (2.52b)$$

If $w_+ = w_-$, then $R = 0$, $D = 1$, that is, the wave does not reflect from the interface of the media with the same wave resistance. This follows from the fact that in media with $w = const$ the direct wave (2.49a) propagates without generating the back one (2.49b), which is true for any $\varepsilon(z)$ and $\mu(z)$, in particular, in the case when these functions are discontinuous.

The energy flux falling onto the boundary $z = 0$ from the medium $z < 0$, created by a sum of the direct and back waves, is equal (with the accuracy to the factor $c/(8\pi)$) to $\text{Re}[(1 + R)(1 - R^*)/w_-^*]$. If the medium $z < 0$ has no losses, that is, $\text{Im}\, w_- = 0$, then this flux is equal to $(1 - |R|^2)/w_-$, that is, to the difference of fluxes, transferred by the direct and back waves apart. This property is not inherent in media with losses, that is, with $\text{Im}\, w_- \neq 0$.

Note a specific case when the dielectric permittivity is negative in the medium at $z > 0$, for instance, as in plasma at low frequencies (see (1.25)). Then w_+ is imaginary, and according to (2.52a), $|R| = 1$. The wave completely reflects from the medium with $\varepsilon < 0$. However, the field penetrates into this medium. In this case $D \neq 0$, but since γ_+ is imaginary, the field is fast decreasing, as $\exp(-|\gamma_+|)$. The energy flux outgoing into this medium equals zero, because, according to (2.49a), E_x and H_y are in-quadrature. If ε is complex at $z > 0$, that is, the losses exist in the medium, then $|R| < 1$, and the energy flux through the boundary $z = 0$ is not equal to zero.

2.2.2
The slowly varying wave resistance

If the wave resistance is a "slow" function, then an approximate solution to system (2.3) can be obtained in an explicit form. The solution should be close to the solution at $dw/dz = 0$ when $\vec{E}$ and $\vec{H}$ are superpositions of the waves (2.49a) and (2.49b), that is, of the waves propagating in both directions. Taking into account the symmetry of equations with respect to $\vec{E}$ and $\vec{H}$, we introduce two new functions $A(z)$ and $B(z)$ by the formulas

$$E(z) = \sqrt{w}[A(z)\exp(-i\gamma) + B(z)\exp(i\gamma)], \tag{2.53a}$$

$$H(z) = \frac{1}{\sqrt{w}}[A(z)\exp(-i\gamma) - B(z)\exp(i\gamma)]. \tag{2.53b}$$

Here $E(z)$, $H(z)$ denote the components E_x, H_y, respectively. Substituting (2.53) into (2.3) gives the following differential equation system for $A(z)$, $B(z)$:

$$\frac{dA}{dz} = -\frac{1}{2w}\frac{dw}{dz}B\exp(2i\gamma), \tag{2.54a}$$

$$\frac{dB}{dz} = -\frac{1}{2w}\frac{dw}{dz}A\exp(-2i\gamma). \tag{2.54b}$$

The equations are exact, equivalent to (2.3).

System (2.54) is constructed in such a way that the small function dw/dz is a factor at dA/dz and dB/dz; this fact causes the form of definitions (2.53). An iterative method is convenient for obtaining approximate solutions to equations (2.54). For the zeroth approximation we take $dA/dz = 0$, $dB/dz = 0$. In this order, E and H are given by (2.53) at $A = const$, $B = const$. This approximate solution to system (2.3), or which is the same as that to the wave equations, so-called *WKB approximation* (Wenzel, Kramer, Brillouin) is often used in different branches of the mathematical physics. It can also be obtained when considering the medium with variable $\varepsilon(z)$, $\mu(z)$ as a limit of an aggregate of sequential layers with constant values of w, different in different layers. The thickness Δz of each layer is small. When passing through each layer the phase obtains the additional summand $k\sqrt{\varepsilon\mu}\Delta z$, and the total phase accumulation is integral (2.50). The amplitude of the transmitted wave obtains the additional factor $D(z+\Delta z)/D(z)$, where D is given in (2.52b). At $w_+ = w_- + \Delta w$, this factor is equal to $1 + \Delta w/(2w)$ (with the accuracy to the terms of the order $(\Delta w)^2$), that is (with the same accuracy), to $\sqrt{w_+/w_-}$. This is in accordance with both formulas (2.53).

The solution $A = const$, $B = const$ obtained in the zeroth approximation does not describe the appearance of a reflected wave. The coefficients A and B are independent in this solution; for instance, the solution $A = 1$, $B = 0$ is possible. The reflection is an effect of the first (not zeroth) order. System (2.54) allows us to give an analytical (but, of course, also approximate) description of this process. Let, for instance, a wave of the unit amplitude fall onto a certain layer of a finite thickness L, in which dw/dz is small, and there are no jumps of w on the layer boundaries, that is, the reflection from the boundaries does not occur. At the layer input $z = 0$, we have $A = 1$, $B = 0$. Substituting the latter equality into (2.54a), we obtain $A(z) = 1$ in this approximation. Alteration of $A(z)$ through the layer thickness is not accounted in this order, more precisely, in this step of the iterative procedure. Alteration of the phase of incident wave is accounted by the factor $\exp(-i\gamma)$. Formula (2.54b) gives an explicit expression for dB/dz, integrating which gives

$$B(z) = -\frac{1}{2} \int_L^z \frac{dw(\xi)}{d\xi} \frac{1}{w(\xi)} \exp(-2i\gamma(\xi)) \, d\xi. \tag{2.55}$$

Here it is assumed that at $z = L$ there are no incoming waves from the side $z > L$, that is, $B(L) = 0$. The reflection factor of the layer is $R = B(0)$, that is,

$$R = \frac{1}{2} \int_0^L \frac{dw(z)}{dz(z)} \frac{1}{w(z)} \exp(-2i\gamma(z)) \, dz. \tag{2.56}$$

The approximation $A = const$ ceases to be valid at very large values of L, and formula (2.56) becomes unapplicable. It is possible to obtain the next approximation, determining the function $A(z)$ from equation (2.54a), into which function (2.55) is substituted. This iterative procedure can be continued; it can also be generalized for the case when w has jumps on the layer boundaries, etc.

Similarly to the zeroth approximation, formula (2.56) of the first approximation can also be obtained from imaging the layer with continuous distribution of $w(z)$ as a limit of an aggregate of thin homogeneous plates. As it follows from (2.52a), when a wave falls onto the boundary where $w(z)$ jumps by the value Δw, the reflected wave has the amplitude $\Delta w/(2w)$. Replacing Δw by $dw/dz \cdot \Delta z$ we get the value of this local reflection coefficient as $dw/dz \cdot 1/(2w) \cdot \Delta z$. The wave reflected from the entire layer is a sum of the local reflected waves. The wave falling onto the thin plate has the phase $\gamma(z)$ (at $A(z) = 1$). After passing through the layer the local reflected wave obtains an addition phase which is also equal to $\gamma(z)$. Therefore, the total reflected wave has the amplitude (2.56) in the first approximation (i. e., without accounting additional re-reflections).

We describe another way of obtaining formula (2.56), equivalent, in fact, to introducing the functions $A(z)$, $B(z)$, but a little simpler. Introduce a new function $\rho(z)$ by the formula

$$\rho(z) = \frac{B(z)}{A(z)} \exp(2i\gamma), \tag{2.57}$$

which has a meaning of the "local reflection factor." It is easy to check that, according to (2.54), $\rho(z)$ satisfies the first-order nonlinear differential equation of Riccati

$$\frac{d\rho}{dz} - 2ik\sqrt{\varepsilon\mu}\rho + \frac{1}{w}\frac{dw}{dz}(1 - \rho^2) = 0. \tag{2.58}$$

The equation is exact, equivalent to (2.54) (and (2.3)).

If dw/dz and $|\rho|$ are small (the latter follows from the expression for $\rho(z)$, which will be obtained below), then we can drop the quadratic terms in (2.58). The obtained linear differential equation has an explicit solution

$$\rho(z) = \frac{1}{2}\exp(2i\gamma)\int_L^z \frac{1}{w}\frac{dw(\varsigma)}{d\varsigma}\exp(-2i\gamma)\,d\varsigma + \rho(L). \tag{2.59}$$

The last summand differs from zero only if there is a wave incoming from the side $z > 0$. If there is no such wave, that is, $\rho(L) = 0$ and therefore $R = \rho(0)$, then we obtain expression (2.56) for the total reflection coefficient again.

2.2.3
The periodical layer

The formation of the wave reflected from the periodical layer will be considered below under the assumption providing the validity of (2.56). Let the medium parameters be periodical functions of z:

$$\varepsilon(z) = \varepsilon(z + l), \qquad \mu(z) = \mu(z + l). \tag{2.60}$$

The form of the functions is nonessential in our approximation. For simplicity we assume that the layer contains an integer number of periods, that is, $L = Nl$ with integer N.

If z is placed in the nth layer, $z = nl + \xi$, $0 < \xi < l$, then expression (2.50) for the phase can be written as a sum:

$$\gamma(z) = k\sum_{m=1}^{n}\int_{(m-1)l}^{ml}\sqrt{\varepsilon(\varsigma)\mu(\varsigma)}\,d\varsigma + k\int_{nl}^{z}\sqrt{\varepsilon(\varsigma)\mu(\varsigma)}\,d\varsigma. \tag{2.61}$$

All the terms in the first sum are identical owing to the periodicity condition (2.60), and they are equal to the so-called *optical thickness*

$$a = \int_0^l \sqrt{\varepsilon(\varsigma)\mu(\varsigma)}\,d\varsigma \tag{2.62}$$

of the layers. Under this definition

$$\gamma(z) = nka + k \int_0^{\varsigma} \sqrt{\varepsilon(\varsigma)\mu(\varsigma)}\, d\varsigma. \tag{2.63}$$

Expression (2.56) for the reflection factor R can also be written as a sum

$$R = \frac{1}{2} \sum_{n=0}^{N-1} \int_{nl}^{(n+1)l} \left[\frac{1}{w}\frac{dw}{dz} \exp(-2inka) \exp\left(-2ik \int_0^{\varsigma} \sqrt{\varepsilon(\varsigma)\mu(\varsigma)}\, d\varsigma\right) \right] dz. \tag{2.64}$$

Since the function $w^{-1}\, dw/dz$ is also periodical with the period l, the integrals in (2.64) differ from each other by the multiplier $\exp(-2ina)$. Introduce R_0 as a reflection factor of the layer containing only one period:

$$R_0 = \frac{1}{2} \int_0^{l} \frac{1}{w}\frac{dw}{dz} \exp(-2i\gamma(z))\, dz. \tag{2.65}$$

Then the reflection factor of the sequence of N identical layers is

$$R = R_0 \sum_{n=0}^{N-1} \exp(-2ikan). \tag{2.66}$$

The phase of each term in the sum in (2.66) is equal to the phase of the wave reflected from the nth layer (partial wave) when it reaches the input of the layer system.

With the accuracy to a nonessential factor of the unit modulus, the sum in (2.66) is

$$\sin(Nka)/\sin(ka). \tag{2.67}$$

This factor determines, in principle, the frequency dependence of the reflection factor of the periodical layer. If the product ka (proportional to the frequency) is not close to π (or $2\pi, 3\pi, \ldots$), that is, if the thickness of each layer is not close to the length of the half-wave (or its multiple), then the factor (2.67) has the order R_0. If ka is close to π, more precisely, if $|ka - \pi| < 1/N$, then the factor (2.67) is large. At $ka = \pi$, all the partial reflected waves are in-phase, and hence $R = NR_0$. After reflection from a thick periodical layer a nonsmall reflected wave ($|R| \gg |R_0|$) may arise in narrow frequency bands close to $k = m\pi/a$, $m = 1, 2, \ldots$ even if dw/dz is small, that is, if the reflection factor from one layer is small ($|R_0| \ll 1$).

2.2.4
Nonreflecting coating

A dielectric layer with finite thickness may serve as a nonreflecting coating placed on a certain medium, which possesses the property that the wave falling onto the medium does not reflect; it transmits completely into the layer, and deeper into the medium. Such a layer does not protect the medium against the penetration of the field into it, but provides the absence of the reflected signal, that is, its radar invisibility. The layer matches the medium with free space from which the wave comes.

The simplest matching layer is a homogeneous plate. Denote the parameters of the medium located at $z > L$ by ε_0, μ_0 and the parameters of the plate placed at $0 < z < L$, by ε_1, μ_1. We set $\varepsilon = 1$, $\mu = 1$ in free space from which the wave comes. The parameters ε_1, μ_1, L should be found from the demand that the reflection factor equals zero. Similar to (2.53), the field in the plate is a sum of the direct and back waves; the field in the medium is a direct wave (as (2.51b)); the field in free space is the sum of the direct wave of unit amplitude and the reflected wave (as (2.51a)). Four unknown coefficients participate in the formulas for these fields: A and B at the respective waves in the plate, D at the outgoing wave in the medium, and the reflection factor R. Equating the values of tangential components on both sides of the interfaces $z = 0$ and $z = L$, we obtain four nonhomogeneous algebraic equations for these coefficients. The demand $R = 0$ gives the conditions on the plate parameters providing the complete antiradar protection of the medium. Elementary, but a little cumbersome, derivations give

$$w_1 = \sqrt{w_0}, \tag{2.68a}$$
$$\sqrt{\varepsilon_1 \mu_1} L = \frac{\pi}{2k}. \tag{2.68b}$$

Since $w = 1$ in free space, equality (2.68a) means that the wave resistance of the plate material, completely matching two media, is the geometric mean between those of the media. The optical thickness (2.68b) of the plate should provide the phase difference of the waves, reflected from both sides, to be π, that is, the optimal conditions of the mutual suppression of the waves.

The matching is narrow-band; the deviation of the frequency from the value corresponding to (2.68b) violates the condition necessary for the complete suppression. The larger the wave resistance of the protected medium is, the larger w_1 should be and the narrower is the frequency band in which $|R|$ is essentially smaller than the reflection factor of the nonprotected medium. At $\varepsilon_0 \to \infty$ the band width tends to zero.

The frequency band can be expanded if ε_1 and μ_1 also depend on the frequency in a way that w_1 remains independent of the frequency and the quantity $\sqrt{\varepsilon_1 \mu_1}$ decreases as $1/k$ with increasing frequency.

If we want to match free space with metal, then the material of the matching layer should have losses; otherwise $|R| = 1$ at any values of the parameters. The wider the frequency band, where the condition $|R| \ll 1$ should be fulfilled, the smoother the transition from free space to the metal must be. The smoothness can be reached either by varying ε, μ of one plate or by using a set of thin plates with constant ε, μ in each of them.

Using the model of the continuous medium with $\varepsilon = \varepsilon(z)$, $\mu = \mu(z)$, we may solve either the nonlinear equation (2.58) with the end condition $\rho(L) = 1$ or system (2.54) with the end conditions

$$A(0) = 1, \qquad B(L) = -A(L)\exp(-2i\gamma(L)). \qquad (2.69)$$

The latter provides the condition $E(L) = 0$.

These conditions are stated at both ends of the segment $0 \le z \le L$, which leads to certain difficulties when solving problem (2.54), (2.69) numerically. There exists a technique which allows us to avoid them. Instead of solving the above problem, we consider two auxiliary problems for the function pairs $A^{(1)}(z)$, $B^{(1)}(z)$ and $A^{(2)}(z)$, $B^{(2)}(z)$, respectively. They satisfy the same equation system (2.54) and the end conditions

$$A^{(1)}(0) = 1, \qquad B^{(1)}(0) = 0, \qquad (2.70a)$$
$$A^{(2)}(0) = 0, \qquad B^{(2)}(0) = 1. \qquad (2.70b)$$

The end conditions for these functions are stated at the same end of the segment. Each of problems (2.54), (2.70a) and (2.54), (2.70b) is the Cauchy problem and it can be solved simpler than the initial one (2.54), (2.69). After the values of the auxiliary functions are found at $z = L$, their linear combination is constructed, which satisfies conditions (2.69) and hence is the sought solution.

Considering the model of the layer medium, we express the fields $E_n(z)$, $H_n(z)$, $n = 0, \ldots, N$, in each of the N layers by the two numbers A_n, B_n as follows:

$$E_n = A_n \exp(-i\gamma_n) + B_n \exp(i\gamma_n), \qquad (2.71a)$$
$$H_n = \frac{1}{w_n}\left[A_n \exp(-i\gamma_n) - B_n \exp(i\gamma_n)\right], \qquad (2.71b)$$

where $\gamma_n = k\sqrt{\varepsilon_n\mu_n}(z - z_{n-1})$, $w_n = \sqrt{\mu_n/\varepsilon_n}$, and the numbers ε_n, μ_n are the parameters of the nth layer, $\varepsilon_0 = 1$, $\mu_0 = 1$. The nth layer occupies the segment $z_{n-1} < z < z_n$. The wave of the unit amplitude falls onto the medium, $A_0 = 1$. On metal $E = 0$, so that, similarly as in (2.69), $B_N = -A_N \exp(-2i\gamma_N)$. The continuity conditions for the fields $\vec{E}$, $\vec{H}$ should be satisfied on the N interfaces $z = z_{n-1}$, $n = 1, 2, \ldots, N$. A system of $2N$ nonhomogeneous linear algebraic equations is obtained for $2N$ coefficients $A_1, \ldots, A_N, B_0, \ldots, B_{N-1}$. The reflection factor $R = B_0$ is found from this system. Here not only the numbers

$\gamma_n(z_n)$ depend on the frequency (they are proportional to k) but also the parameters ε_n, μ_n can depend on it.

2.2.5
The oblique incidence

In the medium with parameters independent of x and y, a solution of the Maxwell equation is possible, such that the fields satisfy only the condition $\partial/\partial y \equiv 0$. This condition is not so strong as (2.1); it results in either

$$E_y = 0, \quad H_x = 0, \quad H_z = 0, \tag{2.72a}$$

or

$$H_y = 0, \quad E_x = 0, \quad E_z = 0. \tag{2.72b}$$

For the fields fulfilling (2.72a) the components E_x, E_z are expressed by H_y:

$$E_x = -\frac{1}{ik\varepsilon}\frac{\partial H_y}{\partial z}, \tag{2.73a}$$

$$E_z = -\frac{1}{k\varepsilon}\frac{\partial H_y}{\partial x}, \tag{2.73b}$$

with H_y satisfying the partial differential wave equation

$$\varepsilon\frac{\partial}{\partial z}\left(\frac{1}{\varepsilon}\frac{\partial H_y}{\partial z}\right) + k^2\varepsilon\mu H_y + \frac{\partial^2 H_y}{\partial x^2} = 0, \tag{2.74}$$

generalizing the ordinary differential equation (2.47). In a similar way, for the fields fulfilling (2.72b), the components H_x, H_z are expressed by E_y, which, in turn, satisfies the equation of type (2.74).

Solution to equation (2.74) can be found in the form $H_y(x,z) = Z(z)X(x)$. Substituting this function into (2.74) and separating the functions of x and z (the *Fourier method*), we obtain

$$\left[\varepsilon\frac{d}{dz}\left(\frac{1}{\varepsilon}\frac{dZ}{dz}\right) + k^2\varepsilon\mu Z\right]\bigg/ Z = -\frac{d^2 X}{dx^2}\bigg/ X. \tag{2.75}$$

Both sides of this equality must equal a constant number. Denote this number (separation constant) as $k^2\sin^2\alpha$ (physical sense of α will be established further). Then

$$X(x) = \exp(\pm ik\sin\alpha \cdot x), \tag{2.76}$$

and $Z(z)$ satisfies the equation

$$\varepsilon\frac{d}{dz}\left(\frac{1}{\varepsilon}\frac{dZ}{dz}\right) + k^2\left(\varepsilon\mu - \sin^2\alpha\right) Z = 0. \tag{2.77}$$

This equation differs from (2.47) by replacing $\varepsilon\mu$ with $\varepsilon\mu - \sin^2\alpha$.

In the wave with $H_y(x, z) = Z(z) \exp(-ik \sin \alpha \cdot x)$ the fields depend on x as in the plane wave; however, their dependence on z is different. The wave phase is not a linear function of x and z; the equiphase surfaces are not planes. This is valid even in the medium with $w = const$. In the medium with $\varepsilon = \varepsilon(z)$, $\mu \neq \mu(z)$, there is no solution to the Maxwell equations in the form of "incline plane wave."

Such a wave exists in the layer with $\varepsilon = const$, $\mu = const$. In this case equation (2.77) has the solution

$$Z(z) = \exp\left(-ik\sqrt{\varepsilon\mu - \sin^2 \alpha} z\right), \tag{2.78}$$

so that the phase of $H_y(x, z)$ is a linear function of x and z. The field can be written as a plane wave

$$H_y(x, z) = \exp\left[-ik\sqrt{\varepsilon\mu}(z \cos \beta + x \sin \beta)\right]; \tag{2.79}$$

the normal to its equiphase planes makes an angle β with the z-axis, where

$$\sin \beta = \frac{\sin \alpha}{\sqrt{\varepsilon\mu}}. \tag{2.80}$$

If $\varepsilon(z)\mu(z) < \sin^2 \alpha$ in a certain part of the medium, then $\sin \beta > 1$, $\cos^2 \beta < 0$, that is, $\cos \beta$ is imaginary, and the field (2.79) has a decreasing multiplier

$$Z(z) = \exp(-k\sqrt{\varepsilon\mu}|\cos \beta|z). \tag{2.81}$$

This fact is seen from formula (2.78), which contains a negative value under radical. We will return to this situation later.

If the medium consists of the layers with constant $\varepsilon = \varepsilon_n$, $\mu = \mu_n$ in the nth layer, then, according to (2.80), β has different values β_n in each layer. When passing the interface between the nth and $(n + 1)$th layers, the direction of the wave propagation (i.e., the direction of the normal to equiphase planes) changes; the angles β_n and β_{n+1} are connected by the relation (*refraction law*)

$$\frac{\sin \beta_{n+1}}{\sin \beta_n} = \sqrt{\frac{\varepsilon_n \mu_n}{\varepsilon_{n+1}\mu_{n+1}}}. \tag{2.82}$$

If $\varepsilon_n = 1$, $\mu_n = 1$ in the nth layer, then $\beta_n = \alpha$. The physical meaning of the angle α participating in the separation constant of equation (2.75) consists in the fact that the field in such a layer is a plane wave propagating in the direction which makes the angle α with the z-axis.

In formula (2.76) for $X(x)$ (and, correspondingly, in (2.79) for $H_y(x, z)$), we can replace $-i$ with i. Four waves propagating in the directions $\pm x$, $\pm z$, respectively, may exist. In media with variable ε, μ, in particular, on interfaces

of the layers, two waves corresponding to the directions $\pm z$ are mutually connected. This connection exists even in the case when $w = const$ at $\alpha \neq 0$,
although these waves are independent at $w = const$, $\alpha = 0$. In contrast, the
dependence on x is the same in the whole nonhomogeneous medium: if the
medium properties do not depend on x, then the waves propagating in the directions $\pm x$ are also independent. For instance, if a plane wave falls onto such
a nonhomogeneous medium, then the field in the medium and the reflected
wave have the same dependence on x as the incident wave.

If the wave falls from the nth layer with constant parameters onto the
$(n+1)$th one with other constant parameters, then the reflection and transmission factors are

$$R = \frac{w_n \cos \beta_n - w_{n+1} \cos \beta_{n+1}}{w_n \cos \beta_n + w_{n+1} \cos \beta_{n+1}}, \tag{2.83a}$$

$$D = \frac{2 w_n \cos \beta_n}{w_n \cos \beta_n + w_{n+1} \cos \beta_{n+1}} \tag{2.83b}$$

(the *Fresnel formulas*), where, according to (2.80), $\sin \beta_n = \sin \alpha / \sqrt{\varepsilon_n \mu_n}$. For
the second possible polarization ($H_y = 0$, $E_y = 0$, $E_z = 0$) it is required
to replace w with w^{-1}. In contrast to the normal incidence, $R \neq 0$ even at
$w_{n+1} = w_n$ in the case considered.

Let the wave fall slantwise from an "optically more dense" medium onto
the interface with an "optically less dense" one, that is, $\varepsilon_n \mu_n > \varepsilon_{n+1} \mu_{n+1}$, and
the incidence angle β_n is larger than β_{cr}, where β_{cr} for these two media is
defined by the equality

$$\sin \beta_{cr} = \sqrt{\frac{\varepsilon_{n+1} \mu_{n+1}}{\varepsilon_n \mu_n}}. \tag{2.84}$$

According to (2.82), $\sin \beta_{n+1} > 1$ at $\beta_n > \beta_{cr}$. In this case, $\cos \beta_{n+1}$ is imaginary (or complex when the media have losses). The wave damps in the op-
tically less dense medium in the direction normal to the interface (not in the
direction defined by the angle β_n).

The waves decreasing in the direction perpendicular to the propagating one
are called the *nonhomogeneous waves*. They arise in the cases when in some
medium a wave propagates with the phase velocity smaller than the light velocity in this medium. At $\varepsilon = 1$, $\mu = 1$, the velocity of the wave (2.79) in the
x-direction equals $\omega / (k \sin \beta)$, that is, $c / \sin \beta$. If $\sin \beta > 1$, then the velocity
is smaller than c. In Section 5.1 we will show that the velocity of waves in
the dielectric waveguides is also smaller than c, and the field of these waves
decreases in the direction perpendicular to the guide axis. If the spherical
wave is expressed as an integral by the plane waves of all directions, then the
nonhomogeneous waves are among them.

If both media have no losses, then, according to (2.83a), $|R| = 1$. The so-
called *complete internal reflection* occurs. The field penetrates into the optically

less dense medium, but exponentially decreases in it. The component E_x is in-quadrature with H_y, and, according to (2.73a), the Poynting vector equals zero. If there are losses in the $(n + 1)$th medium, then the difference between the phases of E_x and H_y does not equal $\pi/2$, and the energy flux into this medium differs from zero, and $|R| < 1$.

At $\beta_n > \beta_{cr}$ the reflection is incomplete also in the case when the optically less dense $(n + 1)$th medium has a finite thickness, and the optical density $\varepsilon_{n+2}\mu_{n+2}$ of the $(n + 2)$th medium is so large that $\sin \beta_{n+2} < 1$. In this medium the wave propagates without damping. It partially leaks through the $(n + 1)$th medium as through a "slot," losing a part of its energy for the reflection. In this case formula (2.83a) for the energy factor is not valid; it is true only for the case when the $(n + 1)$th medium is infinitely thick. After passing through the $(n + 1)$th medium, the wave propagates forward into the medium with the large enough optical density. This effect also takes place in the case when ε and μ are not constant in the "slot," but their product is smaller than $\varepsilon_n\mu_n \sin^2 \beta_n$. It is called a *tunneling* or *under-barrier transmission*.

The situation is similar to that at the normal incidence onto the layer with $\varepsilon < 0$, for instance, onto the plasma layer. The wave completely reflects from the very thick layer in the absence of the losses. If the thickness is not very large, then the wave partially leaks through the slot. At the normal incidence the layer with $\varepsilon < 0$ behaves in the same manner as the layer of an optically small density at the incidence from the direction making a large angle with the normal to the interface. In the first case, the factor at the second summand in (2.46) is negative. In the second case, the second summand in (2.74) is negative at $\beta > \beta_{cr}$.

2.2.6
Plano-parallel chiral medium

The waves in which the fields do not depend on x, y, that is, satisfy conditions (2.1), can also propagate in the medium where all three constitutive constants $\varepsilon, \mu, \varkappa$ are functions of z. As follows from the Maxwell equations, these *waves* are also *transversal*, that is, conditions (2.2) hold. Four other equations are obtained from (2.20) by substituting the left-hand sides by the corresponding derivatives. The first equation is

$$-\frac{dH_y}{dz} = ik(\varepsilon E_x - i\varkappa H_x); \tag{2.85}$$

the others are not shown here. Introducing the auxiliary functions $p_{rg}(z), \ldots,$ instead of the fields, by formulas (5.21), we obtain the system of ordinary differential equations

$$\frac{dp_{rg}}{dz} = k(\mu g_{rg} + i\varkappa p_{rg}), \qquad \frac{dg_{rg}}{dz} = -k(\varepsilon p_{rg} - i\varkappa g_{rg}), \qquad (2.86a)$$

$$\frac{dp_{lf}}{dz} = -k(\mu g_{lf} + i\varkappa p_{lf}), \qquad \frac{dg_{lf}}{dz} = k(\varepsilon p_{lf} - i\varkappa g_{lf}). \qquad (2.86b)$$

System (2.86) consists of the two independent subsystems for $p_{rg}(z)$, $g_{rg}(z)$ and $p_{lf}(z)$, $g_{lf}(z)$, respectively. Propagation of the right circularly polarized waves (with the only nonzero "fields" $p_{rg}(z)$, $g_{rg}(z)$) does not cause the appearance of the left circularly polarized waves (with the only nonzero "fields" $p_{lf}(z)$, $g_{lf}(z)$). Waves of two different polarizations, propagating in the z-direction in a plano-layered chiral medium, are not connected, similarly as such waves are not connected at any propagation direction in the homogeneous chiral medium.

System (2.86a) can be immediately rewritten for the field components of the right circularly polarized waves:

$$\frac{dE_x}{dz} = k(\mu H_x + i\varkappa E_x), \qquad \frac{dH_x}{dz} = -k(\varepsilon E_x - i\varkappa H_x). \qquad (2.87)$$

The system for the left circularly polarized waves is similar.

If the wave resistance $w = \sqrt{\mu/\varepsilon}$ does not depend on z, then the system (2.87) has an explicit solution

$$E_x(z) = E_x(0) \exp(-i\gamma), \qquad H_x(z) = H_x(0) \exp(-i\gamma), \qquad (2.88)$$

where

$$\gamma = \int_0^z h(\varsigma)\, d\varsigma. \qquad (2.89)$$

According to (2.23), $h = k\sqrt{\varepsilon\mu} - k\varkappa$. At $w = const$ there exists a connection (2.24) between the components E_x, H_x, similarly as it takes place in the homogeneous medium. For the waves propagating in the opposite direction $-z$ it is required to change the sign at i in (2.24).

The existence of solution (2.88) means that the propagation of the direct wave is not accompanied by the appearance of the opposite back wave. If the wave number of the plano-layer medium is constant, then the propagation of the wave in the direction normal to the interface does not cause the reflection. It occurs only if $dw/dz \neq 0$. If this derivative is small, then approximate methods can be applied to system (2.87), similar to those applied to the system describing the waves in the nonchiral plano-layered medium.

On an elementary example, we illustrate the absence of the reflection at $w = const$ and normal incidence. Let the right circularly polarized wave fall onto the interface $z = 0$ between the half-space $z < 0$ filled by a homogeneous medium with parameters ε_-, μ_-, $\varkappa_-$ and the half-space $z > 0$ filled by a homogeneous medium with parameters ε_+, μ_+, $\varkappa_+$. Then the reflected and transmitted waves arise. The reflection and transmission factors R and D are to be determined.

All four tangential components of the field must be continuous at the interface. Since $p_{\mathrm{lf}} = 0$, $g_{\mathrm{lf}} = 0$ in the whole space, only the continuity of the components E_x, H_x should be provided. These components are

$$\left.\begin{aligned} E_x &= \exp(-ih_-z) + R\exp(ih_-z), \\ H_x &= -\tfrac{i}{w_-}\left[\exp(-ih_-z) - R\exp(ih_-z)\right] \end{aligned}\right\}, \quad z < 0; \tag{2.90a}$$

$$E_x = D\exp(-ih_+z), \quad H_x = -\frac{iD}{w_+}\exp(-ih_+z), \quad z > 0. \tag{2.90b}$$

The field does not contain the parameters $h_\pm$ at $z = 0$, and the boundary conditions are the same as those in the problem for the nonchiral media:

$$1 + R = D, \qquad \frac{1 - R}{w_-} = \frac{D}{w_+}. \tag{2.91}$$

Solution to this system coincides with (2.52). In particular, $R = 0$, $D = 1$ at $w_- = w_+$.

As in the direct wave, $E_y = iE_x$, $H_y = iH_x$ in the reflected wave, that is, the vectors $\vec{E}$, $\vec{H}$ are turning from the x-axis to the y-axis. Formally, this fact is often explained as passing from the one circular polarization to another, although in fact only the propagation direction is changing in this case (see the text below (2.19)).

2.2.7
Media with $\varepsilon < 0$, $\mu < 0$

As was noted in Subsection 1.2.4, there exist media in which ε, μ are the real scalar negative numbers (so-called *metamaterials*). Here we briefly describe properties of the wave propagation in such exotic media. We suppose that the media are homogeneous, that is, $\varepsilon = const < 0$, $\mu = const < 0$.

At $\varepsilon < 0$, $\mu < 0$, the value $\sqrt{\varepsilon\mu}$ is real; we accept that this value is positive, similarly as at $\varepsilon > 0$, $\mu > 0$. If the plane wave propagates in the z-direction (the phase velocity is z-directed), then all the components depend on z as $\exp(-ik\sqrt{\varepsilon\mu}z)$. In such a medium the Poynting vector flux $\vec{S}$ (1.54) is $(-z)$-directed, that is, opposite to the phase velocity direction. If the polarization is such that the wave contains only the components E_x, H_y, then, according to (2.3),

$$E_x = A\exp(-ik\sqrt{\varepsilon\mu}z), \qquad H_y = \frac{\sqrt{\varepsilon\mu}}{\mu}A\exp(-ik\sqrt{\varepsilon\mu}z). \tag{2.92}$$

The only component of the Poynting vector is $S_z = (c/(8\pi))|A|^2\sqrt{\varepsilon\mu}/\mu$. Since $\mu < 0$, then

$$S_z < 0. \tag{2.93}$$

The same property is inherent in the plane wave of the second polarization. It contains the components E_y, H_z, connected by relation (2.4), and the energy flux is directed opposite to the phase velocity as well.

In the media with scalar ε, μ, the energy flux is directed in the same way as the group velocity. The energy is transferred in the same way as the impulse signal. In the medium considered here, the phase and group velocities are "antiparallel." There also exist other devices in which waves have this property, but such devices are not homogeneous and cannot be characterized by any values of ε and μ.

Consider the problem about an oblique incidence of the plane wave from the vacuum ($\varepsilon = 1$, $\mu = 1$) half-space $z < 0$ onto the interface with the half-space $z > 0$ in which $\varepsilon < 0$, $\mu < 0$ (Fig. 2.1). Denote the incidence angle by α. The reflected wave outgoes in the direction making the same angle α with normal to the interface. All the components of the incident wave are proportional to the factor $\exp(-ikx\sin\alpha)$. Since properties of the media do not depend on x, the same factor defines the x-dependence on the fields in the whole plane.

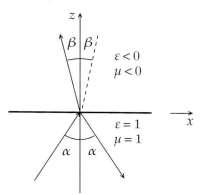

Fig. 2.1 The plane wave refraction on the medium with $\varepsilon < 0$, $\mu < 0$

In the upper half-space, the field components satisfy the wave equation (1.41)

$$\left(\frac{\partial^2}{\partial x^2} + \frac{\partial^2}{\partial z^2} + k^2\varepsilon\mu\right)u = 0. \tag{2.94}$$

According to this equation, the z-dependence of the components is defined by the factor

$$\exp(\mp ipz), \qquad p = k\sqrt{\varepsilon\mu - \sin^2\alpha}. \tag{2.95}$$

To exclude the total internal reflection, we assume that $\varepsilon\mu > 1$. Introduce the angle β by the relation $p = k\sqrt{\varepsilon\mu}\cos\beta$. It is connected with the incidence angle by the condition $\sin^2\beta = \sin^2\alpha/(\varepsilon\mu)$.

The upper sign in (2.95) corresponds to the case $\varepsilon > 0$, $\mu > 0$. In this case the field dependences on x and z are described by the factors $\exp(-ikx\sin\alpha)$ and $\exp(-ipz)$, respectively. The refracted wave outgoes from the interface at the angle β such that $\sin\beta = \sin\alpha/\sqrt{\varepsilon\mu}$. This is the usual refraction law. The Poynting vector flux and the group velocity of this wave are also directed from the interface into the medium (dashed line in Fig. 2.1).

However, in the medium with $\varepsilon < 0$, $\mu < 0$, the energy flux of such a wave would be directed from the medium to the interface. This fact contradicts the energy conservation law, according to which, div $\vec{S} = 0$ in the medium without losses and currents (see (1.56)). The integral form of this law demands that the integrals of S_z over the planes $z = +0$ and $z = -0$ are equal. This means that the energy flux in the domain $z > 0$ must be z-directed. Therefore, the phase velocity must be $(-z)$-directed, that is, to the interface. To fulfill this demand in the case $\varepsilon < 0$, $\mu < 0$, we must take the lower sign in the exponent (2.95) of the z-dependence of the refracted wave.

The field in the refracted wave is described by the factor

$$\exp(-ik\sqrt{\varepsilon\mu}\sin\beta \cdot x)\exp(ik\sqrt{\varepsilon\mu}\cos\beta \cdot z). \tag{2.96}$$

This is the wave propagating from the medium to the interface. The z-component of its phase velocity is negative. The refracted ray in the medium with $\varepsilon < 0$, $\mu < 0$ is mirror-directed to the normal to the ray in the medium with $\varepsilon > 0$, $\mu > 0$. The arrow on the solid line at $z > 0$ in Fig. 2.1 shows the direction of the energy flux.

The second factor in (2.96), dependent on z, can be written in the form $\exp(-iknz\cos\beta)$, where the quantity n is defined in the medium with $\varepsilon < 0$, $\mu < 0$ as

$$n = -\sqrt{\varepsilon\mu}, \qquad n < 0. \tag{2.97}$$

In this notation, the refraction law can be written in the usual form

$$\sin\beta = \frac{\sin\alpha}{n}. \tag{2.98}$$

The angle β is defined to be negative for the rays obtained by anticlockwise turning of the normal to the interface, similarly as for the refracted ray in Fig. 2.1. In usual media ($\varepsilon > 0$, $\mu > 0$) the factor n is positive, $n = \sqrt{\varepsilon\mu}$, and $\beta > 0$. The medium with $\varepsilon < 0$, $\mu < 0$ can be described as a medium with the negative refraction factor.

If the medium with $\varepsilon < 0$, $\mu < 0$ spreads "to infinity," then the asymptotical conditions (1.33) would contain the factor $\exp(ikR)$ instead of $\exp(-ikR)$.

At the interface between media with $n = 1$ and $n = -1$, the refracted ray is the mirror reflection of the incident ray by the interface. The plano-parallel plate of the material with $n = -1$ acts as a converging lens if its depth is larger than the distance to the source. All rays outgoing from point A (Fig. 2.2) converge at point B. The arrows show the direction of the vector $\vec{S}$. Inside the plate the phase velocity is directed oppositely.

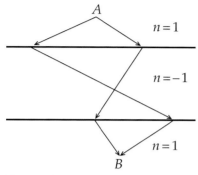

Fig. 2.2 The ray propagation through the layer with $n = -1$

The *Fermat principle* (see Subsection 5.1.5), according to which the integral $\int_{s_1}^{s_2} n ds$ (5.32) is extremal if the integral is taken over the ray connecting points s_1 and s_2, is valid also in the presence of domain with $\varepsilon < 0$, $\mu < 0$ if n is chosen to be negative in this domain (see (2.97)).

3
Closed Waveguides

3.1
Eigenmodes in nonfilled waveguides

3.1.1
Eigenmodes

In this chapter the structure of the electromagnetic waves in the waveguides is investigated. Only one type of waveguides, namely, the waveguides in the form of metallic tubes, is considered. The cross-section of the waveguide does not vary throughout it. The cross-section can be arbitrary; the waveguides of rectangular and circular cross-sections are usually used in practice.

Similarly as in majority of the lines designed for transmitting the field to the distances much larger than the wavelength, the field in the waveguides is convenient to express as a superposition of the eigenwaves (*eigen modes*). Similar to the eigenwaves in the homogeneous medium, the eigenmode is the field which can propagate without changing its structure in the infinitely large waveguide in the absence of other fields. The structure of the eigenmode, that is, the relation between the components of the fields $\vec{E}$, $\vec{H}$, depends on cross-section of the waveguide, its filling, and properties of its walls. Almost everywhere in the closed lines, any field can be expressed as an infinite sum of the eigenmodes with coefficients depending on the excitation type. In the majority of the transmission lines of the usual technical applications, the field contains only one eigenmode.

The field of the eigenmode satisfies the following conditions:

$$\vec{E}^n(x,y,z) = \vec{E}^n(x,y,0)\exp(-ih_nz), \tag{3.1a}$$
$$\vec{H}^n(x,y,z) = \vec{H}^n(x,y,0)\exp(-ih_nz), \tag{3.1b}$$

where z is the coordinate directed along the waveguide, and n is the mode number. The fields $\vec{E}^n(x,y,0)$, $\vec{H}^n(x,y,0)$, as well as the *wave numbers* h_n (*propagation constants*), $n = 1, 2, \ldots$, are different for different eigenmodes (although there exist exceptions from this rule). The number of the eigenmodes in the closed waveguide is infinite.

High-frequency Electrodynamics. Boris Z. Katsenelenbaum
Copyright © 2006 WILEY-VCH Verlag GmbH & Co. KGaA, Weinheim
ISBN: 3-527-40529-1

According to (3.1), $\partial/\partial z \equiv -ih$ for the fields of eigenmodes (the number n is omitted here and in some formulas below). Performing the differentiation with respect to z in the four Maxwell equations involving this derivative, we obtain

$$ihH_y - ik\varepsilon E_x = -\frac{\partial H_z}{\partial y},$$

$$ihE_y + ik\mu H_x = -\frac{\partial E_z}{\partial y},$$

$$-ihH_x - ik\varepsilon E_y = \frac{\partial H_z}{\partial x},$$

$$-ihE_x + ik\mu H_y = \frac{\partial E_z}{\partial x}.$$

(3.2)

Here ε and μ are the parameters of the medium filling the waveguide. In this chapter they are assumed to be independent of z (*regular waveguide*) and x, y (*homogeneous waveguide*).

The transverse components E_x, E_y, H_x, H_y can be expressed by derivatives of the longitudinal components E_z, H_z from equations (3.2). The components E_z, H_z can play the role of scalar potentials. However, it is expedient to introduce the potentials in another way. The determinant of the matrix of system (3.2) equals $k^2\varepsilon\mu(k^2\varepsilon\mu - h^2)$. As we will show further, in the waveguides with multiple-connected cross-section, the modes with $h^2 = k^2\varepsilon\mu$ exist. In these modes, $E_z = 0$, $H_z = 0$, and the transverse components are not expressed by the longitudinal ones. Instead of these components, it is more convenient to introduce the two functions $\chi(x,y)$ and $\psi(x,y)$ proportional to E_z, H_z, as follows:

$$E_z = (k^2\varepsilon\mu - h^2)\chi(x,y),$$ (3.3a)
$$H_z = (k^2\varepsilon\mu - h^2)\psi(x,y).$$ (3.3b)

Here and henceforth, we drop the factor $\exp(-ihz)$, common for all the components. According to (3.2), the transverse components are expressed by the potentials χ and ψ, as

$$E_x = -ih\frac{\partial\chi}{\partial x} - ik\mu\frac{\partial\psi}{\partial y},$$

$$E_y = -ih\frac{\partial\chi}{\partial y} + ik\mu\frac{\partial\psi}{\partial x},$$

$$H_x = ik\varepsilon\frac{\partial\chi}{\partial y} - ih\frac{\partial\psi}{\partial x},$$

$$H_y = -ik\varepsilon\frac{\partial\chi}{\partial x} - ih\frac{\partial\psi}{\partial y}.$$

(3.4)

Equations for the potentials are found from the remaining two Maxwell equations. Substituting expressions (3.4) into these equations, we get the same two wave equations

$$\nabla^2\chi + (k^2\varepsilon\mu - h^2)\chi = 0, \tag{3.5a}$$
$$\nabla^2\psi + (k^2\varepsilon\mu - h^2)\psi = 0, \tag{3.5b}$$

where ∇^2 is the two-dimensional Laplace operator

$$\nabla^2 = \frac{\partial^2}{\partial x^2} + \frac{\partial^2}{\partial y^2}. \tag{3.6}$$

In the cylindrical coordinate system (r, φ, z), formulas (3.4) are

$$\begin{aligned}
E_r &= -ih\frac{\partial\chi}{\partial r} - ik\mu\frac{\partial\psi}{r\partial\varphi}, \\
E_\varphi &= -ih\frac{\partial\chi}{r\partial\varphi} + ik\mu\frac{\partial\psi}{\partial r}, \\
H_r &= ik\varepsilon\frac{\partial\chi}{r\partial\varphi} - ih\frac{\partial\psi}{\partial r}, \\
H_\varphi &= -ik\varepsilon\frac{\partial\chi}{\partial r} - ih\frac{\partial\psi}{r\partial\varphi},
\end{aligned} \tag{3.7}$$

and the two-dimensional Laplace operator is of the form

$$\nabla^2 = \frac{\partial^2}{\partial r^2} + \frac{1}{r}\frac{\partial}{\partial r} + \frac{1}{r^2}\cdot\frac{\partial^2}{\partial\varphi^2}. \tag{3.8}$$

The functions $\chi(x, y)$ and $\psi(x, y)$ should satisfy the boundary conditions on the cross-section contour, which provide fulfilling the boundary conditions for the tangential components of the fields. Further in this section we consider the waveguide with the ideal-conducting walls. Both tangential components of the vector $\vec{E}$ are zero on them. The first tangential component is E_z. The second one lying in the plane of the cross-section is tangential to its contour. Denote this contour by C, and the two unit vectors in this plane, tangential and normal to C, by $\vec{s}$ and $\vec{N}$, respectively. The z-axis and vectors $\vec{N}, \vec{s}$ make up the right-hand triple. It follows from (3.4) that the tangential component E_s is

$$E_s = -ih\frac{\partial\chi}{\partial s} + ik\mu\frac{\partial\psi}{\partial N}. \tag{3.9}$$

Consequently, if the functions χ and ψ satisfy the conditions

$$\chi = 0, \tag{3.10a}$$
$$\frac{\partial\psi}{\partial N} = 0 \tag{3.10b}$$

on the contour C, then the field $\vec{E}$ satisfies condition (1.79) on the ideal-conducting walls of the waveguide.

3.1.2
The TM-, TE-, and TEM-modes

Equation (3.5a) with the boundary condition (3.10a), on one hand, and equation (3.5b) with the boundary condition (3.10b), on the other hand, are independent. Consequently, there exist the eigenmodes with $\psi \equiv 0$ and modes with $\chi \equiv 0$. In modes of the first type, $H_z \equiv 0$ and, according to (3.3b), the only longitudinal component different from zero is E_z. Such modes are called the *transverse magnetic (TM) modes*. The modes of the second type are called the *transverse electric (TE) modes*. In such modes, $E_z \equiv 0$ and the only longitudinal component different from zero is H_z. In other terminology, the TM- and TE- modes are called the *electric* and *magnetic modes*, respectively.

The boundary value problem (3.5a), (3.10a) is homogeneous; it has "nontrivial" (different from zero) solutions only at certain values of the number h. These values are the eigenvalues of the problem; they are denoted by h_n, $n = 1, 2, \ldots$. One eigenfunction $\chi_n(x, y)$ (or several linearly independent ones) corresponds to each eigenvalue h_n. Problem (3.5b), (3.10b) has the solutions $\psi_n(x, y)$, corresponding to the eigenvalues $h = h_n$, which differ (with some exceptions) from those in the first problem. The eigenfunctions are also different. The notations *TM* and *TE* indicate that the magnetic field is transversal in electric modes, and the electric field is transversal in magnetic modes.

Compare problem (3.5a), (3.10a) with the problem

$$\nabla^2 \chi + \alpha^2 \chi = 0, \tag{3.11a}$$

$$\chi|_C = 0. \tag{3.11b}$$

Denote its eigenvalues by α_n, $n = 1, 2, \ldots$. These numbers are defined only by the contour C. The eigenvalues h_n of the electric modes are expressed by α_n as

$$h_n^2 = k^2 \varepsilon \mu - \alpha_n^2. \tag{3.12}$$

Similarly, the problem

$$\nabla^2 \psi + \beta^2 \psi = 0, \tag{3.13a}$$

$$\left. \frac{\partial \psi}{\partial N} \right|_C = 0 \tag{3.13b}$$

is compared with problem (3.5b), (3.10b) for the magnetic modes, and its eigenvalues are expressed by β_n:

$$h_n^2 = k^2 \varepsilon \mu - \beta_n^2. \tag{3.14}$$

Formulas (3.12), (3.14) are main formulas of the theory of waveguides with homogeneous filling.

The quantities α_n^2, β_n^2 are nonnegative numbers, infinitely increasing as n increases. If ε and μ are real, that is, the homogeneous filling of the waveguide has no losses, then the eigenvalues h_n are real if $k^2 \geq \alpha_n^2/\varepsilon\mu$, $k^2 \geq \beta_n^2/\varepsilon\mu$ and imaginary at smaller frequencies. Further, for simplicity, we put $\varepsilon = 1$, $\mu = 1$.

The frequencies

$$k_n^{cut} = \alpha_n,\qquad\qquad (3.15a)$$

$$k_n^{cut} = \beta_n \qquad\qquad (3.15b)$$

are called the *cutoff frequencies* for the nth modes of electric (3.15a) and magnetic (3.15b) types, respectively. At $k < k_n^{cut}$, the eigenvalue h_n is imaginary, and the factor $\exp(-ih_n z)$ describes the damping wave. According to (3.4), the transverse components of the fields $\vec{E}^n$, $\vec{H}^n$ are in-quadrature for both electric ($\psi \equiv 0$) and magnetic ($\chi \equiv 0$) modes at the "low" frequencies, and the mode does not transmit the energy. At the "high" frequencies, when $k > k_n^{cut}$, the eigenvalue h_n is real, $\vec{E}^n$, $\vec{H}^n$ are in-phase, and the mode propagates without damping and transmits the energy. At each frequency, there are only a finite number of modes having real h_n; they can be absent at all. We take that $h_n = -|h_n|$ for the damping modes.

For the multiple-connected cross-sections there exist the solutions $\chi(x,y)$ of problem (3.11), for which $\alpha_n = 0$. We will consider this case in Subsection 3.1.7. Here we assume that the cross-section is single-connected. Then all the numbers α_n, β_n, for which $\chi \neq const$, $\psi \neq const$, differ from zero. For the solutions $\beta = 0$, $\psi = const$, and $\alpha = 0$, $\chi = const$, the fields are zero.

The frequency dependence of the wave numbers h_n are defined by the general formula

$$h_n^2 = k^2 - \left(k_n^{cut}\right)^2. \qquad\qquad (3.16)$$

The phase velocity $v_n = \omega/h_n$ of each eigenmode is larger than c. According to (2.36), the group velocity $v_{n,gr} = h_n/k$ is smaller than c. The relation

$$v_{n,gr}v_n = c^2 \qquad\qquad (3.17)$$

exists for these quantities; it is valid only for the waveguides with the ideal walls, because (3.16) holds only for such waveguides.

3.1.3
The functional orthogonality of the fields of eigenmodes

There exists the relation

$$\int_S \left(\vec{E}^n \times \vec{H}^m\right)_z dS = 0, \qquad n \neq m \qquad\qquad (3.18)$$

between transverse components of the fields of two different eigenmodes. It is called the *orthogonality condition*, implying not the orthogonality of two vectors, the angle between which is $\pi/2$, but the orthogonality as a notion of functional analysis. Indices n, m in (3.18) are the mode numbers; each mode may belong to either TM- or TE-type. Integration is made over the cross-section, that is, over the domain bounded by the contour C. It is assumed that $h_n \neq h_m$. If this condition violates, that is, if the modes correspond to the same eigenvalue, but their fields are different, then two linear combinations can be composed (which are also eigenmodes) for which (3.18) holds.

We prove the property (3.18) using expressions (3.4) for the field components in (3.18) by the scalar functions $\chi(x,y)$, $\psi(x,y)$, together with the equations and boundary conditions (3.11), (3.13) for these functions. The integrand in (3.18) takes the form of one of the following three expressions:

$$- k h_n (\nabla \chi_n \nabla \chi_m), \tag{3.19a}$$

$$- k h_m (\nabla \psi_n \nabla \psi_m), \tag{3.19b}$$

$$- h_n h_m (\nabla \psi_n \times \nabla \psi_m)_z, \tag{3.19c}$$

depending on the type of modes involved in condition (3.18): both modes are electric (a), magnetic (b), or the nth mode is electric and the mth one is magnetic. We show that the integral of each of expressions (3.19) over the cross-section equals zero.

Expression (3.19a) is equal to

$$-\frac{k h_n}{\alpha_n^2 - \alpha_m^2} \operatorname{div} \left(\alpha_n^2 \chi_n \nabla \chi_m - \alpha_m^2 \chi_m \nabla \chi_n \right). \tag{3.20}$$

This assertion is easily checked by unbracketing and then using the identity $\operatorname{div}(a\vec{A}) = a \operatorname{div} \vec{A} + \nabla a \cdot \vec{A}$ and equation (3.5a). According to the *Gauss theorem*, the integral of (3.20) over the cross-section equals the contour integral over C of the normal component of the vector under the divergence; this component vanishes owing to the boundary condition (3.10a). In the same way, using equation (3.5b) and the boundary condition (3.10b), we can prove that the integral of expression (3.19b) equals zero. Expression (3.19c) equals $-h_n h_m \operatorname{rot}_z(\chi_n \nabla \psi_m)$. According to the *Stokes theorem*, the integral of expression (3.19c) over the cross-section equals the contour integral over C of the tangential component of the vector $\chi_n \nabla \chi_m$. Consequently,

$$\int_S \operatorname{rot}_z (\chi_n \nabla \psi_m)\, dS = \int_C (\chi_n \nabla \psi_m)_s\, ds = \int_C \chi_n \frac{\partial \psi_m}{\partial s}\, ds. \tag{3.21}$$

The last integral equals zero owing to the boundary condition (3.10a) for χ_n.

We use the normalization condition

$$\int_S \left(\vec{E}^n \times \vec{H}^n \right)_z dS = -k h_n \tag{3.22}$$

for the fields of eigenmodes. It is easily seen that, according to (3.19a), (3.19b), condition (3.22) leads to the following normalizations:

$$\int\limits_S (\nabla \chi_n)^2 \, dS = 1, \qquad \int\limits_S (\nabla \psi_n)^2 \, dS = 1 \qquad (3.23a)$$

for the potential functions χ_n, ψ_n. They are equivalent to the normalizations

$$\alpha_n^2 \int\limits_S \chi_n^2 \, dS = 1, \qquad \beta_n^2 \int\limits_S \psi_n^2 \, dS = 1. \qquad (3.23b)$$

Hence, the functions $\chi(x, y)$, $\psi(x, y)$ are dimensionless, and the field components have the dimension k^2.

Introducing a certain normalization for the fields, we can use the notion of the *amplitude* of the *eigenmodes*. For instance, the Poynting vector flux transmitted by the mode of the amplitude A is $A^2 \omega / (8\pi) \, \mathrm{Re}\, h_n$.

The system of transverse components of the eigenmodes of waveguide with ideal walls is complete, that is, any transverse component of arbitrary field can be expressed as a series by the corresponding components of the modes. If the field satisfies condition (1.79) on C, then the series converges in the whole cross-section. Further, it will be shown that the set of eigenmodes of the waveguide with impedance walls, or of the waveguide filled with a material nonhomogeneous in the cross-section, may be incomplete. The orthogonality condition for two eigenmodes of such waveguides is more complicated than (3.18).

3.1.4
The electric and magnetic Hertz vectors

The potential functions $\chi(x, y)$, $\psi(x, y)$ introduced by (3.3) are longitudinal components of the *electric* $\vec{\Pi}^{(e)}$ and *magnetic* $\vec{\Pi}^{(m)}$ *Hertz vectors*, respectively. In general form, the electric Hertz vector was introduced in (1.42), (1.43). The fields $\vec{E}$ and $\vec{H}$ appear symmetrically in the Maxwell equations (with accuracy to replacing ε, i by μ, $-i$), but asymmetrically in (1.43). Introduce the vector $\vec{\Pi}^{(m)}(x, y, z)$ satisfying equation (1.42) and express the fields by $\vec{\Pi}^{(m)}$ using the formulas similar to (1.43), as

$$\vec{E} = -ik\mu \, \mathrm{rot}\, \vec{\Pi}^{(m)}, \qquad \vec{H} = k^2 \varepsilon \mu \vec{\Pi}^{(m)} + \mathrm{grad}\, \mathrm{div}\, \vec{\Pi}^{(m)}. \qquad (3.24)$$

Such fields satisfy the Maxwell equations, and the symmetry is renewed.

One can introduce the vectors $\vec{\Pi}^{(e)}$ and $\vec{\Pi}^{(m)}$ simultaneously and express the fields $\vec{E}$, $\vec{H}$ by them. The expressions are "superfluous," because they involve six potential functions (scalar components of the potentials), whereas only two are sufficient for expressing all the six components of any fields $\vec{E}$ and $\vec{H}$, satisfying the Maxwell equations.

As has been already shown, only two scalar potentials are sufficient for describing the eigenmodes in the waveguide. Choosing the longitudinal components of the Hertz vectors as the potentials, we obtain the same functions $\chi(x,y)$, $\psi(x,y)$. Comparing the expressions for the fields by these functions and by the Hertz vectors gives

$$\Pi_z^{(e)}(x,y,z) = \chi(x,y)\exp(-ihz), \tag{3.25a}$$

$$\Pi_z^{(m)}(x,y,z) = \psi(x,y)\exp(-ihz). \tag{3.25b}$$

The wave equations for χ, ψ follow from those for the Hertz vectors (that is, from equations (1.42) without the right-hand sides).

3.1.5
Rectangular waveguides

Apply the general technique based on formulas (3.11)–(3.14) to waveguides with a rectangular cross-section (Fig. 3.1). Locate the origin in the lower-left corner of the section and direct the x-axis along the larger rectangle side of length a and the y-axis along the smaller one of length b.

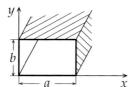

Fig. 3.1 Rectangular waveguide

First, we consider the magnetic modes. We find the function $\psi(x,y)$ from problem (3.13). Expressing it in the form $\psi(x,y) = X(x)Y(y)$ and substituting into (3.13), we get two ordinary differential equations

$$\frac{d^2X}{dx^2} + \beta_x^2 X = 0, \qquad \frac{d^2Y}{dy^2} + \beta_y^2 Y = 0 \tag{3.26}$$

for the functions $X(x)$ and $Y(y)$, respectively; here β_x, β_y are the separation constants connected by the condition $\beta_x^2 + \beta_y^2 = \beta^2$. The boundary condition (3.13b) gives

$$\left.\frac{dX}{dx}\right|_{x=0;a} = 0, \qquad \left.\frac{dY}{dy}\right|_{y=0;b} = 0. \tag{3.27}$$

Only one of the two linearly independent solutions to each equation (3.26) satisfies the corresponding boundary conditions, so that

$$\psi(x,y) = P\cos(\beta_x x)\cos(\beta_y y), \tag{3.28}$$

where the normalizing factor P is calculated from condition (3.23), and β_x, β_y are determined from the boundary conditions (3.27):

$$\beta_x = \frac{\pi m}{a}, \qquad \beta_y = \frac{\pi q}{b}. \tag{3.29}$$

Each of the quantities m, q may take the values $0, 1, 2, \ldots$, but function (3.28) does not correspond to any mode at $m = q = 0$, since in this case $h_n = k$, $\psi = const$ and all field components are zero. The mode number n introduced before is a pair of the integers $\{m, q\}$. The mode with such indices is denoted by TE_{mq} (or H_{mq}); its cutoff frequency is

$$k_{mq}^{cut} = \sqrt{\frac{\pi^2 m^2}{a^2} + \frac{\pi^2 q^2}{b^2}}. \tag{3.30}$$

The simplest of these modes, TE_{10}, corresponding to $m = 1, q = 0$ is mostly used. Its field does not depend on y and has the three non-zero components

$$E_y = Pik\frac{\pi}{a}\sin\frac{\pi x}{a}; \quad H_x = Pih\frac{\pi}{a}\sin\frac{\pi x}{a}; \quad H_z = P\frac{\pi^2}{a^2}\cos\frac{\pi x}{a}. \tag{3.31}$$

The cutoff frequency of the TE_{10}-mode equals $k_{10}^{cut} = \pi/a$, which corresponds to the wavelength $\lambda_{10} = 2a$. All other eigenmodes of the rectangular waveguide have smaller cutoff wavelengths, so that there exists a frequency range, higher than π/a, at which only one (the *main*) mode propagates without dumping. If $a > 2b$, then the next mode is H_{20}, for which $\lambda_{20} = a$; in this case, the one-mode range is $a < \lambda < 2a$. If $a < 2b$, then the cutoff wavelength closest to λ_{10} is λ_{01}, which corresponds to the TE_{01}-mode. The one-mode range is $2b < \lambda < 2a$.

We pass to modes of electric type, denoted as TM_{mq} or E_{mq}. The function $\chi(x.y)$ is found from the same equations (3.26) with the boundary conditions

$$X|_{x=0;a} = 0, \qquad Y|_{y=0;b} = 0, \tag{3.32}$$

following from (3.11b). It is

$$\chi(x, y) = P\sin\frac{\pi m x}{a}\sin\frac{\pi q y}{b}, \tag{3.33}$$

where indices m, q take only positive values $1, 2, \ldots$.

At fixed m, q the TE_{mq}-mode has the same cutoff frequency (3.30) as the TH_{mq}-mode. The propagation constants of these two modes coincide, that is, the modes are degenerated. Any of their superpositions is also the eigenmode of the rectangular waveguide. In particular, a superposition can be combined such that the component E_y, depending on the coordinates as $E_y = \sin(\pi m x/a)\cos(\pi q y/b)$ in both the modes, is not present in the field.

The entire theory of rectangular waveguides can be described in terms of the *longitudinal* modes (in the field of which either $H_y \equiv 0$ or $E_y \equiv 0$), instead of the *transverse* ones (in which either $H_z \equiv 0$ or $E_z \equiv 0$). Such modes make up the complete system as well. They can be formed not by a combination of the two transversal modes, but, in a more natural way, immediately expressing the fields $\vec{E}$, $\vec{H}$ by the Hertz vectors $\vec{\Pi}^{(e)}$ and $\vec{\Pi}^{(m)}$. Choosing the component $\Pi_y^{(m)}(x, y, z)$ as the scalar potential, we can express the field of the mode with $E_y \equiv 0$ by it using formula (3.24). If $\vec{\Pi}^{(m)}$ has only one component $\Pi_y^{(m)}$, then $E_y \equiv 0$; similarly, if $\vec{\Pi}^{(e)}$ has only $\Pi_y^{(e)}$, then $H_y \equiv 0$.

The concept of such longitudinal modes is convenient to apply when investigating the bendings of the rectangular waveguide in the plane of one of its sides. Such a bending mutually connects the eigenmodes TH_n and TM_n of the usual classification, which requires the two scalar potentials. However, it does not connect the longitudinal modes of different types. For instance, if the waveguide is bent in the x-plane (i. e., the bending axis is parallel to the y-axis) and a wave with $E_y \equiv 0$ falls on the bending, then all arisen modes belong to the same type, that is, the total field does not have the component E_y and it can be described by one potential $\Pi_y^{(m)}(x, y, z)$ only.

Return to the usual meaning of transverse modes. The eigenmodes in the rectangular waveguide can be treated as a superposition of several plane waves in free space. We explain such a concept by the simplest example of the TE_{10}-mode. It has only one field component E_y, which, according to (3.31), can be written (with another normalization) as

$$E_y = \exp(-ihz - i\beta x) - \exp(-ihz + i\beta x), \tag{3.34}$$

where $\beta = \pi/a$ and $h^2 + \beta^2 = k^2$. Introduce an angle γ by the equalities

$$\cos \gamma = \frac{h_n}{k}, \qquad \sin \gamma = \frac{\beta_n}{k}. \tag{3.35}$$

Both terms in (3.34) have the form $\exp(-ik(z \cos \gamma \pm x \sin \gamma))$. They describe plane waves the normals to which make the angle γ with the z-axis and the angles $\pi/2 \mp \gamma$ with the x-axis. When interfering, these two waves make up the zero ("nodal") planes $x = 0$ and $x = \pi/\beta$. The field of the plane waves is not disturbed if these planes are metallized. The same effect takes place when introducing the two metallic planes $y = 0$, $y = b$ perpendicular to the electric field. On these planes the condition $E_t = 0$ is fulfilled. The field (3.31) can be treated as a result of interference of two plane waves sequentially reflected from the walls $x = 0$, $x = a$. The plane waves the superposition of which makes up an eigenmode of the waveguide are called the *Brillouin waves*.

The eigenmodes of the rectangular waveguide, the fields of which depend not only on z and x but also on y, can be treated as a sum of four plane waves sequentially reflected from the four walls of the waveguide. All these waves propagate in the directions making the angle γ, $\cos\gamma = h_n/k$, with the z-axis. The phase velocity in these directions is $v = c/\cos\gamma$, and the group velocity is $v_{gr} = c \cdot \cos\gamma$. The relation $vv_{gr} = c^2$, common for all homogeneous waveguides with the ideal-conducted walls, holds.

At high frequencies, $k \gg k_n^{cut}$, the angle γ is small, and the Brillouin waves propagate in the directions close to the z-axis. The mode structure is close to that of the homogeneous plane wave. If the frequency decreases and approaches k_n^{cut} from above ($k > k_n^{cut}$), then the angle γ increases up to $\pi/2$ at $k - k_n^{cut} \ll k_n^{cut}$. At the limit, the Brillouin waves propagate almost perpendicular to the z-axis. The energy flux in the z-direction becomes small at the same amplitude of the mode. According to (3.35), at $k < k_n^{cut}$ we have $\sin\gamma > 1$, the angle γ becomes imaginary, and the mode does not propagate in the z-direction. At $k = k_n^{cut}$ the field corresponds to the wave resonating in the cross-section plane.

If the waveguide walls have the losses, then each reflection of the plane wave is associated with the energy losses. This qualitative conclusion based on consideration of the waveguide mode as a superposition of the plane waves will be confirmed in Section 3.2.

3.1.6
Circular waveguides

The theory of waveguides with the circular cross-section is constructed by the same scheme as the theory of rectangular cross-sections. Solutions to the wave equations (3.11a), (3.13a) with the operator ∇^2 written in cylindrical coordinates in (3.8) are found as a product of a cylindrical function of z and trigonometric function of φ. The Bessel function J_m should be chosen among the cylindrical functions because only it has no singularity at $r = 0$. The argument of the trigonometric function is $m\varphi$, where m takes the values $0, 1, 2, \dots$. At any other m the function would be either ambiguous or nonanalytical at certain φ. The potentials $\psi(r, \varphi)$ and $\chi(r, \varphi)$ of magnetic and electric modes, respectively, are of the form

$$P_{mq}J_m(\beta_{mq}r)\begin{Bmatrix} \cos \\ \sin \end{Bmatrix}(m\varphi). \tag{3.36}$$

All the modes with $m \neq 0$ are polarizationally degenerated, their angular dependences may be $\cos(m\varphi)$, $\sin(m\varphi)$, or any their superposition which can be written as $\cos(m\varphi + \varphi_0)$ with any φ_0. The propagation constant is the same for all φ_0.

For the magnetic modes, the boundary condition (3.13b) gives $J_m'(\beta_{mq}a) = 0$, where a is the cross-section radius, and J_m' is the derivative of the Bessel function with respect to the argument. For the electric modes, replacing β_{mq} with

α_{mq} and using the boundary condition (3.11b), we obtain $J_m(\alpha_{mq}a) = 0$. Introduce the quantities μ_{mq}, ν_{mq} such that $J'_m(\mu_{mq}) = 0$, $J_m(\nu_{mq}) = 0$. For the cutoff frequency of the magnetic mode TE_{mq} (H_{mq}) we have $k^{cut}_{mq} = \mu_{mq}/a$, and for the electric mode TM_{mq} (E_{mq}), $k^{cut}_{mq} = \nu_{mq}/a$.

Among all modes of circular waveguide, the TE_{11}-mode has the largest value of the cutoff wavelength $\lambda_{11} = 3.4a$, which corresponds to $k^{cut}_{11} = 1.84/a$. The structure of its field is very complicated; it has five nonzero components (only $E_z \equiv 0$). It is the main mode, and it is usually applied in the waveguide channels. The closest to the TE_{11}-mode is the symmetrical electric one, TM_{01} (E_{01}). The first zero of the Bessel function J_0 is $\nu_{01} = 2.40$. The cutoff wavelength of this mode is $\lambda_{01} = 2.6a$. In the range $2.6a < \lambda < 3.4a$ only the main mode can propagate in the circular waveguide.

The TM_{01}-mode has the property that its component E_z proportional to $J_0(2.40r/a)$ does not equal zero at $r = 0$. This mode (like similar modes in more complicated waveguides) is applied in electronic devices in which the axial electron flow interacts with the high-frequency electromagnetic field.

We mention two more eigenmodes of the circular waveguide. The field of the mode TE_{01} (H_{01}), as well as of all the TE_{0q}-modes, does not have the component H_φ; hence, the z-component is not present in the current on the walls: $I_z = 0$. Only the azimuth current flows in the walls, although the energy propagates in the z-direction. The charges do not appear on the walls, either, because $E_r = 0$ and, according to (1.81), the two-dimensional divergence of the surface current is zero: $\partial I_\varphi / \partial \varphi = 0$. The absence of the longitudinal current means that the field of this mode is slightly disturbed by the transversal slots; such slots do not overcut the currents. For this reason, the TE_{01}-mode is used in the channels which have turning junctions. As will be shown later, this property of the TE_{01}-mode leads to the fact that its damping caused by the finite conductivity of the walls is small at high frequencies.

Since $\nu_{11} = \mu_{01}$, the mode TM_{11} (E_{11}) has the same cutoff frequency $k^{cut}_n = 3.83/a$ as TE_{01}. These two modes are degenerated. For this reason, they are strongly connected in the long irregular waveguides. This fact will be considered in Section 3.4.

In the waveguides with the elliptic cross-section, the field of the modes is usually expressed in elliptic coordinates. The functions ψ and χ are written as a product of the radial Mathieu function (an analog of the Bessel function in (3.36)) and the angular Mathieu function (an analog of the trigonometric function). The boundary condition on C determines the coefficients in arguments of these functions, that is, the numbers analogous to μ_{mq}, ν_{mq}, and then the cutoff frequencies of the eigenmodes by these coefficients. The field structure of these modes is similar to that of the circular waveguide. However, there is no polarization degeneracy, as well as the degeneracy between the modes analogous to TM_{11} and TE_{01}, in elliptical waveguides. The longitudinal current exists on the walls in the mode analogous to TE_{01}.

The waveguides of more complicated cross-sections are used in the case when certain field structure should be formed. In this case the functions ψ and χ are found from the two-dimensional scalar boundary value problems by special (as a rule, numerical) methods. The fields to be found are expressed by these functions with the use of the universal formulas (3.4).

3.1.7
Multiple-connected cross-sections

If the cross-section contour is not single-connected, that is, if it consists of several closed subcontours having no common points, then there exist eigenmodes in the waveguide, the fields of which are transversal, $E_z \equiv 0$, $H_z \equiv 0$. The propagation constant of these modes is equal to that in free space, that is, their phase velocity is independent of the frequency (it equals c); the same is true for the group velocity.

At $\alpha = 0$, (3.5a) becomes the Laplace equation

$$\nabla^2 \chi = 0. \tag{3.37}$$

It is known that this equation has no nonzero solutions, satisfying the boundary condition (3.10a). However, for the multiple-connected domain condition (3.10a), which follows from the condition

$$E_s|_C = 0 \tag{3.38}$$

and formula (3.9), can be weakened at $h = k$ (it is assumed that $\varepsilon = 1$, $\mu = 1$). In this case, equation (3.38) is satisfied if $\psi \equiv 0$ everywhere and $\chi = const$ on certain subcontours. The values of χ may be different on different subcontours. Then equation (3.37) has nontrivial solutions, that is, there exist the functions satisfying (3.37) and having constant values on the subcontours. Under above assumptions the condition $E_z = 0$ is automatically fulfilled owing to (3.3a).

The above modes are called the TEM-modes (*cable modes*). Similarly as in the plane wave, the field of these modes is transversal. With accuracy to a nonessential common factor, the field components of the TEM-mode are

$$E_x = H_y = -\frac{\partial \chi}{\partial x}\exp(-ikz); \qquad E_y = -H_x = -\frac{\partial \chi}{\partial y}\exp(-ikz). \tag{3.39}$$

In contrast to the homogeneous plane wave, the fields of which are also expressed by these formulas, where χ is a linear function of x and y (then (3.37) is satisfied), in the TEM-mode the function χ is a more complicated solution to (3.37).

Formally, the cutoff frequency of the *TEM*-mode equals zero; hence any frequency is larger than it and the mode can propagate at all frequencies.

A *two-conductor line* consisting of an exterior shell and interior wire is of the greatest interest among the closed lines with multiple-connected cross-sections. The *TEM*-mode of such a line may be described in terms of the long line theory. Since $H_z \equiv 0$ in this mode, the current flows only in the longitudinal direction along both conductors: shell and interior wire. One can introduce the notion of the total current, equal to the integral of the surface current density I_z, taken over the contour of a single conductor. The currents on the conductors differ only in sign. According to (1.80), the current density equals $c/(4\pi)H_s$, so that

$$J = \frac{c}{4\pi} \oint \frac{\partial \chi}{\partial N} \, ds, \tag{3.40}$$

where ds is an element of the contour arc. If V_1, V_2 are the values (constants) of the function $\chi(x,y)$ on the interior (wire) and exterior (shell) contours, respectively, then, according to (3.39), the integral of E_τ over any line connecting the contours (τ is the direction tangential to this line) equals $V_1 - V_2$, because $\int_{P_1}^{P_2} E_\tau \, d\tau = \int_{P_1}^{P_2} \partial \chi / \partial \tau d\tau$. Here P_1 and P_2 are the points on the two conductors. The integral does not depend on the integration path, that is, the field $\vec{E}$ is the potential in cross-section, and $V_1 - V_2$ is the difference of the potentials (in terms of the long line theory) on the central wire and the shell.

The ratio

$$W = \frac{1}{c} \cdot \frac{4\pi \, (V_1 - V_2)}{\oint \partial \chi / \partial N ds} \tag{3.41}$$

between the potential difference and the current is called the *wave resistance* of the line. When connecting two lines, it plays the same role as the quantity $w = (\mu/\varepsilon)^{1/2}$ does in the theory of plane waves in the plano-layered medium. The smaller the difference between the wave resistances of the lines, the easier they can be matched, that is, the smaller the reflection factor of the *TEM*-mode from the connection is.

If the exterior and interior conductors are circular cylinders, then the solution to equation (3.37) has the form

$$\chi = \frac{V_1 - V_2}{\ln(a/b)} \ln r + const, \tag{3.42}$$

where a and b are the radii of the conductors, $a > b$. Such a line is called *co-axial*. According to (3.41), its wave resistance is $W = (2/c) \ln(a/b)$. The wave resistance has the same dimensionality as the usual resistance in the theory of direct currents, and it is convenient to express it in the practical system of units, that is, in ohms. In the Gauss system the resistance unit is 9×10^{11} times larger than an ohm. Introducing this coefficient into (3.41) and substituting the light velocity value there are equivalent to replacing the factor $1/c$ with the number 30. For the co-axial cable we have $W = 60 \ln(a/b)$ ohm.

There exists an analogy between the double-conductor transmission line and the long condenser, that is, an electrostatic object. The field $\vec{E} = -\nabla \chi(x, y)$, with χ satisfying the Laplace equation (3.37) and having the values V_1, V_2 on the conductors, coincides with the static field of the long condenser consisting of the same conductors being under the potentials V_1, V_2. With accuracy to the sign, the charges on both the conductors are the same. The value of the charge per unit length is calculated by formula (3.40) (without the factor c), where the integration is made over the contour of any of the two conductors.

The ratio of the charge per unit length to the potential difference is the capacity per unit length of the condenser; it is dimensionless. The condenser is assumed to be so long that the end effects do not influence the most of its parts, and the charge per unit length does not depend on the length itself. The second multiplier in (3.41) is inversely proportional to the charge per unit length.

As in any closed waveguide, in the two-conductor line, the waveguide modes can propagate, in which (in contrast to the TEM-mode) either E_z or H_z differs from zero, the cutoff frequency exists, the phase and group velocities depend on the frequency, and so on. The potential functions of the waveguide modes satisfy not the Laplace equation (3.37), but the wave equations (3.11a), (3.13a).

The potential function for the TE-modes of the co-axial cable has the form

$$\psi(r, \varphi) = \left[A J_m(\beta_{mq} r) + B N_m(\beta_{mq} r) \right] \begin{Bmatrix} \cos \\ \sin \end{Bmatrix} (m\varphi). \tag{3.43}$$

The existence domain for this solution to equation (3.13a) does not include the point $r = 0$; hence, the Neumann function N_m having a singularity at this point should be presented in (3.43). The boundary condition (3.10b) should be fulfilled at $r = a$ and $r = b$. They give two equations

$$A J'_m(\beta_{mq} a) + B N'_m(\beta_{mq} a) = 0, \quad A J'_m(\beta_{mq} b) + B N'_m(\beta_{mq} b) = 0. \tag{3.44}$$

From this equation we obtain the ratio A/B and transcendental equation

$$J'_m(\beta_{mq} a) N'_m(\beta_{mq} b) - J'_m(\beta_{mq} b) N'_m(\beta_{mq} a) = 0 \tag{3.45}$$

for β_{mq}. Of course, at $b = 0$ it transforms into the equation $J'_m(\beta_{mq} a) = 0$ for the TE-modes in the circular waveguide, since $N'(\beta_{mq} b) \to \infty$ as $b \to 0$.

Similar to the case of the single-connected circular waveguide, the smallest value of the cutoff frequency β_{mq}, that is, the smallest root of equation (3.45), corresponds to the TE_{11}-mode. Its cutoff wavelength is $\lambda_{11} = 2\pi/\beta_{11}$; it is approximated (with an accuracy to about 10%) by the formula $\lambda_{11} = \pi(a + b)$. The co-axial cable is a single-mode line for all waves longer than $\pi(a + b)$.

3.2
Waves in waveguides with nonhomogeneous cross-section filling

3.2.1
Nonhomogeneous filling and impedance walls

In this subsection we consider waves in closed waveguides filled with the material of variable permittivity $\varepsilon = \varepsilon(x, y)$ and permeability $\mu = \mu(x, y)$, dependent on the transverse coordinates x, y. In fact, this means that longitudinal rods or plates are introduced into the waveguide. The cross-section as well as ε and μ is also independent of the z-coordinate, that is, similarly as in Section 3.1, the considered waveguide is regular.

The waveguide, the field on the walls of which fulfills the impedance condition (1.67) instead of (1.79), also belongs to this type. Its cross-section boundary on which (1.79) holds is displaced into the wall by several skin layers, though. The electric field is zero at this depth, but the cross-section is filled with a nonhomogeneous material: $\varepsilon \neq 1$ in the layer near the boundary of this extended cross-section. This conventional boundary should not be taken into account; the integrals over the cross-section and its contour imply the integration over the domain on the boundary of which condition (1.79) holds. However, the complications caused by the nonhomogeneous filling also relate to nonfilled waveguides with impedance walls.

In the general case, the eigenmodes in waveguides with nonhomogeneous filling are not divided into two classes – TE- and TM-modes. Both longitudinal components are present in the field of the eigenmodes of such waveguides, $E_z \neq 0$, $H_z \neq 0$ in this field.

We point out two exceptions when TM- and TE-modes exist independently. The first exception concerns the circular waveguide filled with a material having the permittivity and permeability independent of φ, that is, $\varepsilon = \varepsilon(r)$, $\mu = \mu(r)$. Waveguides with the walls of constant impedance belong to this type, as well. Either $E_z \equiv 0$ or $H_z \equiv 0$ in the eigenmodes, fields of which do not depend on φ. If $\partial/\partial\varphi = 0$ and $\partial/\partial z = -ih$ in the homogeneous Maxwell equations (1.32), then the equations split into the two independent groups: for the components E_z, E_r, H_φ and H_z, H_r, E_φ, respectively. From equations of the first group it follows that

$$E_r = \frac{ih}{h^2 - k^2 \varepsilon \mu} \frac{dE_z}{dr}, \qquad H_\varphi = \frac{ik\varepsilon}{h^2 - k^2 \varepsilon \mu} \frac{dE_z}{dr}. \tag{3.46}$$

Similar equations follow from those of the second group. These equations together with the equation for $E_z(r)$ have a solution; hence, the assumption that all the fields do not depend on φ is noncontradictory if ε and μ are independent of φ, as well. The third equation for the components of the first group leads to the second-order ordinary differential equation for E_z, in which h plays a role of the eigenvalue. This equation has the same meaning as (3.5a); however, it is more complicated than (3.11). There is no simple expression similar to (3.12) for the wave numbers h_n. Modes of this group are the TM_{0q}-modes. The theory of the TE_{0q}- modes in such waveguides is developed in a similar way.

A waveguide of another type in which the division onto the TE- and TM-modes is kept is a rectangular waveguide filled with a material the parameters of which do not depend on the Cartesian coordinate y: $\varepsilon = \varepsilon(x)$, $\mu = \mu(x)$. There exist eigenmodes in this waveguide, having the fields which are independent of y, as well. Similar to (3.46),

$$E_y = \frac{-ik\mu}{h^2 - k^2 \varepsilon \mu} \frac{dH_z}{dx}, \qquad H_x = \frac{ih}{h^2 - k^2 \varepsilon \mu} \frac{dH_z}{dx} \tag{3.47}$$

for these modes, so that if ε and μ do not depend on y, then the fields can also be independent of y. In the fields of these modes $E_z \equiv 0$, that is, the modes are of the TE_{0q} type.

3.2.2
Wave number in impedance waveguide

We find an expression for the wave number in the waveguide, on the walls of which conditions (1.67) hold. At $w = 0$ they transform into conditions (1.79), and h is expressed by formulas (3.12), (3.14), (3.16) at $\varepsilon = 1$, $\mu = 1$.

Considering the transformation from (1.79) to (1.67) as a small perturbation with the smallness parameter $|w|$, we expand h^2 into a series by the powers of w. The first term of this series is $h_0^2 = k^2 - \alpha_0^2$ or $h_0^2 = k^2 - \beta_0^2$, where α_0 and β_0 are eigenvalues of problems (3.11), (3.13), correspondingly, for the waveguide of the same cross-section with ideal-conducting walls. We find the second term in h^2, proportional to w.

The local Cartesian coordinate system mentioned after (1.67) is connected with the system (r, s, z) used here, as $x = -s$, $y = z$, $z = -r$. According to (1.67) the boundary conditions on the cylinder surface are

$$E_s = wH_z, \qquad E_z = -wH_s. \tag{3.48}$$

We express the tangential components in this condition in terms of the functions χ and ψ, using formulas (3.9) (and a similar formula for H_S) and (3.3). The boundary conditions for these functions are

$$-ih\frac{\partial\chi}{\partial s} + ik\frac{\partial\psi}{\partial N} = w\beta^2\psi; \qquad \alpha^2\chi = w\left(ik\frac{\partial\chi}{\partial N} + ih\frac{\partial\psi}{\partial s}\right). \tag{3.49}$$

If the fields of the eigenmodes do not depend on s, then the functions χ and ψ are not connected by the boundary conditions. In waveguides having such eigenmodes, the division onto the TM- and TE-modes is kept also for the case of impedance walls. Of course, this is in accordance with the first example of the preceding subsection. If $\partial/\partial s \neq 0$, then the division is impossible.

The functions χ and ψ satisfy the same equations (3.11a), (3.13a) and α^2, β^2 in (3.49) are eigenvalues of these equations. They are expanded into series by powers of w, as

$$\alpha^2 = \alpha_0^2 + w(\alpha^2)_1 + \cdots, \qquad \beta^2 = \beta_0^2 + w(\beta^2)_1 + \cdots. \tag{3.50}$$

The squared wave number is also expanded in such a series

$$h^2 = h_0^2 - w(\alpha^2)_1 + \cdots \quad \text{or} \quad h^2 = h_0^2 - w(\beta^2)_1 + \cdots \tag{3.51}$$

(the lower indices are not the mode numbers here).

However, the expansions of the type $h = h_0 - w(\alpha^2)_1/2h_0 + \cdots$, following from (3.51), are not applicable at $h_0 = 0$ and, in general, if $|h_0|$ is not large in comparison with $w(\alpha^2)_1$, that is, at the waveguide cutoff frequency or in its neighborhood. According to (3.49), both the functions also cannot be expanded into series by w applicable at any frequency.

The disturbance caused by passing from $w = 0$ to $w \neq 0$ should be investigated for the three different cases: (a) the nondisturbed mode is the TE-mode ($\chi_0 \equiv 0$); (b) it is the TM-mode ($\psi_0 \equiv 0$); (c) the degeneration occurs ($\psi_0 \neq 0$, $\chi_0 \neq 0$). In the first case, besides disturbing the field of the TE-mode, the TM-mode of a small amplitude appears. In the second case, the small TE-mode appears. In the third case, the disturbance, generally speaking, cancels the degeneracy.

We begin with the disturbance of the TE-mode. Multiply the second boundary condition in (3.49) by h and introduce the function $\Xi = h\chi$. Then the first order of h does not take part in (3.49) and, therefore, all the quantities involved in this condition are expanded into series by w, existing at all frequencies.

Substituting expansions (3.50) into (3.13a) and (3.49) yields

$$\psi = \psi_0 + w\psi_1, \qquad \Xi = w\chi_1. \tag{3.52}$$

In the zeroth order we have problem (3.13); in the first one we obtain the equation

$$\nabla^2\psi_1 + \beta_0\psi_1 = -\left(\beta^2\right)_1\psi_0 \tag{3.53}$$

and boundary conditions

$$\frac{\partial \psi_1}{dN} = \frac{-i\beta_0^2}{k}\psi_0 + \frac{1}{k}\frac{\partial \Xi_1}{\partial s}, \qquad \Xi_1 = \frac{ih_0^2}{\beta_0^2}\frac{\partial \psi_0}{\partial s}. \tag{3.54}$$

Eliminating Ξ_1 from the first condition in (3.54) with the use of the second one, we obtain the boundary condition for ψ_1, containing the quantities of the zeroth order only:

$$\frac{\partial \psi_1}{\partial N} = \frac{-i\beta_0^2}{k}\psi_0 + \frac{ih_0^2}{k\beta_0^2}\frac{\partial^2 \psi_0}{\partial s^2}. \tag{3.55}$$

Problem (3.53), (3.55) has a solution only if the corresponding homogeneous one (3.13) has a nonzero solution. It is possible only if a certain connection exists between the right-hand sides of both the boundary condition (3.55) and equation (3.53). This condition gives an equation for $(\beta^2)_1$. In order to obtain this equation, the expression $\psi_1 \nabla^2 \psi_0 - \psi_0 \nabla^2 \psi_1$, which equals $\operatorname{div}(\psi_1 \nabla \psi_0 - \psi_0 \nabla \psi_1)$, should be integrated over the cross-section and transformed into the contour integral using the Green formula, and then the boundary conditions (3.55) for ψ_0 and ψ_1 should be used. In this way the sought expression

$$(\beta^2)_1 = \frac{i}{k}\left[\beta_0^4 \oint \psi_0^2\, ds + h_0^2 \oint \left(\frac{\partial \psi_0}{\partial s}\right)^2 ds\right] \tag{3.56}$$

is obtained for $(\beta^2)_1$.

The squared wave number of the eigenmode in the waveguide with impedance walls, close to the magnetic mode in the same waveguide with ideal-conducting walls, can be written in the following form:

$$h^2 = h_0^2 + (1-i)\,M, \tag{3.57}$$

where, according to (3.56),

$$M = \frac{\mu d}{2}\left[\beta_0^4 \oint \psi_0^2\, ds + h_0^2 \oint \left(\frac{\partial \psi_0}{\partial s}\right)^2 ds\right]. \tag{3.58}$$

Here d is the thickness of the skin layer (1.77), and expression (1.78) for w together with normalizing condition (3.22) for ψ_0 is used. According to (3.57), h' and h'' are obtained from the following system of two linear algebraic equations:

$$\left(h'\right)^2 - \left(h''\right)^2 = h_0^2 + M, \qquad h'h'' = \frac{1}{2}M. \tag{3.59}$$

Far away from the cutoff frequency, h differs from h_0 by a value of the order $k|w|$. In the immediate neighborhood of the cutoff frequency, when $h_0 = 0$, $|h|$ is of the order $k|w|^{1/2}$.

The minimal value of $|h|$ is reached not at $k = \beta_0$, but at $k = \{\beta_0^2 - M\}^{1/2}$. It equals $\sqrt{M}$, where M is calculated by (3.58) at $\beta_0 = k$ and $h_0 = 0$. At the frequency much larger than the cutoff one, that is, at $h_0^2 \gg M$ the damping factor equals $h'' = M/2h_0$, the phase velocity is

$$v = v_0 \left(1 - M/2h_0^2\right). \tag{3.60}$$

It is smaller than v_0, that is, smaller than that in the same waveguide with ideal-conducting walls. It seems as though the finite conductivity enlarges the cross-section, it moves the cross-section contour to the distance proportional to d. The relative decrease of β^2 caused by the enlargement of the cross-section linear dimensions by $(1 + M/2\beta_0^2)$ times would lead to the same relative decrease of the phase velocity.

At the frequency much smaller than the cutoff one, that is, at $h_0^2 < 0$, $|h_0|^2 \gg M$, we have from (3.57)

$$h = -i \left(|h_0| - \frac{M}{2\,|h_0|} \right) + \frac{M}{2\,|h_0|}. \tag{3.61}$$

The energy flux is proportional to h', that is, to the second term in (3.61).

The ratio E_z/H_z, equal to zero in the field of the TE-mode propagating in the waveguide with ideal-conducting walls, has the order $|w|$ far from the cutoff frequency and $|w|^{3/2}$ near it for the waveguide with impedance walls.

The disturbance of the TM-modes, caused by passing to the impedance walls, is investigated by the same scheme. First, the function $\Psi = h\psi$ is introduced; then the first of conditions (3.49) is multiplied by h. Equation (3.11) and conditions (3.49) are expanded into series by w. The expansion of α^2 is given in (3.50), and the functions Ψ and χ are expanded as follows:

$$\Psi = w\Psi_1 + \cdots, \qquad \chi = \chi_0 + w\chi_1. \tag{3.62}$$

The functions χ_1 should solve the problem

$$\nabla^2 \chi_1 + \alpha_0^2 \chi_1 = -(\alpha^2)_1 \chi_0, \qquad \chi_1 = \left. \frac{ik}{\alpha_0^2} \frac{\partial \chi_0}{\partial N} \right|_C. \tag{3.63}$$

Similar to (3.56), the solvability condition for this problem has a form of the explicit expression:

$$(\alpha^2)_1 = ik \oint \left(\frac{\partial \chi_0}{\partial N} \right) ds. \tag{3.64}$$

Formulas (3.59)–(3.61), describing the disturbance of the modes of magnetic type, are transferred to the case of modes of the electric type, by replacing expression (3.58) with

$$M = \frac{\mu k^2 d}{2} \oint \left(\frac{\partial \chi_0}{\partial N} \right)^2 ds. \tag{3.65}$$

Consider briefly the third case when two eigenmodes of a waveguide with ideal-conducting walls are degenerated. Then the main result of the replacement of ideal-conducting walls by the impedance ones consists in the fact that the eigenmodes of the "nondisturbed" waveguide cannot propagate without a strong distortion of their structure, that is, they cease to be eigenmodes. Two linear combinations of both the waves exist, being close to the eigenmodes of the impedance waveguide. Only these combinations are weekly disturbed by the transition from the ideal-conductive walls to the impedance ones.

If $\psi_0 \neq 0$, $\chi_0 \neq 0$ at $a_0 = \beta_0$ (as, for instance, for the TE_{mq}- and TM_{mq}-modes in the rectangular waveguide), then both systems (3.11) and (3.13) have solutions. At $w \neq 0$ there must exist solutions to the nonhomogeneous equation (3.53), together with a similar equation for χ_1 and nonhomogeneous boundary conditions for χ_1 and ψ_1 obtained from (3.49) after replacing all the quantities on the right-hand side by the corresponding nondisturbed values. This demand is fulfilled only for the two values of the ratio between amplitudes involved in the boundary condition at ψ_0 and χ_0. Only the waves corresponding to these ratios are propagating without distortion, that is, they are the eigenmodes of the impedance waveguide. Both the components E_z and H_z are present in the field of these *hybrid modes*. Their propagation constants differ by a value of the order $\Delta h \approx k|w|$.

The situation is almost the same as in the case when the disturbance is caused by the appearance of the chirality in dielectric. The disturbance removes the polarization degeneracy of the plane waves. However, this effect has no great practical importance in the waveguide engineering. If the field of a degenerated mode is created at the waveguide entrance, then two eigenmodes propagate in it, but their misphasing occurs only at the distance of the order $1/\Delta h$, that is, λ^2/d. This is about hundreds of meters, whereas the rectangular waveguides are usually no more than several meters long. The field structure created at the waveguide entrance does not practically change at this distance.

3.2.3
Losses in impedance waveguide

If material, filling the waveguide, has losses and $\varepsilon'' < 0$ or $\mu'' < 0$, then the eigenmodes propagate with damping at all frequencies, $h = h' + ih''$, $h'' < 0$, and h^2 is complex. In this case, the notion of the cutoff frequency k_n does not have such an exact meaning as in the nonfilled waveguides with ideal-conducting walls for which h^2 is real and either $h'' = 0$ or $h' = 0$. However, if the losses are not very large, then there exists a high-frequency range, in which $|h''| \ll h'$ and propagation of the eigenmode is accompanied by not very large damping (the damping per wavelength is small, $|h''| \ll k$), and a low-frequency range in which $|h''|$ is large. At the high frequencies the energy flux along the waveguide is not zero, $h'' \neq 0$. A part of the energy transforms into the heat and it should be compensated by the energy influx. Between these two frequency ranges there exists an intermediate range in which h' and $|h''|$ are of the same order.

The formulas of the preceding subsection give the values of h' and h'' for the impedance waveguide. The most significant influence of the nonideal walls on the field consists in the appearance of damping. Far from the cutoff frequency (in the high-frequency range) the damping coefficient is

$$h'' = -\frac{M}{2h_0},$$ (3.66)

where M is given in (3.58) for TE-modes and in (3.65) for TM-modes.

The above formula can be obtained using the energetic considerations. The energy flux, transferred by the mode, decreases as $\exp(2h''z)$, $h'' < 0$. Consider a waveguide segment between the cross-sections z and $z + \Delta z$. A part of the transferred energy, equal to $2|h''|\Delta z$, is lost in the segment. It is the energy outgoing into the impedance wall. According to (1.68), this energy equals $\Delta z \cdot c/(8\pi)w \int (H_z^2 + H_s^2)\, ds$, where the integral is taken over the cross-section contour. Equating these two expressions and substituting the tangential components of the magnetic field in the waveguide with ideal-conducting walls, that is, according to (3.4), (3.8), $H_s = ih\partial\psi/\partial s$, $H_z = \beta^2\psi$ (for TE-modes) and $H_s = ik\partial\psi/\partial N$ (for TM-modes), we obtain (3.56) and (3.58), (3.65).

The energy losses occur in the boundary layer, the cross-sectional area of which has the order dl, where l is the length of the contour C. The energy is transferred in the z-direction through the cross-section of the area proportional to l^2. The relative energy losses have the order of the ratio of these two areas, that is, h'' has the order kd/l. As any surface effect, the damping has the order of the ratio of the cross-sectional area of the boundary layer and the area of the entire cross-section.

At the frequency a little larger than the cutoff one, h'' mainly depends on the frequency just as $1/h_0$ does, that is, it is fast decreasing when moving away from the cutoff frequency. At the high frequency, h'' increases fast with increasing frequency, which is described by the factor dh_0 or dk^2/h_0 in the expression for h''. At the high frequency $h_0 \approx k$, $d \sim k^{-1/2}$, and h'' increases as $k^{1/2}$.

There exists an exception from the rule "the value h'' increases with increasing frequency," which concerns the TE_{01}-mode (and, in general, all TE_{0q}-modes) in the waveguide with circular cross-section. For this wave $\partial\psi/\partial s \equiv 0$, and $h'' \sim d/h_0$, that is, $h'' \sim k^{-3/2}$; the second term in (3.58) is not present. As has already been noticed, there are no longitudinal currents I_z in the TE_{01}-mode and the electric field has the only component E_s there. The losses increase in all other waveguides is caused just by the longitudinal currents.

In the frequently used waveguide channels, formulas (3.56), (3.58), and (3.65) lead to the following expressions for h'':

$$h'' = \frac{d}{h_0} \left(\frac{k^2}{2b} + \frac{\pi^2}{a^3} \right), \tag{3.67a}$$

$$h'' = \frac{d}{2(\mu^2 - 1)ah_0} \left[k^2 + \frac{1}{a^2}\mu^2 \left(\mu^2 - 1 \right) \right], \qquad \mu = 1.84; \tag{3.67b}$$

$$h'' = kd \frac{1/a + 1/b}{2\ln a/b} \tag{3.67c}$$

for the TE_{10}-modes in the rectangular waveguide (3.67a), the TE_{11}-mode in the circular waveguide (3.67b), and the TEM-mode in the coaxial cable (3.67c).

In the decimeter band, the losses have the order of hundredth of decibel per meter. For instance, for the standard waveguide with $a = 0.072$ m, $b = 0.034$ m, the damping coefficient is $h'' = 1.8 \times 10^{-3}$ m^{-1}, that is, 0.015 dB/m at $\lambda = 0.1$ m and $d = 1\,\mu k$. In the centimeter range, h'' is much larger, particularly due to the fact that the cross-section must be smaller; otherwise the waveguide ceases to be a single-mode waveguide. For instance, at $\lambda \sim 0.002$ m the damping is $h'' \sim 5$ dB/m. Mainly due to the above, the application of the closed waveguides is limited in the millimeter range.

The losses caused by the material nonideality can be decreased by placing dielectric layers of finite thickness onto the walls. The decrease is essentially large at high frequencies, that is, when the wavelength is much smaller than the linear sizes of the cross-section. However, even for the standard waveguide and $\lambda = 0.1$ m, the damping decreases 3–4 times when the layer of thickness $\lambda/(4(\varepsilon - 1)^{1/2})$ is placed onto the side walls.

3.2.4
Orthogonality condition

Relation (3.18) between the fields of the two different eigenmodes (the *orthogonality condition*) was proven using the properties of the two scalar functions $\psi(x,y)$ and $\chi(x,y)$. However, only for the nonfilled waveguides the fields of all eigenmodes are expressed by the two functions satisfying equations (3.11a), (3.13a) and only for the waveguides with ideal-conducting walls these functions fulfill the simple boundary conditions (3.11b), (3.13b). For the more complicated waveguides, the orthogonality condition has a more general form, and (3.18) is only its particular case. Derivation of this general condition is directly based on the Maxwell equations.

Using formula (1.45) and the homogenous Maxwell equations, we can verify that the fields of two eigenmodes are connected by the equality

$$\text{div} \left(\vec{E}^n \times \vec{H}^m - \vec{E}^m \times \vec{H}^n \right) = 0, \tag{3.68}$$

similar to (1.49). We integrate this equality over the domain bounded by the waveguide segment $z_1 < z < z_2$ lying between the two cross-sections $z = z_1$ and $z = z_2$, and then transform the volume integral of the divergence into the surface one. The surface integral is a sum of the integral

$$\int_{z_1}^{z_2} \int_C (E_z^n H_s^m - E_s^n H_z^m - E_z^m H_s^n + E_s^m H_z^n)\, ds\, dz \tag{3.69}$$

of the transverse components of the vector located under the divergence sign in (3.68), taken over the waveguide surface, and the integral taken over the cross-sections $z = z_1$, $z = z_2$ of the z-component of the vector under divergence in (3.68). In (3.69) indices z, s correspond to the two unit vectors $\vec{z}$ and $\vec{s}$ tangential to the wall, and ds is an element of the contour. The integral (3.69) equals zero either for the waveguide with ideal-conducting walls or for the impedance waveguide, on walls of which the boundary conditions (3.48) hold. The integrals taken over the cross-sections $z = z_1$ and $z = z_2$, respectively, are equal to each other and opposite in sign. Therefore, the normals to the planes $z = z_1$ and $z = z_2$ have opposite directions. The dependence of the components of this vector on the z-coordinate is given by the multiplier $\exp[-i(h_n + h_m)z]$. Thus, the quantity

$$\exp\left[-i\left(h_n + h_m\right)z\right] \int_S \left[\left(\vec{E}^n \times \vec{H}^m\right)_z - \left(\vec{E}^m \times \vec{H}^n\right)_z\right] dS \tag{3.70}$$

is the same at $z = z_1$ and $z = z_2$; here dS is an element of the cross-section. The values z_1 and z_2 may be chosen arbitrarily so that at $h_n + h_m \neq 0$ the product (3.70) is independent of z. This means that

$$\int_S \left[\left(\vec{E}^n \times \vec{H}^m\right)_z - \left(\vec{E}^m \times \vec{H}^n\right)_z\right] dS = 0. \tag{3.71}$$

This is the general form of the sought functional orthogonality condition for the components of the fields of two eigenmodes. It involves only the transverse components of these fields. It is valid for the modes for which $h_m \neq -h_n$.

Consider a special case. Let the nth eigenmode have the property that if the sign of the wave number h_n is changed, that is, if the direct mode of the nth number is replaced with the back one of the same number, then either the two components E_x, E_y change the sign and H_x, H_y remain the same, or, conversely, the components H_x, H_y change the sign and E_x, E_y remain the same. Then the equality

$$\int_S \left[\left(\vec{E}^n \times \vec{H}^m\right)_z + \left(\vec{E}^m \times \vec{H}^n\right)_z\right] dS = 0 \tag{3.72}$$

holds together with (3.71). Hence, the integral of each term from (3.71), (3.72) equals zero for the eigenmodes having the above property. In this case the orthogonality condition holds in the simpler form (3.18).

It follows from (3.2) or (3.4) that the modes with either $E_z = 0$ or $H_z = 0$, that is, the TE- or TM-modes, possess the above property. For instance, if the sign of h changes in (3.4) for the electric waves, then the components E_x, E_y change the sign, and H_x, H_y remain the same. For the more complicated waveguides in which the fields of eigenmodes contain, in general, both the longitudinal components, the orthogonality condition is valid only in the general form (3.71).

Relation (3.71) takes place for any two waves of the same direction or two opposite modes with different numbers. The integral in (3.71) differs from zero only for the two modes of the same number, directed toward each other. We use the following normalization for the modes:

$$\int_S \left[\left(\vec{E}^n \times \vec{H}^{-n} \right)_z - \left(\vec{E}^{-n} \times \vec{H}^n \right)_z \right] dS = -2kh_n, \tag{3.73}$$

where $\vec{E}^{-n}$, $\vec{H}^{-n}$ are the fields of the nth mode propagating in the $(-z)$-direction. It is clear that the previous normalization (3.22) conforms with (3.73). The mode of the amplitude A transmits the energy $\omega/(8\pi) \cdot |A|^2 \operatorname{Re} h_n$.

3.2.5
Complex and associated modes

For the nonfilled waveguides with ideal-conducting walls, the squared wave number h^2 is real, and h is either real or imaginary. If the material filling (fully or partially) the waveguide has losses or there are losses into the walls, then h^2 is complex. The eigenmodes with complex h, that is, with $h' \neq 0, h'' \neq 0$, can also exist in the waveguide with nonhomogeneous filling although if there no losses in the filling material, and the waveguide walls are ideal-conducting without losses. The energy flux through the cross-section equals zero,

$$\int_S \operatorname{Re} \left(\vec{E} \times \vec{H}^* \right)_z dS = 0, \tag{3.74}$$

for such modes. The integrand in (3.74) is not identical zero, as for the eigenmodes for which $h^2 < 0$ at low frequencies, and the transverse components of $\vec{E}$ and $\vec{H}$ are in-quadrature. For the modes with complex h^2 and without losses, the transverse component of the Poynting vector $\omega/(8\pi) \cdot \operatorname{Re}(\vec{E} \times \vec{H}^*)_z$ is positive in one part of the cross-section and negative in another one, so that condition (3.74) holds.

Sometimes, the above assertion is formulated as the existence of the opposite energy fluxes. This formulation is illustrative, but not correct due to the fact that only the integral of the Poynting vector over a closed or infinite surface has a meaning of the energy flux.

For the modes fulfilling condition (3.74), the quantity h'' may differ from zero also in the absence of losses. This fact does not contradict formula (3.66). The circular waveguide with a round dielectric rod placed along its axis is an example of such a waveguide. In a certain frequency range, h^2 is complex for some nonsymmetrical $(\partial/\partial\varphi \neq 0)$ eigenmodes of this waveguide.

The squared wave number h^2 is an eigenvalue of some vector boundary problem for the transverse components of the fields. In the case of nonfilled waveguides with ideal-conducting walls, this problem is reduced to the scalar self-adjoint ones (3.13) and it has the real eigenvalues only. For the waveguides of more complicated form, the problem, in general, is non-self-adjoint; more precisely, a non-self-adjoint operator corresponds to it. Therefore, it can have complex eigenvalues.

The non-self-adjointness of the problem, the eigenfunctions of which are the fields of eigenmodes, is essential for the completeness of the system of transverse components of eigenmodes and for the existence of the so-called *associated* modes.

The completeness of a system of functions implies that the functions of a certain class can be approximated by a finite superposition of functions of this system so that this approximation can be made as precise as wanted by increasing the number of the superposition terms. The inaccuracy can be characterized, for instance, in the L_2 metric, that is, by the value of the integral of the squared modulus of the difference between the given function and its approximation, taken over the function definition domain. The system of the eigenfunctions of the self-adjoint operator is always complete. The completeness may not take place for the non-self-adjoint operator.

If the eigenvalues of the two modes coincide at a certain frequency, then, in general, this leads only to a usual degeneration ($h_n = h_m$). If the fields of these modes also coincide, then the system of the eigenmodes ceases to be complete. In this case the system of eigenfunctions supplemented by the associated mode is complete. It can be shown that the fields of the associated mode can be obtained by differentiating the coincident fields with respect to the parameter h. According to (3.1), this means that the electric field of the associated mode equals

$$\left\{ \frac{\partial \vec{E}^n(x,y,z)}{\partial h_n} - iz\vec{E}^n(x,y,z) \right\} \exp\left(-ih_n z\right). \tag{3.75}$$

A similar formula expresses the magnetic field of the associated mode by the magnetic field of the eigenmode, the fields of which coincide with those of another wave at the same frequency. If the fields of three eigenmodes coincide, then two associated modes appear which together with all the eigenmodes make up a complete system.

The existence of the modes (3.75) dependent on the z-coordinate in a different way than the eigenmodes leads to some nontypical resonance phenomena. In certain cases the field near the source with given excitation current $\vec{j}^{\text{ext}}$ turns to be formally infinite, as in a more simpler case of the waveguide excitation at the cutoff frequency of an eigenmode when its amplitude turns to be infinite, either. This means that some idealizations fail to be consistent: they become mutually exclusive. For the above simple case, we consider this question in the next section.

3.3
Excitation of closed waveguides

3.3.1
Excitation by extrinsic current

Fields of the modes considered in Subsection 3.1.8 satisfy the homogeneous Maxwell equations. These modes are used for constructing solutions of the problem about the fields created in the waveguide by certain sources located inside it or on its walls. The problem of determining the mode amplitudes is a typical problem of diffraction theory. The given are sources, equations, and boundary conditions, which the solution must satisfy. In this subsection we find the solution for the case of sources given in the simplest form of the extrinsic current with density $\vec{j}^{\text{ext}}$.

One of the most general methods for solving the diffraction problems is based on the use of the Green function. In the vector problems the Green function describes a field created by a certain simple source, and subjected to certain boundary conditions on certain surfaces. These surfaces and boundary conditions are usually chosen such that the Green function can be written in the explicit form. The flexibility of the Green function method consists in the fact that the surfaces and conditions on them can be agreed with the diffraction problem for which the Green function is introduced.

Application of this method to the Maxwell equations is based on formula (1.49) connecting the fields $\vec{E}^{(1)}$, $\vec{H}^{(1)}$ and $\vec{E}^{(2)}$, $\vec{H}^{(2)}$, created by the extrinsic currents with densities $\vec{j}^{(1)}$ and $\vec{j}^{(2)}$, respectively. An integral form of this equality is

$$\int_S \left[\left(\vec{E} \times \vec{H}^g \right)_N - \left(\vec{E}^g \times \vec{H} \right)_N \right] dS = \frac{4\pi}{c} \int_V \left(\vec{E}^g \vec{j}^{\text{ext}} - \vec{E} \vec{j}^g \right) dV. \qquad (3.76)$$

Here $\vec{E}$, $\vec{H}$ are the sought fields, and $\vec{j}^{\text{ext}}$ is a current creating them; $\vec{E}^g$, $\vec{H}^g$

are the fields of the Green function and $\vec{j}^g$ is a current creating them. The surface integral is taken over any surfaces enclosing all the currents $\vec{j}^{ext}$; N is the normal to this surface, and V is the volume bounded by S. The surface integral involves only the tangential components of the fields.

In the waveguide problems, the Green function should be chosen as one of the eigenmodes of the unit amplitude. The electric field $\vec{E}^g$ must satisfy the same condition as the sought field $\vec{E}$. The source $\vec{j}^g$ is as if moved into "infinity," it lies outside the integration domain of (3.76), and, therefore, $\vec{j}^g$ should be set equal to zero in this formula. The existence of the source of the Green function manifests itself by the fact that this function violates the radiation condition at $z = \infty$ or $z = -\infty$: it does not represent an outgoing wave at one of the two "infinities."

We apply the Green function method to the problem about field created by a current $\vec{j}^{ext}$. As the Green function, we take a field of an eigenmode propagating in the back direction (into $z = -\infty$). First, we suppose that the waveguide walls are ideal conducting and $\{\vec{E}^g, \vec{H}^g\}$ is a field of an eigenmode in such a waveguide. Then $E_t = 0$, $E_t^g = 0$ on the waveguide walls, and the integral in (3.76) taken over the walls is zero. Similar to deriving the orthogonality condition (3.71), here we choose the waveguide segment between the cross-sections $z = z_1$ and $z = z_2$ ($z_1 < z_2$), as an integration domain in (3.76). The currents j^{ext} lie in this domain. The integrals over the cross-sections are the same as those involved in the derivation of (3.71). Since at $z = z_1$, the field $\{\vec{E}, \vec{H}\}$ consists of modes of the same direction as the mode $\{\vec{E}^g, \vec{H}^g\}$, the integral over the cross-section $z = z_1$ is zero. At $z = z_2$ the field $\{\vec{E}, \vec{H}\}$ contains only modes opposite to the mode $\{\vec{E}^g, \vec{H}^g\}$. Only the term proportional to the amplitude of the mode, having the same number as in the Green function with opposite sign, remains under the integral over $z = z_2$. Denote the field of this mode by $\{\vec{E}_n, \vec{H}_n\}$. Then the field $\{\vec{E}^g, \vec{H}^g\}$ can be denoted as $\{\vec{E}_{-n}, \vec{H}_{-n}\}$. According to the normalizing condition (3.73), the integral over the cross-section $z = z_2$ equals $-2kh_n \cdot A_n$, where A_n is the sought amplitude of the nth mode created by the extrinsic current with density $\vec{j}^{ext}$. Thus,

$$A_n = \frac{2\pi}{\omega} \cdot \frac{1}{h_n} \int_V \vec{j}^{ext} \vec{E}_{-n} \, dV. \tag{3.77}$$

When finding the amplitudes of modes propagating in the back direction, a direct mode must be chosen as the Green function. Then the integral over the cross-section $z = z_2$ contains only the fields of the same-directed modes and, therefore, equals zero. The term containing the product of the two fields of opposite modes differs from zero under the integral over the cross-section $z = z_1$. The same formula (3.77) with n replaced by $-n$ (recall that $h_{-n} = -h_n$) can be obtained for the amplitude of the back mode created by the extrinsic current $\vec{j}^{ext}$.

Formula (3.77) remains valid also in the case of the waveguide with impedance walls. The Green function must fulfill the same impedance conditions and satisfy the same equations (3.71), (3.73). The above result also takes place for the waveguide filled with nonhomogeneous dielectric, $\varepsilon = \varepsilon(x,y)$, $\mu = \mu(x,y)$. It is clear that $\vec{E}^g$, $\vec{H}^g$ must be chosen as the fields of the eigenmode in this waveguide. Using the same scheme, the amplitude of the associated wave could be found in the waveguides with an incomplete system of eigenmodes.

According to (3.77), in order to excite a mode with a noticeable amplitude, the excitation currents must be located along the electric field of this mode. For instance, the current located in the cross-section of the rectangular waveguide parallel to the narrow wall excites the main TE_{10}-mode with amplitude increasing as the current location approaches the cross-section center where the field of this mode is maximal. The longitudinal current located on the axis of the circular waveguide excites only the TH_{0q}-modes, the only modes in the field of which $E_z \neq 0$ on the axis. The longitudinal currents ($j_{x,y} = 0$) do not excite the TE-modes. The circular current located in the middle of the cross-section of the circular waveguide excites the TE_{0q}-modes.

If the currents are located in several cross-sections, then the modes arisen in these cross-sections are summed with the corresponding phase incursions. The phase incursion of the direct mode is described by the factor $\exp(ih_n z)$ involved in $\vec{E}_{-n}$. The fields of the current elements located at $z = -L$ and $z = 0$ appear in A_n, $n > 0$, with factors $\exp(-ih_n L)$ and 1, respectively. The difference between the phases of these two summands is equal to the electric ("optical") distance between the two elementary sources.

The amplitude A_n contains the wave number h_n in the denominator. Near the cutoff frequency of the mode excited by a given current, $A_n \to \infty$. The transverse components of either electric or magnetic fields tend to zero, as well. However, the flux of the energy carried away by this mode is proportional to $|A_n|^2$, that is, it tends to infinity. In order that a finite (not infinitely large) current radiates the infinite energy, it should be excited by an infinite field. The formulation of the problem about the field of a given current has no physical sense near the cutoff frequency. We must give the field exciting the current, but not the current itself. At the small h_n ($h_n^2 > 0$) the arisen current is small and the energy carried away into infinity is finite if the given field is finite. No infinities appear in the case when the inverse influence of the excited field onto the current is taken into account, that is, when the idealization "$j^{\text{ext}} = const$" is abandoned. Since the inverse influence is not strong at the nonsmall h_n, the multiplier $1/h_n$ properly describes the dependence of A_n on the frequency.

In contrast to the transverse components of the field, the longitudinal ones cannot be, in general, represented as a sum of the longitudinal components of the fields of the excited eigenmodes with the coefficients (3.77). The longitudinal components must be found from the nonhomogeneous Maxwell equations using the transverse components. It follows from the equations that

$$E_z = \frac{1}{ik} \left(\mathrm{rot}\, \vec{H} \right)_z - \frac{4\pi}{i\omega} j_z^{\mathrm{ext}} \qquad (3.78)$$

in the domain with extrinsic currents. The first term in (3.78) contains only the derivatives of the transverse components with respect to x, y, and equals the sum of the longitudinal components of the arisen eigenmodes. The second term is present only for the domains where the longitudinal extrinsic currents exist. It is essential in the theory of electron devices for describing the interaction between the electromagnetic field and electron beam.

The appearance of the second term in (3.78) can be simply explained. The first term describes the longitudinal field which would be created in some cross-section $z = \tilde{z}$ by the currents located at $z < \tilde{z}$ and $z > \tilde{z}$. This field would appear in the narrow slot between the cross-sections $z = \tilde{z} - 0$ and $z = \tilde{z} + 0$. When cutting a slot, the currents would become discontinuous and the surface charges ρ_{surf}, equal to $\pm 1/i\omega \cdot j_z$ by (1.18), would arise on both the cross-sections. According to (1.17), these charges would create the field $E_z = 4\pi\rho_{\mathrm{surf}}$ in the slot. The field (3.78) is a field in the absence of such a slot. The second term in (3.78) compensates this additional field, which would appear only in the "virtual" experiment with the slot cutting and which does not exist in the absence of the slot.

3.3.2
The slot excitation and end-plane excitation

The waveguide may be excited not by a current $\vec{j}^{\mathrm{ext}}$ located inside it, but by a field created on the hole in the waveguide wall by an external source (Fig. 3.2(a)). In this case the volume integral is not present in formula (3.76), since $\vec{j}^g$ equals zero. Similarly as in Subsection 3.3.1, the field $\{\vec{E}^g, \vec{H}^g\}$ is a field of the $(-n)$th mode with unit amplitude. Both the cross-sections must be chosen such that the hole in the wall lies between them. Similar to the above, the integral over the cross-section $z = z_1$ is zero, whereas the integral over the cross-section $z = z_2$ $(z_2 > z_1)$ equals $2kh_n A_n$, where A_n $(n > 0)$ is the sought amplitude of the direct mode.

Only the integral over the hole remains from the integral over the side wall in (3.76). Since the tangential component $E_{\mathrm{tan}}^g = 0$ on both the wall and the hole, the second term under the surface integral is zero, as well. Hence,

$$A_n = \frac{1}{2kh_n} \int\limits_S \left(\vec{E} \times \vec{H}^g \right)_N dS, \qquad (3.79)$$

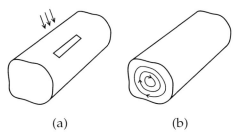

(a)   (b)

Fig. 3.2 Slot excitation and end-plane excitation

where the integral is taken over the hole and $\vec{E}$ is a field created on it by the external sources. This formula can be rewritten in another form by replacing the tangential components of $\vec{H}^g$ with the surface current density $\vec{I}^g$ by (1.80). Then (3.79) gives

$$A_n = \frac{2\pi}{\omega}\frac{1}{h_n}\int\limits_S \vec{E}\cdot\vec{I}^g\,dS, \tag{3.80}$$

where the integral is taken over the hole, $\vec{I}^g$ is a current flowing in the waveguide without the hole when the back mode of the nth number and unit amplitude propagates in it. The field $\vec{E}$ in (3.80) plays a role of the current $\vec{j}^{\text{ext}}$ in (3.77); the current $\vec{I}^g$ in (3.80) plays a role of the field $\vec{E}_{-n}$, that is, the Green function.

If the hole has the form of the long narrow slot, then the electric field $\vec{E}$ on it is directed perpendicular to the longer side of the slot. According to (3.80), in order to excite a mode, the direction of this field must coincide with the direction of the current flowing in the waveguide without the hole. Hence, the slot should cut these currents. The most intense field interaction in the slot, cut in the metallic surface, occurs if the slot cuts the currents flowing in the metal.

The efficiency of applying the Green function in the problems connected with finding the field created either by E_{tan} given on some surface, or by a given current, is explained by the unnecessity to find the currents induced on the metal. Since the electric field E_{tan}^g of the Green function is zero on the metal, the currents do not appear in the main formula (3.76). The usual in the wire antenna theory approach for finding the field with use of the vector potential or the Hertz vector is not convenient for application in the problems involving the metallic surfaces on which the induced currents arise. Recall that all the currents including the induced ones appear in the wave equation (1.42) for the Hertz vector.

If a waveguide end is closed by a plane metallic wall (diaphragm), then in order to find the amplitudes A_n, the field $\{\vec{E}^g, \vec{H}^g\}$ must be subjected to the condition $E_{\text{tan}}^g = 0$ on this wall. The cross-section $z = z_1$ should be made coincident with the diaphragm plane. The currents induced on it do not appear in the expression for A_n. The field $\{\vec{E}^g, \vec{H}^g\}$ should be chosen as a field of the standing wave, incoming from $z = \infty$ and reflected from the diaphragm. The integral over the cross-section $z = z_1$ is zero, whereas the integral over $z = z_2$ is calculated in a similar way as for the waveguide without the diaphragm. The amplitude A_n is expressed by the same formula (3.80) where I^g is a current flowing in the waveguide with diaphragm (without slot).

The waveguide can be excited by the electric field created in the cross-section (*end-plane excitation*) (Fig. 3.2(b)). The expression for amplitudes of arisen modes is derived using the same scheme as in the case of the excitation through the wall hole. The same standing wave as in the presence of the metallic diaphragm must be chosen as the Green function and the cross-section must be made coincident with the end plane. The integral of the electric field given on the end plane, multiplied by the magnetic field of the standing wave on the end plane, appears in the expression for A_n, and the formula for A_n coincides with (3.79).

In the problem on the end-plane excitation, the magnetic field can be given on the end plane, instead of the electric one. Then, similar to the above, the standing wave for which the magnetic field (instead of the electric one) equals zero on the end plane should be chosen as the Green function. In analogy with the ideal conductor, on which $E_{\text{tan}} = 0$ and which can be represented as a dielectric with $|\varepsilon| = \infty$, the material with $H_{\text{tan}} = 0$ on its surface is called the *ideal magnetic*, since this condition would hold on the material with $|\mu| = \infty$. The Green function which should be used when the magnetic field is given on the end plane satisfies the conditions different on the waveguide walls ($E_{\text{tan}} = 0$) and on its cross-section ($H_{\text{tan}} = 0$). If the conditions for the Green function are chosen in this way, then neither the values H_{tan} on the walls nor E_{tan} on the end plane appear in the expression for amplitudes A_n. The standing wave with $H_{\text{tan}} = 0$ in the cross-section $z = 0$ differs from the wave with $E_{\text{tan}} = 0$ in this cross-section in displacement along the z-axis by a quarter of the wavelength in the waveguide.

It is impossible to prescribe arbitrarily both E_{tan} and H_{tan} in any cross-section at the same time. Prescribing one field completely defines the other, and, in particular, its values in this cross-section.

3.3.3
Integration in the plane of a complex variable

In this subsection we consider in detail a simple problem about excitation of the circular waveguide by an elementary dipole located on its axis and directed along it. In the general formula (3.77) we should take $j_z = \delta(R)$, $R = \sqrt{r^2 + z^2}$. Only the TM_{0q}-modes arise.

The component E_z equals $\alpha_{0q}^2 \chi$ in such modes, where, according to (3.11), $\chi = P_{0q} J_0(\alpha_{0q} r)$, and $\alpha_{0q} = \nu_{0q}/a$, $J_0(\nu_{0q}) = 0$. According to (3.22), the normalizing factor is $P_{0q} = [\sqrt{\pi} \nu_{0q} J_0'(\nu_{0q})]^{-1}$. Substituting these values into (3.77) gives

$$A_{0q} = \frac{2\sqrt{\pi}}{\omega} \frac{1}{h_{0q} \cdot a^2} \frac{\nu_{0q}}{J_0'(\nu_{0q})}, \tag{3.81}$$

where a is the waveguide radius.

Below we derive this formula once more using the method of integration in the plane of the complex variable instead of the Green function method. The considerations of this subsection have a methodical purpose only. They make possible to pass to some problems considered in the next chapters, for which only this method is applicable.

The excitation of the waveguide by a given current $\vec{j}^{\,\text{ext}}$ can be treated as a process of diffraction of the field, created by this current in vacuum, on the waveguide walls. Further this field is called the *incident field*. The secondary (*diffraction*) field is found from the requirement that the sum of these two fields must satisfy the boundary condition on the waveguide metallic surface $r = a$. Inside the waveguide the only nonzero component of the field $\vec{E}$, tangential to the surface, is E_z. We also assume that the waveguide walls are ideal-conducting, that is, $E_z = 0$ at $r = a$.

If the incident field were a cylindrical wave, then $E_z(r, z)$ would be proportional to the function $H_0^{(2)}(\alpha r) \exp(-ihz)$, $\alpha = \sqrt{k^2 - h^2}$, where the parameter h may take any value. Then the diffraction field could be easily constructed. It would also be a cylindrical wave and be proportional to the function $J_0(\alpha r) \exp(-ihz)$. The proportionality factor would be found from the requirement that $E_z(a, z) = 0$, and the total field would be

$$E_z(r, z) = \left[H_0^{(2)}(\alpha r) - \frac{H_0^{(2)}(\alpha a)}{J_0(\alpha a)} J_0(\alpha r) \right] \exp(-ihz). \tag{3.82}$$

At any value of h, the functions $H_0^{(2)}(\alpha r) \exp(-ihz)$ and $J_0(\alpha r) \exp(-ihz)$ satisfy the wave equation just as the Cartesian component E_z of the field should do. The first summand in (3.82) represents the outgoing cylindrical wave. It is proportional to $\exp(-i\alpha r - ihz)/\sqrt{\alpha r}$ at $|\alpha r| \gg 1$, according to the asymptotic of $H_0^{(2)}(\alpha r)$. This function has a singularity at $r \to 0$, which is agreed with the fact that such a field can be created only by the current distributed along the axis $r = 0$. The second summand has no singularities. It describes the field created by the currents induced on the cylindrical surface $r = a$.

However, a dipole located at the origin creates not a cylindrical but a spherical wave in free space. In order to find the diffraction field, the incident field is convenient to be preliminarily represented in the form consistent with the shape of the surface on which the diffraction occurs. Such a consistency is the first stage of solving diffraction theory problems. In our problem the representation of the spherical wave as a superposition of cylindrical ones is also a preparatory stage in the method of integration in the complex variable plane.

We write the Cartesian component of the field of the elementary dipole $j_z^{ext} = \delta(R)$ in Cartesian coordinates instead of spherical ones convenient for finding the spherical components of the field. Similar to the current, the electric Hertz vector $\vec{\Pi}^{(e)}$ has only the z-component, which must satisfy the non-homogeneous wave equation

$$\Delta\Pi_z^{(e)} + k^2\Pi_z^{(e)} = \frac{4\pi i}{\omega}\delta(R) \tag{3.83}$$

(see (1.42)) and the radiation condition. The component E_z is expressed by $\Pi_z^{(e)}$ using the formula

$$E_z = k^2\Pi_z^{(e)} + \frac{\partial^2}{\partial z^2}\Pi_z^{(e)}, \tag{3.84}$$

obtained from (1.43) at $\Pi_x^{(e)} = \Pi_y^{(e)} = 0$. Since in the fields of the eigenmodes, $\Pi_z^{(e)}$ is connected with $\chi(x,y)$ by formula (3.25a), then (3.84) takes the form (3.3). However, E_z cannot be expressed by χ for the spherical wave.

The function $C\exp(-ikR)/R$ is a solution to equation (3.83). The factor C is determined by the value of the singularity at $R = 0$, that is, by the factor $4\pi i/\omega$ in (3.83). Integrating (3.83) over a sphere of small radius ε ($k\varepsilon \ll 1$) centered at $R = 0$, considering that $\Delta \equiv \text{div grad}$ and using the Gauss theorem, we get the surface integral $\int (\partial\Pi^{(e)}/\partial R)\,dS$ over the sphere to the left from the first term. The second term does not give a finite contribution at $\varepsilon \to 0$. In the surface integral $dS = \varepsilon^2\,d\Omega$, where Ω is the entire solid angle. Substituting $\Pi^{(e)} = C\exp(-ikR)/R$, we obtain $-4\pi C$ on the left-hand side. According to the δ-function property, the value $4\pi i/\omega$ appears on the right-hand side. Hence, the solution to (3.83), satisfying the radiation condition, is

$$\Pi_z^{(e)} = -\frac{i}{\omega}\exp(-ikR)/R. \tag{3.85}$$

Formula (3.84) expresses the z-component of the incident field in terms of the Hertz vector. The same formula is valid for the diffraction field if we replace $\Pi_z^{(e)}$ with $\Pi_z^{(e)diff}$ in it. Then the boundary condition $E_z(a,z) = 0$ for the sum of the incident field and the sought diffraction one holds if

$$\Pi_z^{in}(a,z) + \Pi_z^{diff}(a,z) = 0 \tag{3.86}$$

for all z. Here Π_z^{in} redenotes $\Pi_z^{(e)}$ given by (3.85); the index (e) is omitted.

The representation of function (3.85) describing a spherical wave as a super-position of the cylindrical ones is given by the table integral

$$\frac{\exp\left(-ik\sqrt{r^2+z^2}\right)}{\sqrt{r^2+z^2}} = \frac{i}{2}\int\limits_{-\infty}^{\infty} H_0^{(2)}(\alpha r)\exp(-ihz)\,dh, \qquad \alpha = \sqrt{k^2-h^2}.$$

$$(3.87)$$

It is implied in this integral that $\operatorname{Re}\alpha > 0$ if $-k < h < k$, that is, if $\alpha^2 > 0$, and $\operatorname{Im}\alpha < 0$ if $-\infty < h < -k$ or $k < h < \infty$, that is, if $\alpha^2 < 0$. In the first interval, that is, at real α, the integrand describes the outgoing wave. In the remaining two intervals (at the imaginary α) it represents the field decreasing when r increases. This conforms with the fact that the left-hand side of (3.87) describes the outgoing wave. The condition $\operatorname{Im}\alpha < 0$ at $h \to \pm\infty$ provides the convergence of the integral.

Hence, according to (3.84), (3.82), and (3.87), the solution of the diffraction problem considered here is given by the formula

$$E_z(r,z) = \frac{i}{2\omega}\int\limits_{-\infty}^{\infty}\left[H_0^{(2)}(\alpha r) - \frac{H_0^{(2)}(\alpha a)}{J_0(\alpha a)}J_0(\alpha r)\right]\alpha^2\exp(-ihz)\,dh. \qquad (3.88)$$

In the problem of this subsection the finding of the solution in the form of the Fourier integral is elementary. This stage is the most difficult in the more complicated problems. The second stage consists of introduction of the complex plane $h = h' + ih''$ and transformation of the integral (3.88) over the real axis of this plane, by deforming the integration contour. In this procedure, the investigation of the integrand behavior when continuing it onto the whole complex plane of the variable h is decisive.

The integration contour should be deformed into the infinitely remote half-circle located in the lower or upper half-plane of h. The choice of the complex half-plane to which the contour is displaced depends on the points at which the field is calculated, namely, on the sign of the z-coordinate.

We carry out the transformation allowing us to calculate the field at points $z > 0$. In this case the contour is displaced into the lower half-plane. Since, at $z > 0$, the multiplier $\exp(-ihz) = \exp(-ih'z) \cdot \exp(h''z)$ tends to zero as $h \to -\infty \cdot i$, the integral over the infinitely remote half-circle tends to zero, either. When displacing the contour and using the Cauchy theorem, it is essential that the integrand is single-valued, so that its only singularities lying between the initial contour ($h'' = 0$) and the final one ($h'' = -\infty$) are the poles, that is, the points at which the function is infinite.

We show that the expression in the square brackets under the integral in (3.88) is single-valued, although it contains the Hankel function, an ambiguous function of its argument, and an ambiguous function α. The ambiguity of

α is caused by the fact that the function $\alpha(h)$ has the branching points $h = k$ and $h = -k$, at which $\alpha = 0$. When a point on the complex plane passes around the branching point and returns to its initial position, then α changes the sign. Consider the neighborhood of the branching point $h = k$. Let the point $h = k + \rho \exp(i\psi)$ pass around the point $h = k$ along a small circle, $\rho \ll k$. Then $\alpha = \sqrt{2\pi\rho i} \exp(i\psi/2)$. When passing, ψ varies from $\psi = 0$ to $\psi = 2\pi$ and α varies from $\sqrt{2\pi\rho i}$ to $-\sqrt{2\pi\rho i}$, that is, it changes the sign. The same relates to the neighborhood of the branching point $h = -k$.

However, the entire integrand in (3.88) is single-valued; it does not change the sign after the passing. At small arguments, the functions $H_0^{(2)}(\alpha r)$ and $H_0^{(2)}(\alpha a)$ are equal (with accuracy to a nonessential factor) to $J_0(\alpha r) \ln(\alpha r)$ and $J_0(\alpha a) \ln(\alpha a)$, respectively. Hence, the integrand is proportional to the difference $\ln(\alpha r) - \ln(\alpha a)$, that is, to $\ln(r/a)$, at small $|\alpha|$. It does not depend on α near the points $\alpha = 0$, and there are no other points on the plane of h where the functions are ambiguous. Therefore, expression (3.88) equals the sum of residues at the poles, lying between the initial and final integration contours.

The poles are zeros of the denominator, that is, the values of h, for which $J_0(\alpha a) = 0$. Denote these values by h_n. Applying the *theorem of residues*

$$\int \frac{f(h)}{F(h)} dh = 2\pi i \sum_n \frac{f(h_n)}{dF/dh|_{h=h_n}}, \tag{3.89}$$

where $F(h_n) = 0$, and using the fact that the first term under the integral in (3.88) does not have the poles, we transform (3.88) into the sum of the residues of the second term

$$E_z(r, z) = \frac{\pi i}{\omega} \sum_n \frac{H_0^{(2)}(\alpha_n a)}{a J_0'(\alpha_n a)} \frac{\alpha_n^2}{d\alpha_n/dh} J_0(\alpha_n r) \exp(-ih_n z), \tag{3.90}$$

where $h_n^2 = k^2 - \alpha_n^2$, $J_0(\alpha_n a) = 0$. The sum contains all the roots of this equation, so that, replacing the number n by $\{0, q\}$, yields

$$A_n = \frac{1}{P_{oq}} \frac{\pi i}{\omega} \frac{H_0^{(2)}(\nu_{0q}) \alpha_{0q}}{J_0'(\nu_{0q}) h_n a}. \tag{3.91}$$

Using formulas of the cylindrical function theory, namely, $J_0'(x) = -J_1(x)$ and $H_0^{(2)}(\nu_{0q}) = -2i/[\pi \nu_{0q} J_1(\nu_{0q})]$, and substituting the value P_n, we get the same expression (3.81) for A_n as that obtained by the Green function method.

A finite number of poles h_n can be located in the interval $-k < h < k$ of the real axis; they correspond to the roots of the equation $J_0(\alpha_n a) = 0$ for which $\alpha_n < k$. The roots lying on the imaginary half-axis $h' = 0$, $h'' < 0$ correspond to the larger values of α_n. The real h_n relates to the *traveling modes*, and the imaginary h_n to the *damping* ones. If there exist roots lying on the real axis, that is, if k is not smaller than the cutoff frequency of the TM_{01}-mode, then the initial integration contour in (3.88) cannot coincide with the real axis of the complex plane everywhere. The contour should pass around the poles leaving them over or under itself. In (3.88), the location of the contour near the poles on the real axis should be chosen such that this expression satisfies the unused demand about the absence of the eigenmodes incoming from infinity. This means that at $z > 0$ only the modes can arise, for which $\mathrm{Re}\, h_n \geq 0$. Consequently, the poles lying on the half-axis $h'' = 0$, $h' < 0$ must not be located inside the domain which the initial contour covers when transforming into the infinite half-circle lying in the lower half-plane. The contour in (3.88) must be located under the poles lying on the negative part of the real axis of the complex plane h. In a similar way, investigating the field at $z < 0$ and deforming for this end the integration contour into the infinite half-circle lying in the upper half-plane, we can show that the poles lying on the positive part of the real axis must be located under the integration contour (see Fig. 3.3). In the space domain $z > 0$ these poles give the eigenmodes in (3.90) propagating straightforward, that is, in the direction $z = \infty$.

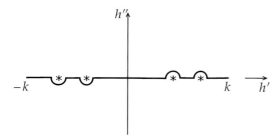

Fig. 3.3 The integration contour for the problem of closed waveguide excitation

If the integrand in (3.88) were not single-valued, then the lines of the cut would be drawn on the plane and after this the integration contour would be deformed in such a way that the lines would not be intersected by the contour. Then the integral in (3.88) would be equal not only to the sum of residues, but it would also contain the integrals over both the sides (banks) of the cut. This would mean that the field consists not only of the sum of fields of the eigenmodes of the *discrete spectrum* (we mean formula (3.90)), but also of the integral of modes of the *continuous spectrum*. The latter notion will be explained further when considering the problem in which such an additive integral exists. Since in the closed waveguide the system of the eigenmodes of field is complete (with adding the associated modes in certain cases and, in particular, for more complicated waveguides), the integral term does not appear when transforming expression (3.88).

A connection exists between the completeness of the system of eigenmodes of a physical object (closed waveguide in our case) and single-valuedness of the expression under the Fourier integral providing a formal solution to the diffraction problem on excitation of this object by a given current. In this case the left-hand side of the equation for the wave numbers of the eigenmodes (equation $J_0(\sqrt{k^2 - h^2}a) = 0$ in our problem) coincides with the denominator of this expression.

3.4
Nonregular closed waveguides

3.4.1
The cross-section method

Each eigenmode may propagate in the waveguide independently, without affecting and generating other modes. If the waveguide is nonregular, that is, some of its parameters are different in different cross-sections, then there cannot exist only one eigenmode. When propagating in a nonhomogeneous segment of the waveguide, the eigenmode generates other modes and, besides, its amplitude varies. The nonhomogeneity may be of the following types: the varying shape or size of the cross-section (expansion or contraction of the waveguide, junction of the waveguides of different cross-sections), varying the orientation of the cross-section (twisted waveguides), or varying the direction of the axis (bending). The impedance of the waveguide walls or the shape and permittivity of the material filling (fully or partly) the waveguide may also vary. In all these cases, the field can be described by the infinite system of the first-order ordinary differential equations for amplitudes of arisen eigenmodes of certain auxiliary waveguides.

As an eigenmode in a cross-section of the nonregular waveguide we call the eigenmode of a certain auxiliary regular waveguide (the *reference waveguide*), with the size, shape and filling, same as those of the actual cross-section of the nonregular waveguide. The reference waveguides are different for different cross-sections of the nonregular waveguide. The system of their eigenmodes depends on the cross-section of the nonregular waveguide to which it corresponds.

For simplicity, when deriving the system of equations we confine ourselves to the waveguides without associated waves. In any cross-section of the nonregular waveguide the transverse components of the fields $\vec{E}$ and $\vec{H}$ (that is, the components lying in the cross-section) can be expanded into series by the complete system of the transverse components of the eigenmodes in the reference waveguide. The expansion coefficients in the series for E_x and for E_y

are the same, as well as those in the series for H_x and for H_y, but the coefficients in the series for $\vec{E}$ and $\vec{H}$ are different. However, one can "double the number" of the expansion functions, using the fields of waves propagating in both the directions (the quotation marks mean that the notion "number" has, formally, no sense for the elements of infinite systems). Below the mode index n will take positive and negative values, $n > 0$ for the direct waves (i. e., the waves propagating or damping in the direction $z \to \infty$) and $n < 0$ for the back waves. Then the coefficients in the expansions of the transverse components of $\vec{E}$ and $\vec{H}$ are the same. From the Maxwell equations it follows that in the domains free of the extrinsic currents, the longitudinal components of the fields are expanded into the series with the same coefficients.

The expansions have the form

$$\vec{E}(x,y,z) = \sum_{n=-\infty}^{\infty} A_n(z)\, \vec{E}^n(x,y,z), \tag{3.92a}$$

$$\vec{H}(x,y,z) = \sum_{n=-\infty}^{\infty} A_n(z)\, \vec{H}^n(x,y,z). \tag{3.92b}$$

Here $\vec{E}^n(x,y,z)$ and $\vec{H}^n(x,y,z)$ are the factors at $\exp(-ih_n z)$ in the fields of the eigenmodes. In a regular waveguide these quantities do not depend on z; the dependence on z is contained in the functions $A_n(z)$, which are $\exp(-ih_n z)$. In a nonregular waveguide $\vec{E}^n$, $\vec{H}^n$ depend on the z-coordinate of the cross-section to which the reference waveguide corresponds, and $A_n(z)$ is a more complicated function depending on the amplitudes of the other modes. The propagation constants also depend on z, $h_n = h_n(z)$.

The equation system for $A_n(z)$ has the form

$$\frac{dA_n}{dz} + ih_n A_n = \sum_{m=-\infty}^{\infty} S_{nm} A_m \qquad n = \ldots - 2, -1, 1, 2, \ldots \tag{3.93}$$

In order to obtain this system and expressions for the coefficients S_{nm} (the *coupling coefficients*), expressions (3.92) should be substituted into the Maxwell equations and both the orthogonality conditions (3.71) and the normalization ones (3.73) should be used. However, this way cannot be applied immediately.

The left-hand side in series (3.92a), that is, the field $\vec{E}$ in a nonregular waveguide does not fulfill the boundary conditions which the fields $\vec{E}^n$ (i. e., all terms of the sum) fulfill. In the simplest case of the waveguide with ideal-conducting walls, the reference waveguides have the ideal-conducting walls, as well, and the longitudinal component E_z^n equals zero on the wall for all eigenmodes. Hence, all terms of the series for E_z (3.92a) are zero on the wall. The component E_z of the field in a nonregular waveguide is not zero on the wall, since it does not coincide with the tangential component not lying in the cross-section plane. Thus, series (3.92a) for E_z does not converge on the wall: all its terms are zero but the quantity represented by it, does not equal zero. The series, divergent on the domain boundary, converges nonuniformly inside the domain, that is, the number of terms needed for obtaining the result with a given accuracy infinitely increases when the point at which the series is calculated approaches the boundary. The derivative of the nonuniformly convergent series does not equal the series of derivatives of its terms. The direct substitution of the vector series (3.92) into the Maxwell equations as well as their termwise differentiation is not allowed.

Several artificial methods are used for obtaining system (3.93), which allow us to avoid the differentiation of series (3.92a) for E_z with respect to x, y. Of course, all they lead to the same form of this system and to the same expression for S_{nm}.

The derivation can be carried out in the way that only the x- and y-components of the vectorial series (3.92) are used. The boundary condition $E_s = 0$ (unit vector $\vec{s}$ lies in the cross-section plane) for these components is valid both for all terms of the series and for the field in the nonregular waveguide. Series for the components are uniformly convergent and they can be differentiated termwise. Another method consists in treating the nonregular waveguide as a limit of sequentially linked short segments of the regular waveguides and using the small perturbation method for investigation of the mutual mode transformation at the segment junctions. From this point of view, the waveguide bending is a subsequence of small sharp bendings whereas the cross-section shape variation is a limit of the small step sequence.

Justification of formula (3.93) (and (3.94) below) is not based on the presentation of the nonregular waveguide as a limiting case of the staircase one; it can be carried out in other ways. For instance, the rigorous method mentioned above can be used. However, the staircase approach leads to the proper result. This can probably be explained by the fact that system (3.93) describes a volume (not surface) effect, whereas the difference between continuous and staircase waveguides is localized near the surface. An analog is a known geometrical effect: the length of the staircase line does not tend to the length of its limiting continuous form, but the areas bounded by these two curves are equal at the limit.

The nonregular waveguide (Fig. 3.4(a)) can also be treated as a limiting case of the waveguide with the regular cross-section and nonhomogeneous filling, at $\varepsilon \to -\infty \cdot i$ (Fig. 3.4(b)). At finite $|\varepsilon|$, the fields of the modes of the reference waveguides approximate the field in the nonregular waveguide in the entire cross-section. The series for the waveguides shown in Fig. 3.4(b), analogous to (3.92), can be differentiated termwise, and in this way the coupling coefficients can be calculated. Their limits at $\varepsilon \to -\infty \cdot i$ are the sought coefficients for the waveguide shown in Fig. 3.4. The boundary conditions on metal for all the components of the field $\vec{E}$ are fulfilled without using series (3.92) for E_z, which is divergent on the boundary.

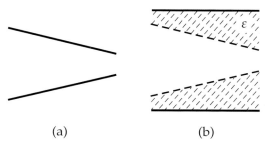

(a) (b)

Fig. 3.4 Nonregular and partially filled waveguides

The larger the coefficient $S_{nm}(z)$ at a certain z, the stronger the influence of the field of the mth mode on the amplitude A_n of the nth mode is in this cross-section. The influence depends on the type and value of the nonregularity at this z. It is proportional to the amplitude A_m of the mth mode.

We give without detailed proof the formulas for S_{nm} in the case of a straight nonfilled waveguide with the variable cross-section. The coefficients S_{nm} can be expressed in terms of the z-derivatives of the fields $\vec{E}^n$, $\vec{H}^n$, participating in series (3.92). However, it is more convenient to express S_{nm} in terms of integrals over the cross-section contour, taken of the potentials χ, ψ (3.3), and their derivatives with respect to s and N, where $\vec{N}$ is a unit vector lying in the cross-section plane normal to the surface of the reference waveguide. Introduce the function $v(s,z)$, $(v \geq 0)$ as the tangent of the angle between the z-axis and the tangent to the surface of the nonregular waveguide, which lies in the plane containing the z-axis, that is, in the plane perpendicular to the cross-section. At a given z, the function $v(s,z)$ describes the degree of the waveguide nonregularity. If the waveguide is regular in a certain interval of z, then $v(s,z) = 0$, and $S_{nm} = 0$ for these z.

The coupling coefficients are equal to

$$S_{nm} = \frac{1}{2h_n (h_n - h_m)} \times$$
$$\oint v(s) \left[\alpha_n^2 \cdot \alpha_m^2 \psi^n \psi^m + \left(h_n h_m - k^2 \right) \frac{\partial \psi^n}{\partial s} \frac{\partial \psi^m}{\partial s} \right] ds \qquad (3.94a)$$

$$S_{nm} = \frac{k^2 - h_n h_m}{2h_n (h_n - h_m)} \oint v(s) \frac{\partial \chi^n}{\partial N} \cdot \frac{\partial \chi^m}{\partial N} ds \qquad (3.94b)$$

$$S_{nm} = -\frac{k}{2h_n} \oint v(s) \frac{\partial \chi^n}{\partial N} \cdot \frac{\partial \psi^m}{\partial s} ds \qquad (3.94c)$$

Formulas (3.94a), (3.94b) refers to the cases, when both the participated modes are of the same type, TE or TM, respectively; (3.94c) gives S_{nm} for the case when the nth is the TE-mode and the mth is the TM-modes. The coupling coefficients have the dimensionality cm^{-1} (the same as k has).

The term $S_{nn}A_n$ involved in series (3.93) for dA_n/dz is proportional to the amplitude of the mode with the same number. The coefficient S_{nn} is proportional to the derivative of the wave number

$$S_{nn} = -\frac{1}{2h_n} \cdot \frac{dh_n}{dz}. \tag{3.95}$$

The derivative dh_n/dz can be expressed by the formulas similar to (3.94),

$$\frac{dh_n}{dz} = \frac{\alpha_n^2}{2h_n} \oint v(s) \left[\alpha_n^2 (\psi^n)^2 - \left(\frac{\partial \psi^n}{\partial s}\right)^2 \right] ds, \tag{3.96a}$$

$$\frac{dh_n}{dz} = \frac{\alpha_n^2}{2h_n} \oint v(s) \left(\frac{\partial \chi^n}{\partial N}\right)^2 ds \tag{3.96b}$$

for the wave number of the magnetic and electric modes, respectively.

The equation system (3.93) should be supplemented by the end conditions at the ends of the nonregular segment. If this segment is located between two regular waveguides, then the reference waveguides corresponding to the input and output cross-sections coincide with these regular waveguides. Therefore, no additional calculations should be made for passing from the amplitudes at the ends of the nonregular segment to the amplitudes in the regular waveguides. The mode amplitudes in the regular segments are equal to $A_n(0)$ and $A_n(L)$, where $z = 0$ and $z = L$ are the z-coordinates of the cross-sections at the input and output of the nonregular segment, respectively. If the first mode falls onto the nonregular segment from the left and there are no other modes incoming, then the end conditions are as follows:

$$A_1(0) = 1, \quad A_n(0) = 0 \quad (n \neq 1, n > 0); \quad A_{-n}(L) = 0 \quad (n > 0). \tag{3.97}$$

The reflected mode with the amplitude $A_{-1}(0)$ and other back modes with the amplitudes $A_{-n}(0)$ $(n > 0)$ outgo into the left waveguide; the direct modes with amplitudes $A_n(L)$ $(n > 0)$ outgo into the right one. The values $A_{-n}(0)$ and $A_n(L)$ are sought. Their aggregate for all modes falling onto the nonregular segment makes up the so-called *scattering matrix*.

System (3.93) supplied with the end conditions (3.97) imposed at the opposite ends of the segment represents a *two-point* problem. Such a problem for the system of two equations was considered in Section 2.2 (see formulas (2.54), (2.69)). Similarly as in that more simpler case, our problem can also be reduced to the same equation system with the end conditions at one end (the Cauchy problem) by means of introducing auxiliary functions following the scheme described by (2.70). However, numerically solving this problem is difficult in the case when the dumping modes are considered. There exists some iterative technique avoiding this difficulty.

3.4.2
Slowly varying parameters

An approximate explicit solution to system (3.93), (3.97) can be found in the case when the coupling of different modes caused by the waveguide nonregularity is weak. For different types of nonregularity, the mode coupling may be treated to be weak in the following cases. For the bent waveguides, the bending radius should be large in comparison with linear sizes of the cross-section. For the nonregular straightline waveguides, the waveguide parameters should be slowly varying in comparison with this size; more precisely, the parameter $v(s, z)$ should be small. Then all the coupling coefficients in (3.97) are small, $|S_{nm}|/k \ll 1$. However, the total variation of the parameters through the entire nonregular segment $0 < z < L$ may be large.

For the solution obtained below to be approximately valid on the large segment, that is, at $kL \gg 1$, it is necessary to pass from $A_n(z)$ to new variables $a_n(z)$ such that the equation for da_n/dz does not contain the term proportional to $a_n(z)$, that is, the matrix of coefficients in the differential equation system does not contain the diagonal elements.

According to (3.95), the term $-(ih_n + 1/2h_n \cdot dh_n/dz)A_n$ is involved in the equation for dA_n/dz in (3.93). Introduce the *reduced amplitudes* $a_n(z)$ using the relation

$$A_n(z) = q_n(z) \exp\left[-i\gamma_n(z)\right] \cdot a_n(z), \tag{3.98}$$

where

$$q_n(z) = \sqrt{\frac{h_n(a)}{h_n(z)}}, \tag{3.99a}$$

$$\gamma_n(z) = \int_0^z h_n(\varsigma)\, d\varsigma. \tag{3.99b}$$

The equation system for $a_n(z)$ contains no diagonal terms

$$\frac{da_n(z)}{dz} = \sum_{m \neq n} \frac{q_m(z)}{q_n(z)} S_{nm}(z) \exp\left[-i(\gamma_m - \gamma_n)\right] a_m. \tag{3.100}$$

The function $\gamma_n(z)$ has the same physical meaning as the similar-denoted function from (2.50) in the theory of waves in a nonhomogeneous medium. The multiplier $\exp[-i\gamma_n(z)]$ describes the main process occurring in the nonregular segment of the waveguide or in the nonhomogeneous medium. This process is known as *phase accumulation*, that is, the summation of the phase incursion on different segments.

The end conditions for the function $a_n(z)$ are the same as those (3.97) for $A_n(z)$.

Similarly as in (3.93), in system (3.100) the small quantities S_{nm} are factors in the expressions for da_n/dz, and the solution is easily found in the first order of the small parameter ν_0, where ν_0 is a certain mean value of the geometric parameter $\nu(s,z)$. Keeping only the reduced amplitude of the incident mode in equation (3.100) for $n = 1$, we get

$$a_1(z) = 1. \tag{3.101}$$

In this approximation, the amplitude of the main mode, transmitted through the nonregular segment, is

$$A_1(L) = q_1(L) \exp\left[-i\gamma_1(L)\right]. \tag{3.102a}$$

The amplitude $A_1(z)$ varies through the nonregular segment in such a way that, according to (3.99) and (3.98), the transmitted energy, proportional to $h_1(z)|A_1(z)|^2$, is constant (see the text after (3.73)).

Substituting the function $a_1(z) = 1$ into right-hand side of equations (3.100) for $n \neq 1$ and dropping the remaining terms (which is possible just owing to the absence of the term proportional to $a_n(z)$ among them), we obtain the sought approximate expressions for both the amplitudes $A_n(L)$ of the direct modes outgoing into the right waveguide and the amplitudes $A_{-n}(0)$ of the back modes outgoing into the left waveguide:

$$A_n(L) = q_n(L) \int_0^L \frac{q_1(z)}{q_n(z)} S_{n1} \exp\left[-i\gamma_1(z)\right] \exp[-i(\gamma_n(L)$$

$$+ i\gamma_n(z))]\, dz, \qquad n = 1, 2, \ldots, \tag{3.102b}$$

$$A_{-n}(0) = -q_n(L) \int_0^L \frac{q_1(z)}{q_n(z)} S_{-n1} \exp\left[-i\gamma_1(L) + i\gamma_{-n}(z)\right] dz,$$

$$n = 1, 2, \ldots \tag{3.102c}$$

The energy carried away by the arisen modes is proportional to $|A_n(L)|^2$ and $|A_{-n}(0)|^2$, that is, it has the order of ν_0^2. The energy of the incident mode does not vary with accuracy to this value, that is, (3.102a) holds.

According to (3.102), the process of appearance of the nth direct ($n > 1$) mode and the $(-n)$th back ($n > 0$) one, when the first ($n = 1$) incident mode is passing, consists in the following: the mode falling onto the nonregular segment acquires the phase $\gamma_1(z)$ on its way to the interval dz. When passing through the dz interval, the mode generates the nth direct and the $(-n)$th back modes with amplitudes $S_{n1}(z)A_1(z)\, dz$ and $S_{-n1}(z)A_1(z)\, dz$, respectively. The arisen direct mode acquires the additional phase $\int_z^L h_n\, dz = \gamma_n(L) - \gamma_n(z)$ when propagating to the end $z = L$ of the nonregular segment (Fig. 3.5, upper

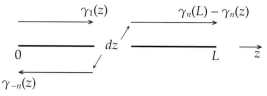

Fig. 3.5 Mode transformation on nonregular waveguide segment

arrows). The back mode acquires the phase $\int_z^0 h_{-n}\, dz$ when propagating to the beginning $z = 0$ of the nonregular segment (Fig. 3.5, lower line).

The appearance of new modes is the interferential process. Modes arisen on different segments of the waveguide are summed together, each with its own phase. The amplitude of the mode scattered by the nonregular segment depends not only on the coupling coefficient between this mode and the incident one, but also on the difference $h_n(z) - h_1(z)$ between their propagation constants. The smaller this difference, the weaker the mutual interferential suppression of the elementary modes arisen in different parts of the nonregular segment is, and the larger the amplitude of the arisen mode is. Thus, the reflection coefficient $A_{-1}(0)$ is, as a rule, small, because for $n = 1$ the phase factor in formula (3.102c) equals $2\gamma_1(z)$ and is large. The most part of the energy of arisen modes is carried out by the direct modes for which this factor is small, equal to $\exp[-i(\gamma_n - \gamma_1)]$.

If there exists a cross-section in the nonregular segment, in which the propagation constant of the nth mode ($n \neq 1$) equals h_1, that is, if $h_n(z_0) = h_1(z_0)$ at $z = z_0$, then in the cross-section neighborhood, the mutual mode suppression described by the exponential factor in (3.102b) is weakened. The nth modes arisen on the interval dz near z_0 reach the end of the segment with close phases. It can be shown that in this case $A_n(L)$ has the order not of v_0 but the smaller one, namely, of $v_0^{1/2}$. However, if the equality $h_n = h_1$ holds for any z, that is, the first and nth modes are degenerated, then the process does not have the interferential nature, and the amplitude $A_n(z)$ is not small. The inverse influence of the nth mode upon the first one causes $a_1(z)$ not to be constant as well as the approximations (3.101) and, consequently, (3.102), not to be valid. Such a situation occurs, for instance, for the bent circular waveguide. The TE_{01}- and TM_{11}-modes are degenerated in such a waveguide; the bending couples these modes together. Regardless of how large the radius of bending is, the field in it consists mainly of the combination of the fields of these two modes which completely transform to each other in cycles. The angle at which the full transformation occurs does not depend on the bending radius. It can be found from the expression for the coupling coefficient caused by the bending. The coupling coefficients in the equations for the bent waveguide play the same role as the coefficients S_{nm} in the equation system for the waveguide with a varying cross-section. The above angle equals $\mu_{01}/(2\sqrt{2}) \cdot \lambda/a = 1.35 \cdot \lambda/a$ radian.

If a mode falls onto the contracting waveguide segment from the wide side, then a situation may occur when in the narrow segment it is already damped, not a traveling one. At a certain value $z = \tilde{z}$ the wave number vanishes, $h_1(\tilde{z}) = 0$, so that h_1 is real at $z < \tilde{z}$ and imaginary at $z > \tilde{z}$. The cross-section $z = \tilde{z}$ is a cutoff one. The mode is reflected on this cross-section; the modulus $|R|$ of the reflection coefficient is close to unity. This situation can occur at as slow as possible contracting, that is, at arbitrarily small v_0. The phase of the reflection coefficient depends on the waveguide shape at $z > \tilde{z}$, that is, in the domain in which the mode is no longer traveling, but into which the field penetrates. The field structure is similar to that of the plane wave in the case of the complete interior reflection. If at a certain cross-section $\tilde{\tilde{z}}$ ($\tilde{\tilde{z}} > \tilde{z}$) the waveguide again widens so that $h_1(z)$ becomes real, then the mode will again propagate with a smaller amplitude and $|R| < 1$, though. The tunneling occurs just as in the case of the disturbance of the complete interior reflection (2.1.4). However, this notion is more descriptive in this case.

From formulas (3.94) it follows that at the cutoff cross-section, $S_{1,-1}$ becomes infinite for nonfilled waveguides. The immediate application of system (3.93) is impossible near this cross-section. Coupling between the modes of numbers $n = 1$ and $n = -1$ is very strong (the complete reflection), separation of the direct and back modes does not correspond to the actual field structure. In this domain the field has nature of a standing wave, and it is expedient to introduce other unknown functions instead of A_n and A_{-n}. Similar transformations are also required in the case when the cross-section is a cutoff one for the arisen mode.

If the nonregularity is not flat (radius of the bending curvature is small or the function $v(s,z)$ is large), then system (3.93), (3.97) must be solved numerically. Although the system is not convenient for the numerical analysis in the case of finite coupling coefficients, the numerical experiments have shown that it provides a valid solution even for such extremely sharp nonregularity as a step between two circular waveguides of different diameters. However, in general, other methods should be applied to such local, not flat, nonregularities.

3.4.3
The diaphragm: equation for field in the slot

One of the examples of a local waveguide nonregularity is a diaphragm, that is, an infinitely thin ideal-conducting plate with a hole, placed perpendicular to the axis. Usually the diaphragm is used in a single-mode rectangular waveguide. Near such a local element occupying a small part of the channel, the field contains many waveguide modes of higher numbers. These waves, however, are damping at some distance from the diaphragm, and the field consists only of the incident mode and the reflected and transmitted ones of the same number. In practice, it is necessary to know only the reflection and transmission coefficients of the main mode. However, formation of the reflected and transmitted modes occurs near the diaphragm, where the field also contains other ("local") modes. Amplitudes of these modes appear in the formulas of the theory of *local nonregularities*.

Let the diaphragm be located in the cross-section $z = 0$ and the main TE_{10}-mode fall on it from the side $z < 0$. The back and direct modes outgo from the diaphragm into the directions $z < 0$ and $z > 0$, respectively.

We consider the so-called *inductive* diaphragm, consisting of two identical metallic strips parallel to the narrow side of the cross-section (and adjacent to walls). There is a slot between the strips (Fig. 3.6). The theory for such a waveguide is a little simpler than that for the *capacitive* diaphragm with strips parallel to the wide side of the cross-section.

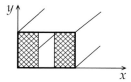

Fig. 3.6 Diaphragm in the waveguide

The field of the incident TE_{01}-mode does not depend on the y-coordinate. Since the parameters of the diaphragm disturbing the field do not depend on y as well, then the diffracted field is also independent of y. Besides the transmitted and reflected modes, the TE_{m0}-modes ($m > 1$) are also radiated from the diaphragm. The cutoff frequencies of these modes (see (3.30)) are $k_{m0}^{\text{cut}} = \pi m / a$, where a is the waveguide width. At $ka < 2\pi$, the waveguide is the single-mode one; all higher modes are damping.

The field of the incident mode contains three components: two transverse E_y and H_x, lying in the diaphragm plane, and one longitudinal H_z. There are no new components appearing at the diffraction on the inductive diaphragm. In the main mode, E_y and H_x are proportional to $\sin(\pi x / a)$; in the higher modes they are proportional to $\sin(\pi m x / a)$.

In the cross-section $z = 0$, E_y vanishes on the metal; E_y and H_x are continuous on the slot. Under these conditions the component H_z is equal to zero on the metal and continuous on the slot automatically.

The electric field E_y is continuous in the whole interval $0 < x < a$, since at both $z = -0$ and $z = +0$ it equals zero on the metal and continuous on the slot. We express the component E_y in the form of the series

$$E_y = e^{-ih_1 z} \sin \frac{\pi x}{a} + R e^{ih_1 z} \sin \frac{\pi x}{a} + \sum_{m>1} A_m e^{ih_m z} \sin \frac{\pi m x}{a}, \quad z \leq 0$$

$$E_y = D e^{-ih_1 z} \sin \frac{\pi x}{a} + \sum_{m>1} A_m e^{-ih_m z} \sin \frac{\pi m x}{a}, \quad z \geq 0. \tag{3.103}$$

Here R and D are the sought reflection and transmission coefficients of the main mode, respectively, and A_m are the amplitudes of local modes.

Since the system of functions $\sin(\pi m x/a)$, $m = 1, 2, \ldots$, is complete in the whole interval $0 < x < a$, it follows from equating the two expressions (3.103) at $z = 0$ that the coefficients A_m are identical in both the sums (which is already considered in (3.103)), and

$$1 + R = D. \tag{3.104}$$

The coefficients in (3.103) can be expressed by the function $E_y(x, z)$. We multiply (3.103) by $\sin(\pi m x/a)$ and integrate over the interval $0 < x < a$. After denoting $e(x) = E_y(x, 0)$, we obtain

$$R = \frac{2}{a} \int_0^a e(\xi) \sin \frac{\pi \xi}{a} \, d\xi - 1, \tag{3.105a}$$

$$A_m = \frac{2}{a} \int_0^a e(\xi) \sin \frac{\pi m \xi}{a} \, d\xi. \tag{3.105b}$$

Relation (3.104) has already been used here. Since $e(x) = 0$ on the metal, the integrals in (3.105) are actually taken over the slot.

The component H_x of the magnetic field equals $1/(ik) \cdot \partial E_y(x, y)/\partial z$, so that, according to (3.103), at $z = \pm 0$ we have

$$H_x = -\frac{h_1}{k} \sin \frac{\pi x}{a} + R \frac{h_1}{k} \sin \frac{\pi x}{a} + \sum_{m>1} A_m \frac{h_m}{k} \sin \frac{\pi m x}{a}, \quad z = -0,$$

$$H_x = -D \frac{h_1}{k} \sin \frac{\pi x}{a} - \sum_{m>1} A_m \frac{h_m}{k} \sin \frac{\pi m x}{a}, \quad z = +0. \tag{3.106}$$

From equating the magnetic field on both sides of the slot and using (3.104), it follows that

$$R h_1 \sin \frac{\pi x}{a} + \sum_{m>1} A_m h_m \sin \frac{\pi m x}{a} = 0. \tag{3.107}$$

Equality (3.107) does not hold in the whole interval $0 < z < a$. It is true only on the slot. The component H_x has a jump on the metal, the current flows on the strips in the y-direction. The system of functions $\sin(\pi m x / a)$ is incomplete on the slot, and, therefore, from (3.107) it does not follow that the coefficients at these functions vanish.

Substituting expressions (3.105) into (3.107) and interchanging the summation over m and integration over ξ, we obtain the following equation for $e(x)$:

$$\int e\left(\xi\right) K\left(x,\xi\right) d\xi = \sin\frac{\pi x}{a}, \tag{3.108}$$

where

$$K(x,\xi) = \frac{2}{a} \sum_{m=1} \frac{h_m}{h_1} \sin\frac{\pi m x}{a} \sin\frac{\pi m \xi}{a}. \tag{3.109}$$

Equation (3.108) as well as equality (3.107) is valid for the points x located on the slot; the integration in (3.108) is performed over the slot. Nevertheless, (3.108) is formally not an integral equation, since its kernel (3.109) has non-integrable singularity at $\xi = x$, caused by the fast increase of the multiplier h_m as m increases, $h_m \rightarrow i\pi/a \cdot m$ as $m \rightarrow \infty$. Interchanging the summation and integration was formally not allowed, since it caused the appearance of the function $K(x,\xi)$ with a strong singularity. However, computational techniques exist which allow us to solve such equations numerically.

Equation (3.108) is an integral equation of the first kind, that is, ill-posed; a small perturbation of the right-hand side may cause a significant distortion of the solution. However, not the function $e(x)$ itself but the coefficients R and D are of interest. According to (3.105a) and (3.104), these coefficients are obtained from $e(x)$ by the integration, that is, by the smoothing operation.

3.4.4
The diaphragm: equation for current on strips

The problem on inductive diaphragm can be solved by another method dual to that considered above. In this method, the jump of the magnetic field on the metallic screen is the sought function instead of the electric field on the slot. The current (1.80) flowing on the strips along the y-axis is proportional to this jump:

$$I\left(x\right) = \frac{c}{4\pi}\left[H_x\left(x+0\right) - H_x\left(x-0\right)\right]. \tag{3.110}$$

The function $I(x)$ differs from zero on the strips and vanishes on the slot. Recall that $e(x)$ possesses the opposite property.

According to (3.110) and (3.104), the function $I(x)$ can be expressed as a series

$$I\left(x\right) = \frac{c}{4\pi}\left[-2R\frac{h_1}{k}\sin\frac{\pi x}{a} - 2\sum_{m>1} A_m \frac{h_m}{k}\sin\frac{\pi m x}{a}\right]. \tag{3.111}$$

This expression is valid in the whole interval $0 < x < a$; therefore, similar to (3.105),

$$R = -\frac{4\pi}{c} \cdot \frac{4k}{h_1} \frac{1}{a} \int_0^a I(\xi) \sin \frac{\pi \xi}{a} \, d\xi, \tag{3.112a}$$

$$A_m = -\frac{4\pi}{c} \cdot \frac{4k}{h_m} \frac{1}{a} \int_0^a I(\xi) \sin \frac{\pi m \xi}{a} \, d\xi. \tag{3.112b}$$

Since $I(\xi) = 0$ on the slot, the integrals are actually taken only over the strips. The electric field vanishes on the strips, so that

$$(1 + R) \sin \frac{\pi x}{a} + \sum_{m>1} A_m \sin \frac{\pi m x}{a} = 0. \tag{3.113}$$

This formula is similar to (3.107). It is also valid only in a part of the interval $0 < x < a$, in which the system of functions $\sin(\pi m x / a)$ is incomplete, and it does not follow from (3.113) that the coefficients at these functions are zero.

Substituting expressions (3.112) into (3.113) and interchanging the integration and summation order gives the first-kind integral equation for $I(x)$,

$$\int I(\xi) \, \mathcal{P}(x, \xi) \, d\xi = \sin \frac{\pi x}{a}. \tag{3.114}$$

The integral is taken over the strips and the equation is valid for the points on the strips. The kernel of the equation has the form

$$\mathcal{P}(x, \xi) = C \cdot \sum_{m=1} \frac{k}{h_m} \sin \frac{\pi m x}{a} \sin \frac{\pi m \xi}{a}, \quad C = \frac{16\pi}{a \cdot c}. \tag{3.115}$$

It has a weak (integrable) singularity at $\xi = x$.

3.4.5
Diffraction on the screen with a hole

The problem about the diaphragm partitioning the waveguide can be treated as a partial case of the diffraction problem on a thin metallic screen with a hole. In this subsection we consider this general problem although it does not relate to the nonregular waveguide theory directly. The same ideas are used for its solving as those used in the waveguide problems in the two preceding subsections. The methods applied to these problems are, in fact, examples of application of more general methods. Two mutually dual methods can also be used for solving this general problem.

In the first method, the sought function is the electric field in the hole, that is, a two-dimensional vector $\vec{e}(x, y)$ lying in the screen plane. The screen is assumed to be planar, although the technique is also applicable to nonplanar screens with holes. The coordinates x, y together with the z-axis, normal to the screen plane, make up the Cartesian system. On the strips $E_{\text{tan}} = 0$, and on the hole $E_{\text{tan}} = \vec{e}$. In order to obtain the integral equation for $\vec{e}$, the magnetic field $\vec{H}$ should be expressed by $\vec{e}$ in the half-spaces $z \leq 0$ and $z \geq 0$, and the equality condition for H_{tan} on both the sides of the hole (at $z = +0$ and $z = -0$) should be considered.

We introduce the field $\{\vec{E}_+^0, \vec{H}_+^0\}$ which would exist in the half-space $z \geq 0$ if the hole were replaced by the metallic surface, and the sources located at $z \geq 0$ were kept. They, in particular, may be the incoming modes. The field $\vec{E}_+^0$ satisfies the condition $E_{\text{tan}} = 0$ in the whole plane $z = 0$. The difference field $\{\vec{E}_+ - \vec{E}_+^0, \vec{H}_+ - \vec{H}_+^0\}$ is created only by the field $\vec{e}$ on the hole; it does not have any other sources. It satisfies the radiation condition.

We use the Green function method for solving the boundary problem for determining the difference field from its value on the boundary. Taking into account that the expression for the magnetic field $\vec{H}_+ - \vec{H}_+^0$ is to be found, we introduce the Green function $\{\vec{E}^g, \vec{H}^g\}$ as a field satisfying certain artificially constructed equation system, different from the Maxwell equations (1.29) by dropping the term $\vec{j}^{\text{ext}}$ in the first equation and introducing the vector denoted as $4\pi/c \cdot \vec{j}^{(m)}$ on the right-hand side of the second one. In analogy with $\vec{j}^{\text{ext}}$ in the equation system (1.29), the vector $\vec{j}^{(m)}$ is called the *magnetic current density*. Being chosen in the form

$$\vec{j}^{(m)} = \vec{a}\delta\left(|\vec{r} - \vec{r}_g|\right), \tag{3.116}$$

it is called the *elementary magnetic dipole*, located at the point $\vec{r} = \vec{r}_g$ and directed along the unit vector $\vec{a}$. The magnetic current does not really exist; it is an auxiliary mathematical notion introduced in order to construct the Green function possessing the desired properties.

The field $\{\vec{E}^g, \vec{H}^g\}$ satisfies the equation system

$$\text{rot}\, \vec{H}^g - ik\varepsilon\vec{E}^g = 0,$$
$$\text{rot}\, \vec{E}^g + ik\mu\vec{H}^g = -\frac{4\pi}{c}\vec{j}^{(m)}, \tag{3.117}$$

where $\vec{j}^{(m)}$ is given in (3.116). A solution to this system can be obtained from the solution of system (1.29) by replacing $\varepsilon \to \mu$, $\mu \to \varepsilon$, $\vec{H} \to \vec{E}$, $\vec{E} \to -\vec{H}$ and $\vec{j}^{\text{ext}} \to \vec{j}^{(m)}$. The Green function is subject to the condition $E_{\text{tan}}^g = 0$ in the whole plane $z = 0$ and to the radiation condition. Similar to the field of the elementary electric dipole satisfying the above conditions, the field of the elementary magnetic one equals the sum of the fields created in free space by the dipole and its mirror reflection in the plane $z = 0$.

Both the fields $\{\vec{E}^g, \vec{H}^g\}$ and $\{\vec{E}_+^0, \vec{H}_+^0\}$ fulfill the simple boundary condition on the simple surface (plane) and are assumed to be known, henceforth.

The fields $\{\vec{E}_+ - \vec{E}^0_+, \vec{H}_+ - \vec{H}^0_+\}$ and $\{\vec{E}^g, \vec{H}^g\}$ are connected by the equality

$$\text{div}\left[\left(\vec{E}_+ - \vec{E}^0_+\right) \times \vec{H}^g - \vec{E}^g \times \left(\vec{H}_+ - \vec{H}^0_+\right)\right]$$
$$= \frac{4\pi}{c}\left(\vec{H}_+ - \vec{H}^0_+\right) \cdot \vec{j}^{(m)}, \quad (3.118)$$

similar to (1.49). The integral form of this relation is

$$\int \left\{\left[\left(\vec{E}_+ - \vec{E}^0_+\right) \times \vec{H}^g\right]_N - \left[\vec{E}^g \times \left(\vec{H}_+ - \vec{H}^0_+\right)\right]_N\right\} dS$$
$$= \frac{4\pi}{c}\int \left(\vec{H}_+ - \vec{H}^0_+\right) \cdot \vec{j}^{(m)} \, dV. \quad (3.119)$$

The surface integral is taken over the infinite half-sphere and the plane $z = 0$, whereas the volume integral is taken over the half-space $z > 0$. The integral over the half-sphere is zero, since the fields satisfy the radiation condition; the proof of this assertion is given after formula (1.49). Since $E_{\text{tan}} = 0$ for all electric fields in the metallic part of the plane $z = 0$, the surface integral is taken over the hole only. The tangential component of the sought field $\vec{E}_+$ differs from zero on the hole. This component is denoted by $\vec{e}$. Keeping in mind the properties of the δ-function, which stands, according to (3.116), under the volume integral, we obtain from (3.119) the relation

$$\int \left(\vec{e} \times \vec{H}^g\right)_N dS = \frac{4\pi}{c}\vec{a} \cdot \left[\vec{H}_+ \left(\vec{r}_g\right) - \vec{H}^0_+ \left(\vec{r}_g\right)\right]. \quad (3.120)$$

From this relation, the magnetic field $\vec{H}_+$ at $z \geq 0$ can be expressed by the electric field $\vec{e}$ on the hole in the plane $z = 0$.

A similar formula can be obtained for $\vec{H}_-$, that is, for the magnetic field in the half-space $z \leq 0$. The continuity of E_{tan} is provided by the fact that the same field $\vec{e}$ is involved in both (3.120) and similar expression for $\vec{H}_-$. Requiring H_{tan} to be the same on both sides $z = +0$ and $z = -0$ of the hole, as well, we obtain the sought integral equation for the vector $\vec{e}$:

$$\int \left(\vec{e} \times \vec{K}\right)_N dS = \left(\vec{H}^0_+ - \vec{H}^0_-\right)\Big|_{z=0} \cdot \vec{a}. \quad (3.121)$$

This vector equation is a system of two scalar equations for the two components of the vector $\vec{e}$ directed along the x- and y-axes, respectively. According to (3.120), the two-dimensional vector $\vec{K}(\vec{r}, \vec{r}_g)$ is constructed from the magnetic fields of the two Green functions created by the magnetic dipoles located on both sides of the metallic screen without hole, that is, at $z_g = +0$ and $z_g = -0$.

The expression on the right-hand side of equation (3.121) describes the current in an auxiliary problem about the screen without hole, which flows in the place on the screen where the hole is cut in the real problem. The larger this current, the larger the electric field in the cut hole (see the text after (3.80)). Equation (3.108) can be treated as a partial form of equation (3.121), in which the half-infinite waveguides are replaced with the half-spaces $z \geq 0$ and $z \leq 0$.

Integration in (3.121) is performed with respect to the point $\vec{r}$. The point $\vec{r}_g$ is a parameter in the integral; the right-hand side of the equation depends on $\vec{r}_g$. Since the plane $z = 0$ where both the magnetic dipole (3.116) and the point at which the field equals $\vec{H}^0$ are located is not bounded, then $\vec{K}$ does not depend on $\vec{r}$ and $\vec{r}_g$ separately, but only on their difference. The similarly denoted kernel (3.109) in equation (3.108) does not possess this property, because the surface on which the electric field of the Green function vanishes is finite and bounded by the cross-section contour.

We describe the second method for solving the same problem on the metallic screen with a hole. This method leads to an integral equation for the current on the screen surface. It generalizes the method leading to equation (3.114) for the case of the diaphragm in waveguide.

Denote the field created by the same sources in the absence of the screen as $\{\vec{E}^0, \vec{H}^0\}$. The difference field $\{\vec{E} - \vec{E}^0, \vec{H} - \vec{H}^0\}$ is created only by the current $\vec{I}$ induced on the screen; it satisfies the radiation condition.

Define the Green functions in the half-spaces $z \geq 0$ and $z \leq 0$ as fields of the electric dipoles located at $z = +0$ and $z = -0$, correspondingly, that is, on both the sides of the plane $z = 0$. We subject them to the condition

$$H^g_{\text{tan}} = 0 \tag{3.122}$$

in this plane. At the end of Subsection 3.3.3 we have also introduced the Green function fulfilling this boundary condition. It was used there when considering the problem about the end-plane excitation of the waveguide with a magnetic field given in the cross-section. Introducing the Green function which satisfies condition (3.122) on a certain surface allows us to construct an expression for the field in the volume, not containing the electric field on this surface. In the problem about diaphragm with a hole, the equality condition for the field on both sides of the hole should be written without knowing the electric field on it. The elementary electric dipoles creating the fields $\{\vec{E}^g_+,\vec{H}^g_+\}$ and $\{\vec{E}^g_-, \vec{H}^g_-\}$ should be located on the screen and oriented tangentially to it, that is, the unit vectors $\vec{a}$ are tangential, $a_z = 0$. Then, the fields $\vec{E}^g_+$ and $\vec{E}^g_-$ coincide on the hole.

According to (1.49), the difference field and the field of the Green function are connected as

$$\int \left\{ \left[\left(\vec{E} - \vec{E}^0 \right) \times \vec{H}^g \right]_N - \left[\vec{E}^g \times \left(\vec{H} - \vec{H}^0 \right) \right]_N \right\} dS$$
$$= \frac{4\pi}{c} \left(\vec{E} - \vec{E}^0 \right) \bigg|_{\vec{r} = \vec{r}_g} \cdot \vec{a}. \quad (3.123)$$

Since both the fields satisfy the same radiation condition, the integral is taken only over the plane $z = 0$. According to (3.122), the first term under the integral equals zero. All the fields in the vector product of the second term are continuous on the hole. Since the normals on both sides of the plane $z = 0$ are opposite-directed, this term is the same in magnitude and different in sign on both sides $z = +0$ and $z = -0$. Summing equalities (3.123) written at $z = +0$ and $z = -0$, on the right-hand side we obtain the integral of the normal component of the vector products $\vec{E}^g \times \vec{H}$ and $\vec{E}^g \times \vec{H}^0$, taken only over the screen. The second sum vanishes since both the multipliers in the integrands are the same on both sides of the screen. The sum of the first integrals gives the normal component of the vector product $\vec{E}^g \times (\vec{H}_+ - \vec{H}_-)$, that is, the scalar product of E_{tan}^g and the jump of the tangential components of the magnetic field on the screen. This jump equals the current (3.110) induced on the screen.

The right-hand side of (3.123) includes the values of the fields $\vec{E}$ and $\vec{E}^0$ at the point $\vec{r} = \vec{r}_g$ lying on the screen where $E_{\text{tan}} = 0$. Thus, we get the sought equation for the current, which can be written as

$$\int \vec{\mathcal{P}} \left(\vec{r}, \vec{r}_g \right) \vec{I} \left(\vec{r} \right) dS_{\vec{r}} = \vec{a} \cdot \vec{E}^0 \left(\vec{r}_g \right). \quad (3.124)$$

The integral is taken over the screen, and the equation is valid for the points on it, so that (3.124) is the sought integral equation (more precisely, the system of two scalar integral equations for I_x and I_y).

The right-hand side of the equation is the tangential component of the electric field created by the same sources in the absence of the screen. The larger this field, the larger the induced current flowing on the screen. This current creates the field on the screen equal (and opposite in direction) to the component E_{tan}^0 existing there in the absence of the screen. Existence of the screen in the domain where the electric field, tangential to it, is small, causes the appearance of the small current and the small disturbance of the field in the domain. If the screen is perpendicular to the electric field $\vec{E}^0$, then the current does not flow on it and the field in the domain remains the same.

The kernel $\vec{\mathcal{P}}$ in equation (3.124), as well as K in (3.121), depends not on both of its arguments, but on their difference only.

3.4.6
Open end of waveguide

The open end of the waveguide can also be considered as a cross-section, where the regularity is violated. Consider a waveguide located only in the domain $z < 0$. The waveguide is broken at $z = 0$; there are no objects outside the waveguide. A direct eigenmode of the waveguide falls onto the open end from $z = -\infty$. A part of energy returns into the waveguide as the reflected mode of the same number together with back modes of other numbers. Another part of energy is radiated into free space (into both the half-space $z > 0$ and $z < 0$). The reflection factor R and amplitudes A_{-p} ($p > 0$) of other back modes, as well as the amplitude and structure of the radiated field, depend on the waveguide shape, the incident mode number, and the frequency.

If the scattered field is known, then the field of the same mode penetrated into the waveguide may be determined from a "reciprocal" problem about a certain wave falling from free space onto the open end of the waveguide. The reciprocity principle should be used for comparing these two problems.

At a high frequency, that is, at $k \gg k_n^{cut}$, where n is the incident mode number, the modules of the reflection factor $|R|$ and the amplitudes $|A_{-p}|$ are small; the most part of energy is radiated into free space. The field in the orifice has approximately the same structure as the incident mode. When the frequency decreases, $|R|$ increases and approaches the unity at k a little larger than k_n^{cut}, so that a small part of energy is radiated or transformed into other modes.

The back modes appear at the waveguide end in a similar way to those at a sharp junction of the waveguide with a horn. In this case the integral in formula (3.102) for A_{-p} can be calculated by parts. Then the integral term becomes of a higher order of smallness, and A_{-p} is proportional to the coupling factor $S_{n,-p}$ at the fracture cross-section. Amplitudes of the modes for which such coefficients are larger (at the same $v(s)$) are larger as well. This assertion is only qualitative because the waveguide break is not a flat disturbance, for which formula (3.102) relates. However, certain "selection rules," related to flat irregularities, are also valid for the diffraction at the open end.

In the circular waveguide, the total field keeps the dependence on the azimuthal coordinate existing in the excited mode. At $m \neq 0$, the TE_{mq}- and TM_{mq}-modes arise when TE_{m1} or TM_{m1} falls. When the magnetic mode TE_{01} falls, only the magnetic ones TE_{0q} arise, but TM_{0q} do not. The modes of electric type do not arise when TE_{0q} falls. The same holds for the circular waveguide with a symmetrical (not necessary small) ledge.

The rigorous theory of diffraction on the open waveguide end is developed only for the circular waveguides and for the waveguide consisting of two infinite planes. This theory uses the technique of integration in the plane of the complex variable. The foundation of this technique is sketched in Section 3.3.

However, in the problem which is considered here, a more complicated variant of the method is applied. This variant is called the *Wiener–Hopf method*, or the *factorization method*. The latter title owes to a procedure used in this method. The procedure consists in expressing a certain function, defined on the real axis ($h'' = 0$) of the complex plane $h = h' + ih''$, as a product of two functions, one of which is analytically continuable into the upper half-plane ($h'' > 0$), so that it has no singularities there and tends to zero as $h'' \to \infty$, whereas another has the same properties in the lower half-plane ($h'' < 0$).

The problem of diffraction at an open end of a half-infinite waveguide, placed in the half-space $-\infty < z < 0$, can be reduced to the integral equation in half-infinite limits

$$\int_{-\infty}^{0} \mathcal{P}(z - \varsigma) I(\varsigma) \, d\varsigma = f(z), \qquad z < 0 \tag{3.125}$$

for the current $I(z)$, with the kernel $\mathcal{P}$ dependent only on the difference between its arguments. The equation has the same form as (3.124) for the current on the screen with a hole. The kernel $\mathcal{P}$ is the Green function. It describes the field at the point ς, created in the infinite waveguide by the dipole placed at the point z. The boundary conditions for this function are different for the waveguides with different cross-sections; the functions $\mathcal{P}$ for them are different as well. They depend not only on z and ς, but also on the transverse coordinates. The function $f(z)$ on the right-hand side of equation (3.125) depends on the field of the mode which falls onto the open end from the waveguide.

The integral equation of the type (3.125) or a system of such equations is obtained not only in the problem about a half-infinite waveguide, but also in the problem of diffraction on a simple half-plane or on an equi-distant array of half-planes, as well as in other electrodynamic problems and problems of other radio-physical branches.

We sketch a way for solving this problem, without a detailed explanation. Denote by $F(z)$ the function equal to $\int_{-\infty}^{0} \mathcal{P}(z - \varsigma) I(\varsigma) \, d\varsigma$ at $z < 0$ and to zero at $z > 0$. Similarly, denote by $\Phi(z)$ the function equal to the same integral at $z > 0$ and to zero at $z < 0$. Hence, by definition,

$$F(z) = 0 \quad \text{at } z > 0; \qquad \Phi(z) = 0 \quad \text{at } z < 0. \tag{3.126}$$

Make up the following function of the variable h:

$$\int_{-\infty}^{0} F(z) \exp(-ihz) \, dz + \int_{0}^{\infty} \Phi(z) \exp(-ihz) \, dz. \tag{3.127}$$

The integrand in first term of (3.127) involves the left-hand side of equation (3.125), so that this term equals $\int_{-\infty}^{0} f(z) \exp(-ihz) \, dz$.

Substitute the functions $F(z)$ and $\Phi(z)$, according to their definitions at $z < 0$ and $z > 0$, respectively, into (3.127), and then interchange the integration order with respect to z and ς. The inner integral of the product $\mathcal{P}(z - \varsigma) \exp(-ihz)$ is taken over the whole z-axis $-\infty < z < \infty$. Introduce the new variable $t = z - \varsigma$ instead of z there. Then expression (3.127) becomes a product of two integrals. Equating it to (3.127) and using equation (3.125) give

$$\int\limits_{-\infty}^{0} I(z) \exp(-ihz)\, dz \cdot \int\limits_{-\infty}^{\infty} \mathcal{P}(t) \exp(-iht)\, dt$$

$$= \int\limits_{-\infty}^{0} f(z) \exp(-ihz)\, dz + \int\limits_{0}^{\infty} \Phi(z) \exp(-ihz)\, dz. \quad (3.128)$$

Formula (3.128), as an equality of two functions of h, is equivalent to the integral equation (3.125), as an equality of two functions of z.

We factorize the second multiplier in the left-hand side of (10.36), that is, express it as a product of two functions of h:

$$\int\limits_{-\infty}^{\infty} \mathcal{P}(t) \exp(-iht)\, dt = P_1(h) P_2(h), \quad (3.129)$$

where $P_1(h)$ is an analytical function tending to zero in the upper half-plane as $h'' \to \infty$, and $P_2(h)$ is an analytical function tending to zero in the lower half-plane as $h'' \to -\infty$. Divide both sides of equality (3.128) by $P_2(h)$ and write

$$\frac{\int_{-\infty}^{0} f(z) \exp(-ihz)\, dz}{P_2(h)} = g_1(h) + g_2(h), \quad (3.130)$$

where $g_1(h)$, $g_2(h)$ have the same properties as $P_1(h)$, $P_2(h)$, respectively. Then, (3.128) becomes

$$\int\limits_{-\infty}^{0} I(z) \exp(-ihz)\, dz \cdot P_1(h) - g_1(h) =$$

$$g_2(h) + \frac{1}{P_2(h)} \int\limits_{0}^{\infty} \Phi(z) \exp(-ihz)\, dz. \quad (3.131)$$

All these procedures are made for expressing equation (3.128) as an equality of two functions of h, one of which is analytical and tends to zero in the upper half-plane ($h'' > 0$), and another has the same properties in the lower half-plane ($h'' < 0$). It is easy to check that (3.131) has these properties. Indeed, $z \leq 0$ in the integrand of the left-hand side, so that $\exp(-ihz) \to 0$ as $h'' \to \infty$. The functions $P_1(h)$ and $g_1(h)$ have this property by definition. In a similar way it may be checked that all functions on the right-hand side have analogous property in the lower half-plane.

Under certain conditions, from the equality on the axis $h'' = 0$ of two functions having the above properties, it follows that these functions equal zero on this axis. This fact leads to an explicit expression for the integral $\int_{-\infty}^{0} I(\varsigma) \exp(-ih\varsigma) \, d\varsigma$. The integration domain can be extended to $+\infty$ since $I(\varsigma) = 0$ at $\varsigma > 0$. The inverse Fourier transformation of this function gives $I(z)$, that is, the sought solution of the integral equation (3.125).

In the waveguide problem the function $\mathcal{P}$ depends not only on the longitudinal coordinates z and ς, but also on the transverse ones x, y. The field at a point with coordinates x, y, z is proportional to $\int_{-\infty}^{0} \mathcal{P}(x, y, z - \varsigma) I(\varsigma) \, d\varsigma$. For points inside the waveguide, this integral is reduced to a sum of residues, analogous to (3.90). The field consists of the incident mode and a series of the back ones; the amplitudes of these modes (in particular, the reflection coefficient) are proportional to values of residues. For exterior points the integral is not reduced to a residual sum; it describes the wave outgoing from the open end of the waveguide.

4
Closed Resonators

4.1
Resonators with ideal-conducting walls

4.1.1
Waveguide resonators without filling

At certain frequencies, in the closed domain bounded by the walls on which the condition $E_{\tan} = 0$ holds, there exist solutions to the homogeneous Maxwell equations (1.29) with $\vec{j}^{\text{ext}} = 0$. This fact was mentioned in Section 1.4 when considering the existence and uniqueness of the solution to these equations. For any domain, such solutions exist only at certain frequencies depending on the domain shape and size. These frequencies make up the countable sequence (the *discrete spectrum*). These frequencies are called the *eigenfrequencies* of the domain whereas the fields corresponding to them are called the eigenoscillation fields. The solutions to the homogeneous equations may be used for solving the nonhomogeneous ones. The field excited by the extrinsic current in the closed domain (*resonator*) is expressed in a series by its eigenoscillations. Further this statement will be specified more exactly. Such an expression exists at any frequency except for the eigenfrequency of the resonator. In general, the nonhomogeneous problem is not solvable at the eigenfrequency. The assumption about the possibility of creating an arbitrary current at this frequency contradicts the accepted idealizations (monochromatic oscillation, ideal-conducting walls).

We investigate the eigenoscillations of a simple-shaped waveguide resonator, namely, the waveguide segment closed by two solid diaphragms at the ends. Assume that the waveguide walls and diaphragms are ideal-conducting. The field in the resonator coincides with the field of one of its eigenmodes, but with a standing, not a propagating, one. The dependence of such fields on the longitudinal coordinate is determined by the multiplier $\cos(h_n z)$ or $\sin(h_n z)$ instead of $\exp(\pm i h_n z)$. Let this mode be a TE-mode for the definiteness. Then the field in the resonator is expressed by the longitudinal component of the magnetic Hertz vector $\Pi_z^{(m)}(x, y, z)$ having in this case

High-frequency Electrodynamics. Boris Z. Katsenelenbaum
Copyright © 2006 WILEY-VCH Verlag GmbH & Co. KGaA, Weinheim
ISBN: 3-527-40529-1

the form $\psi^n(x,y)\sin(h_n z)$, where $\psi^n(x,y)$ is a solution to system (3.13) corresponding to the eigenvalue β_n, $h_n^2 = k^2 - \beta_n^2$. The transverse components of the electric field are

$$E_x = -ik\frac{\partial\psi^n}{\partial y}\sin(h_n z), \qquad E_y = ik\frac{\partial\psi^n}{\partial x}\sin(h_n z). \tag{4.1a}$$

In the planes $z = 0$ and $z = L$, where $L = \pi l/h_n$ ($l = 1, 2, \ldots$), we have $E_x = 0$, $E_y = 0$. The condition $E_{\text{tan}} = 0$ holds on the diaphragms located in these two planes. The field with the components (4.1a) and

$$E_z = 0, \quad H_x = h_n\frac{\partial\psi^n}{\partial x}\cos(h_n z),$$

$$H_y = h_n\frac{\partial\psi^n}{\partial y}\cos(h_n z), \quad H_z = \beta_n^2\psi^n\sin(h_n z) \tag{4.1b}$$

solves the homogeneous Maxwell equations, and fulfills the condition $E_{\text{tan}} = 0$ both on the side walls of the cylinder and on its bases. The field (4.1b) is obtained by formulas (3.4) with replacing the multiplier $-ih_n$ by the z-differentiation.

At a given L, solution (4.1) exists only if $h_n = l\pi/L$, $l = 1, 2, \ldots$ Since $h_n^2 = k^2 - \beta_n^2$, the frequency should be equal to

$$k_N = \sqrt{\left(\frac{\pi l}{L}\right)^2 + \beta_n^2}. \tag{4.2}$$

These frequencies are the eigenfrequencies of the waveguide resonator for the TE-modes. Since the eigenmode number n is a combination of the two numbers m, q (see the text above (3.30)), N is an aggregate of three integer indices: $N = \{m, q, l\}$. For instance, according to (3.30) and (4.2), for the rectangular waveguide resonator, that is, for the rectangular parallelepiped, we have

$$k_N^2 = \frac{\pi^2}{a^2}m^2 + \frac{\pi^2}{\beta^2}q^2 + \frac{\pi^2}{L^2}l^2, \tag{4.3}$$

where a, b, L are the resonator sides. One of the numbers m, q, l in (4.3) may be zero, the two others should be positive. The three parallelepiped sizes appear in k_N equally.

In a similar way as for the waveguides, the oscillations of the TE (with respect to the z-axis) and TM types may be introduced for the waveguide resonators. The fields of the eigen TM-oscillations are expressed by the longitudinal component of the electric Hertz vector $\vec{\Pi}^{(e)}$ (3.25a) of the form $\chi^n(x,y)\cos(h_n z)$ by formulas similar to (3.4). These fields are

$$E_x = -h_n\frac{\partial\chi^n}{\partial x}\sin(h_n z), \quad E_y = -h_n\frac{\partial\chi^n}{\partial y}\sin(h_n z),$$

$$E_z = \alpha_n^2\chi^n\cos(h_n z) \tag{4.4a}$$

$$H_x = ik\frac{\partial\chi^n}{\partial y}\cos(h_n z), \quad H_y = -ik\frac{\partial\chi^n}{\partial x}\cos(h_n z), \quad H_z = 0. \tag{4.4b}$$

The oscillations exist at the frequencies for which $h_n = \pi l/L$ ($l = 0, 1, 2, \ldots$).

The eigenfrequencies are connected to α_n by the relation

$$k_N = \sqrt{\left(\frac{\pi l}{L}\right)^2 + \alpha_n^2}, \tag{4.5}$$

where L is a distance between the diaphragms, that is, a cylinder height, just as in (4.2). For the rectangular waveguide α_n^2 is expressed by the cross-section sides a and b by the same formula (3.30) and the same expression (4.3) is obtained for k_N^2. The eigenfrequency depends only on the geometric parameters of the resonator.

In the circular waveguide resonator, the fields of the eigenoscillations are expressed by formulas similar to (4.1) and (4.4) with replacing the derivatives with respect to x and y by those with respect to r and φ, correspondingly. The functions $\Pi_z^{(e)}$ and $\Pi_z^{(m)}$ have the form

$$J_m(\alpha_n z)\begin{Bmatrix}\cos\\\sin\end{Bmatrix}(m\varphi)\cos(h_n z); \quad J_m(\beta_n r)\begin{Bmatrix}\cos\\\sin\end{Bmatrix}(m\varphi)\sin(h_n z). \tag{4.6}$$

Similarly as in the circular waveguides, there exists the polarization degeneration for the oscillations with $m > 0$ in such resonators.

Usually, the resonators are used at the frequencies close to the frequency of the so-called main eigenoscillations, that is, oscillations with the smallest k_N. In the circular waveguide resonators the main oscillation is either of the magnetic type TE_{111} ($m = 1, q = 1, l = 1$) with the potential function $J_1(1.84r/a)\cos\varphi\sin(h_n z)$ and the eigenfrequency

$$k_{111}^{(m)} = \sqrt{\left(\frac{1.84}{a}\right)^2 + \left(\frac{\pi}{L}\right)^2}, \tag{4.7a}$$

or of the electric type TM_{010} ($m = 0, q = 1, l = 0$) with the potential function $J_0(2.40r/a)$ and the eigenfrequency

$$k_{010}^{(e)} = \frac{2.40}{a}. \tag{4.7b}$$

The field of this oscillations depends neither on φ nor on z. The electric field of it, as well as of any other TM-oscillation with $l = 0$, has almost the same structure as a field in the electrostatic condenser.

For the short resonators with $L < 2.04a$, the main oscillation is TM_{010}; for the resonators with $L > 2.04a$, it is TE_{111}.

We mention two more types of oscillations in the circular waveguide resonator, namely, the magnetic TE_{011}-oscillation and the electric TM_{111}-oscillation. These oscillations are degenerated just like the TE_{01}- and TM_{11}-modes in the circular waveguide. Their eigenfrequencies coincide and equal

$$k_{011}^{(m)} = k_{111}^{(e)} = \sqrt{\left(\frac{3.83}{a}\right)^2 + \frac{\pi^2}{L^2}}. \tag{4.8}$$

According to (4.1), (4.4) all the components of the field $\vec{E}$ as well as those of the field $\vec{H}$ are in-phase, whereas the fields themselves are in-quadrature. During each half-period the energy transfers from the electric field to the magnetic one and back. After passing from the complex amplitudes to the physical quantities by (1.3), this fact makes the fields $\vec{E}$ and $\vec{H}$ proportional to $\cos(wt - \alpha)$ and $\sin(wt - \alpha)$, respectively.

From (4.1) and (4.4) it follows that the energies of the electric and magnetic fields coincide,

$$\int_V \left|\vec{E}\right|^2 dV = \int_V \left|\vec{H}\right|^2 dV. \tag{4.9}$$

Here the integral is taken over the resonator volume; this relates to all volume integrals below. Of course, equality (4.9) conforms with the property mentioned above. In the next subsection it will be shown that both these properties of the eigenoscillations are inherent to the arbitrarily shaped resonators.

4.1.2
Resonators of arbitrary shape

The field of the eigenoscillation in the resonator of arbitrary shape solves the homogeneous equations

$$\text{rot}\,\vec{H} - ik\vec{E} = 0, \qquad \text{rot}\,\vec{E} + ik\vec{H} = 0, \tag{4.10a}$$

with the boundary condition

$$E_{\text{tan}} = 0. \tag{4.10b}$$

The homogeneous system (4.10a) has nontrivial (i. e., different from the identical zero) solution at certain frequencies only. These eigenfrequencies are denoted by k_N, whereas the fields of the corresponding eigenoscillations can be denoted by $\vec{E}^N, \vec{H}^N$.

Consider two oscillations of numbers M and N. In a similar way to Subsection 1.2.2, we integrate the expression $\text{div}(\vec{E}^M \times \vec{H}^N)$ over the volume. Using formula (1.55), equations (4.10a), and the boundary condition (4.10b), we obtain the identity

$$
k_M \int\limits_V \vec{H}^N \vec{H}^M \, dV + k_N \int\limits_V \vec{E}^N \vec{E}^M \, dV = 0.
\tag{4.11}
$$

In particular, if $N = M$, then

$$
\int\limits_V \left(\vec{H}^N \right)^2 dV = - \int\limits_V \left(\vec{E}^N \right)^2 dV.
\tag{4.12}
$$

Exchanging indices N and M in (4.11), we get another identity analogous to (4.11). It follows from both the identities that if the eigenfrequencies do not coincide, then the fields of two eigenoscillations are orthogonal in the following sense:

$$
\int\limits_V \vec{H}^N \vec{H}^M \, dV = 0, \quad \int\limits_V \vec{E}^N \vec{E}^M \, dV = 0, \quad k_N \neq k_M.
\tag{4.13}
$$

Analogously, taking the volume integral of the expression $\mathrm{div}(\vec{E}^N \times \vec{H}^{N*})$ yields the identity

$$
k_N \int\limits_V \left| \vec{H}^N \right|^2 dV = k_N^* \int\limits_V \left| \vec{E}^N \right|^2 dV.
\tag{4.14}
$$

From (4.14) it follows that the eigenfrequencies k_N are real and

$$
\int\limits_V \left| \vec{H}^N \right|^2 dV = \int\limits_V \left| \vec{E}^N \right|^2 dV.
\tag{4.15}
$$

From the reality of k_N and the wave equation (1.37) it follows that if the eigenoscillation is not degenerated, then each of the fields $\vec{H}^N$ and $\vec{E}^N$ has the same phase in the whole domain. Writing, for instance, the first one as $\vec{H}^N = \vec{H}^{N\prime} + i\vec{H}^{N\prime\prime}$, we obtain the equations for $\vec{H}^{N\prime}$ and $\vec{H}^{N\prime\prime}$ with the same eigenvalue k_N^2. Since it is not degenerated, the real and imaginary parts of $\vec{H}^N$ differ only in the constant factor, that is, $\vec{H}^N$ is in-phase in the whole domain. Then it follows from (4.12), (4.15) (or directly from (4.10a)) that the fields $\vec{E}^N$ and $\vec{H}^N$ are in-quadrature.

These conditions may not be satisfied for the degenerated eigenoscillations. For instance, the polarization degeneration exists in the circular waveguide resonator, such that the φ-dependence of fields may be determined either by the multiplier $\cos(m\varphi)$ or by $\sin(m\varphi)$. The eigenoscillation exists in which the fields are proportional to $\exp(im\varphi)$. Such an oscillation represents not a standing but propagating (in the azimuthal direction) wave and the field is not in-phase in it, of course.

The main results, namely, the functional orthogonality of the fields of differ-ent eigenoscillations (4.13), equality of energies (4.15), reality of eigenvalues, equiphaseness of the fields $\vec{H}^N$ and $\vec{E}^N$, and relative time shift of these fields by a quarter of period, are obtained for the resonators with ideal-conducting walls, which do not contain any objects inside. With some specifications these results are valid for the resonators containing any nonchiral objects with-out losses. In such resonators, the Maxwell equations for the fields of the eigenoscillations differ from (4.10a) by the multipliers $\varepsilon(\vec{r})$ and $\mu(\vec{r})$

$$\operatorname{rot} \vec{H}^N - ik_N \varepsilon \vec{E}^N = 0, \qquad \operatorname{rot} \vec{E}^N + ik_N \mu \vec{H}^N = 0. \tag{4.16}$$

Assume that ε and μ are independent of frequency and are real; we will specify this condition in Section 4.2. Repeating the derivations which resulted in (4.13), (4.15), we obtain the orthogonality conditions in the form

$$\int_V \mu \vec{H}^N \vec{H}^M \, dV = 0, \quad \int_V \varepsilon \vec{E}^N \vec{E}^M \, dV = 0 \quad (k_N \neq k_M). \tag{4.17}$$

The condition of equality of energies is

$$\int_V \mu \left| \vec{H}^N \right|^2 dV = \int_V \varepsilon \left| \vec{E}^N \right|^2 dV. \tag{4.18}$$

The reality of eigenfrequencies, synchronism of the fields $\vec{E}^N$, $\vec{H}^N$, and their misphasing by a quarter of period remain to be valid.

4.1.3
Calculation of eigenfrequencies

The volume resonators are used, in particular, for measuring the parameters of materials inserted into the resonator as certain "samples." The measurements are carried out at the frequency close to the eigenfrequency k_N of the resonator. For nonfilled waveguide resonators the explicit formulas (4.2), (4.5) exist for finding k_N; it is sufficient to find an eigenvalue of one of the two-dimensional scalar problems (3.11), (3.13).

If a sample inserted into the waveguide resonator is a homogeneous cylin-der of the same cross-sections as the resonator, then a simple equation can be derived for k_N containing the material parameters ε and μ. Having measured k_N, we can find the parameters from this equation.

We begin the derivation of this equation with the case when the TE_{mql}-oscillation ($l > 0$) exists in the resonator without the sample. After the sample with the shape described above is inserted, the dependence of the field com-ponents on the transverse coordinates remains the same, whereas in the filled part of the resonator the transverse components of the field $\vec{E}$ get the addi-tional multiplier μ in comparison with formulas (4.1). In the filled part of the resonator, the field has the same structure of the standing TE_{mq}-wave as in the nonfilled one; only the propagation constant $h = (k^2 - \beta^2)^{1/2}$ should be replaced by the quantity $h_\varepsilon = (k^2\varepsilon\mu - \beta^2)^{1/2}$. The lower index n is omitted in the above formulas. The components E_x, E_y are proportional to $\sin(h_\varepsilon z)$ and the components H_x, H_y to $\cos(h_\varepsilon z)$. It is assumed that the sample lies at the bottom of the resonator, that is, its base is located at $z = 0$ (Fig. 4.1). Neither new components, nor different dependences on x, y arise, similarly as they do not arise in the eigenmode of the waveguide with the homogeneous dielectric cylinder filling the entire cross-section. According to (4.1), in the nonfilled part, the fields are proportional to the functions $\sin[h(z - L)]$ and $\cos[h(z - L)]$; arguments of the functions are constructed in such a way that the requirement $E_x = 0$, $E_y = 0$ is fulfilled at $z = L$. The equality conditions for the transverse components of the fields $\vec{E}$ and $\vec{H}$ at $z = d - 0$ and $z = d + 0$, where d is the height of the sample, lead to the sought equation

$$\frac{\mu}{h_\varepsilon} \tan(h_\varepsilon d) + \frac{1}{h} \tan[h(L - d)] = 0. \tag{4.19}$$

At $\varepsilon = 1$, $\mu = 1$ this equation becomes $\tan(Lh) = 0$, that is, $h = \pi l / L$, resulting in expression (4.2) for k_N.

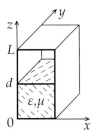

Fig. 4.1 Partially filled resonator

The problem about the TM_{mql}-oscillation in the waveguide resonator with the same inclusion is reduced to the similar equation connecting k_N with parameters of the sample material. In this case the fields are also expressed by formulas similar to (4.4). Analogously to the above, h should be replaced by h_ε in the filled part $0 < z < d$ and the components H_x, H_y get the additional multiplier ε. Repeating the derivations which led to (4.19) for the TE-oscillations, for the TM-oscillations we obtain

$$\frac{h_\varepsilon}{\varepsilon} \tan(h_\varepsilon d) + h \tan[h(L - d)] = 0. \tag{4.20}$$

Inserting the object with $\varepsilon \geq 1$, $\mu \geq 1$ into the resonator causes the field redistribution along the z-axis. The eigenfrequency k_N decreases which follows also from formulas (4.19), (4.20). Increasing ε or μ even in a part of the volume is equivalent to increasing the length of the nonfilled resonator, which, according to (4.2), (4.5), causes the decrease of k_N.

In the nonfilled resonator the field of the TM_{mq0}-oscillation ($l = 0$) is the same at all z, the eigenfrequency does not depend on the length L; it equals α_n (see (4.7b); for this case $\alpha_n = 2.40/a$). If a sample with $\varepsilon\mu > 1$ is introduced, then $h_\varepsilon^2 > 0$, $h^2 < 0$. The field in the free (nonfilled) part of the resonator is proportional not to the trigonometric but to the hyperbolic functions $\sinh[|h|(z - L)]$ and $\cosh[|h|(z - L)]$. The field concentrates near the sample and weakens when approaching the opposite resonator side just as a field in the waveguide at the frequency lower than a cutoff one. This type of oscillations is not used in the measuring devices.

4.1.4
Variational technique

If the resonator is not a segment of the waveguide closed by the two diaphragms at the ends, but an arbitrary volume V bounded by a surface S, then the field of its eigenoscillation cannot be written in the explicit form even if the resonator does not contain any dielectric inclusion. Therefore, there are also neither an explicit expression for its eigenfrequencies k_N, nor any simple equation from which k_N could be found. In this case k_N can be calculated using the variational technique applicable to the arbitrarily shaped resonators with any filling.

We describe the simplest version of this method as follows. Let the eigenfrequency (more exactly, its square) be represented by a functional

$$k^2 = \mathcal{L}(\vec{H}) \tag{4.21}$$

In this subsection we omit the index N in the notation of the eigenfrequency and fields of the eigenoscillations. Substituting the field (4.1b) or (4.4b) into (4.21) yields (4.2) or (4.5), respectively. However, the field $\vec{H}$ is unknown. A peculiar property of the functionals used in the variational method consists in the fact that the inaccuracy of k^2 is much smaller than that of the approximate field $\vec{H}$ substituted into (4.21).

We specify this assertion. Denote exact values of the field $\vec{H}$ and eigenfrequency k by $\vec{H}^0$ and k_0, respectively. By definition, $\mathcal{L}(\vec{H}^0) = k_0^2$. The mentioned feature of the functional $\mathcal{L}$ is called the *stationarity* and means that

$$\mathcal{L}\left(\vec{H}^0 + v\vec{\varphi}\right) = k_0^2 + O\left(v^2\right). \tag{4.22}$$

Here $\vec{\varphi}$ is any vector (this formulation will be specified later), and v is a small real constant. According to (4.22), the inaccuracy of k^2 calculated by (4.21) has the higher order of smallness than that of the field.

For calculating the rough estimate of k_0, formula (4.22) with the approximate field $\vec{H}$ substituted into $\mathcal{L}(\vec{H})$ may be used. However, a more consistent method is known, the so-called *Ritz method*, allowing us to find k_0 with any given accuracy. We express the field $\vec{H}^0$ as an infinite series $\sum_p A_p \vec{h}_p$, where $\{\vec{h}_p\}$ is a complete system of vector functions. Substitute this series into the functional (4.21). It becomes the function $\mathcal{L}(A_1, A_2, \ldots)$ of the coefficients A_p. At certain unknown values of the coefficients A_p^0, the sum of the series equals the field $\vec{H}^0$. According to (4.22), the function $\mathcal{L}(A_1, A_2, \ldots)$ reaches its extreme values at $A_p = A_p^0$ ($p = 1, 2, \ldots$). Consequently, the derivatives of $\mathcal{L}$ with respect to all the coefficients are zero at the stationary point A_p^0:

$$\frac{\partial \mathcal{L}}{\partial A_p} = 0, \qquad p = 1, 2, \ldots, \tag{4.23}$$

which gives an infinite system of algebraic equations. After finding A_p^0 from these equations, we can find the field $\vec{H}^0$ and the extreme value of $\mathcal{L}$, that is, k_0^2.

The functional $\mathcal{L}(\vec{H})$ can be obtained after substituting the expression for $\vec{E}$ from the first equation of system (4.10a) into identity (4.15), which is valid for the fields satisfying (4.10a). Thus,

$$k^2 = \frac{\int_V \left| \operatorname{rot} \vec{H} \right|^2 dV}{\int_V \left| \vec{H} \right|^2 dV}. \tag{4.24}$$

We prove that the functional (4.24) possesses the property (4.22). Substituting $\vec{H} = \vec{H}^0 + v\vec{\varphi}$ into (4.24) and keeping only the terms of the zeroth and first order with respect to v, for the denominator of (4.24) we obtain that

$$\int_V \left(\vec{H}^0 + v\vec{\varphi} \right)^2 dV = \int_V \left(\vec{H}^0 \right)^2 dV + 2v \int_V \vec{H}^0 \vec{\varphi} \, dV, \tag{4.25}$$

where, for simplicity, not only $\vec{H}^0$ but also $\vec{\varphi}$ is assumed to be a real vector. With the same accuracy, the numerator is

$$\int_V \left[\operatorname{rot} \left(\vec{H}^0 + v\vec{\varphi} \right) \right]^2 dV = \int_V \left(\operatorname{rot} \vec{H}^0 \right)^2 dV + 2v \int_V \operatorname{rot} \vec{H}^0 \operatorname{rot} \vec{\varphi} \, dV. \tag{4.26}$$

Substituting $\vec{A} = \vec{\varphi}$, $\vec{B} = \vec{H}^0$ into the table identity

$$\operatorname{rot} \vec{A} \cdot \operatorname{rot} \vec{B} = \vec{A} \operatorname{rot} \operatorname{rot} \vec{B} + \operatorname{div} \left(\vec{A} \times \operatorname{rot} \vec{B} \right) \tag{4.27}$$

yields for the second term in (4.26)

$$\int_V \operatorname{rot} \vec{H}^0 \operatorname{rot} \vec{\varphi} \, dV = \int_V \vec{\varphi} \operatorname{rot} \operatorname{rot} \vec{H}^0 \, dV + \int_S \left(\vec{\varphi} \times \operatorname{rot} \vec{H}^0 \right)_N dS. \tag{4.28}$$

According to the wave equation (1.37), the first integral on the right-hand equals $k_0^2 \int_V \vec{H}^0 \vec{\varphi} \, dV$. The second term is zero owing to the boundary condition (4.10b). Hence, substituting $\vec{H} = \vec{H}^0 + v\vec{\varphi}$ into (4.24) results in the denominator coincident (with the accuracy to a value of the second order of v) with the numerator multiplied by k_0^2. The deviation of the functional (4.24) from k_0^2 has the order v^2, that is, this functional reaches the extremum on the solutions of system (4.10a) with the boundary conditions (4.10b). In this case, there are no conditions imposed on the function $\vec{\varphi}$, that is, on the functions $\vec{h}_p$, except for the condition of differentiability in the whole domain.

Substituting the series with the coefficients A_p for $\vec{H}$ gives the expression for k^2 in the form of the ratio of two bilinear forms of A_p

$$k^2 = \frac{\sum_p \sum_t a_{pt} A_p A_t}{\sum_p \sum_t b_{pt} A_p A_t}. \tag{4.29}$$

For simplicity the vector functions $\vec{h}_p$ and coefficients A_p are assumed to be real. The coefficients a_{pt} and b_{pt} are

$$a_{pt} = \int_V \operatorname{rot}\vec{h}_p \cdot \operatorname{rot}\vec{h}_t \, dV, \quad b_{pt} = \int_V \vec{h}_p \cdot \vec{h}_t \, dV. \tag{4.30}$$

They are known values depending only on the chosen system of functions $\vec{h}_p$ and the shape of the resonator. It is clear that $a_{pt} = a_{tp}$, $b_{pt} = b_{tp}$ (under the above assumption of the reality of $\vec{h}_p$).

Differentiating (4.29) with respect to A_p and equating the results to zero, we obtain the following infinite system of the homogeneous algebraic equations for A_t^0:

$$\sum_t \left(a_{pt} - k_0^2 b_{pt} \right) A_t^0 = 0, \quad p = 1, 2, \ldots \tag{4.31}$$

Here we have considered that the stationary values of the ratio (4.29) equal k_0^2. These values play a role of the eigenvalues in system (4.31). The system has a nontrivial solution only if k_0^2 satisfies the equation

$$\det\{a_{pt} - k_0^2 b_{pt}\} = 0. \tag{4.32}$$

If only a finite number of functions $\vec{h}_n$ are used in the expression for $\vec{H}_0$, then the roots of this algebraic equation tend to the eigenvalues of the resonator as the matrix dimension grows. In this version of the variational method the eigenfrequencies are determined (as the roots of equation (4.32)) before finding the coefficients A_p^0 (from system (4.31)), that is, the fields of the corresponding eigenoscillations.

The similar functional

$$k^2 = \frac{\int_V \left(\mathrm{rot}\, \vec{E} \right)^2 dV}{\int_V \left(\vec{E} \right)^2 dV} \tag{4.33}$$

does not possess the above property if $\vec{\varphi}$ is an arbitrary differentiable vector function. When calculating the linear (with respect to $\vec{\varphi}$) term in the expression $\int_V (\mathrm{rot}\, \vec{E} + v \vec{\varphi})^2 \, dV$ and performing the derivations similar to those made above for the functional (4.24), the term $\int_S (\vec{\varphi} \times \mathrm{rot}\, \vec{E}^0)_N \, dS$ does not vanish for any vector $\vec{\varphi}$, since the component $(\mathrm{rot}\, \vec{E}^0)_{\mathrm{tan}}$ proportional to $\vec{H}^0_{\mathrm{tan}}$ does not vanish on the walls. The demand that the functional (4.33) is stationary on $\vec{E}^0$ is fulfilled only if $\vec{\varphi}$ satisfies the condition $\vec{\varphi}_{\mathrm{tan}} = 0$. This demand should be imposed on the vector functions $\vec{e}_p$ by which the sought field $\vec{E}^0$ is expressed.

An auxiliary way exists allowing us to modify the functional (4.33) so that it becomes stationary for any perturbation of its argument, that is, no demand on the system of functions $\vec{e}_p$ is imposed to satisfy certain boundary conditions on the resonator walls. Using the direct checking, it is easy to show that the functional

$$k^2 = \frac{\int_V (\mathrm{rot}\, \vec{E})^2 \, dV - 2 \int_S \left(\vec{E} \times \mathrm{rot}\, \vec{E} \right)_N \, dS}{\int_V (\vec{E})^2 \, dV} \tag{4.34}$$

possesses the above property.

The functionals of the type (4.24), (4.34) exist also for the eigenvalues of the resonators containing any material objects.

4.1.5
Excitation of resonators

We find the field which exists in the resonator in the presence of the current with density $\vec{j}^{\mathrm{ext}}$. The resonator may have any filling described by the functions $\varepsilon(\vec{r})$ and $\mu(\vec{r})$. The field satisfies the nonhomogeneous equation system

$$\mathrm{rot}\, \vec{H} - ik\varepsilon\vec{E} = \frac{4\pi}{c} \vec{j}^{\mathrm{ext}}, \tag{4.35a}$$

$$\mathrm{rot}\, \vec{E} + ik\mu\vec{H} = 0 \tag{4.35b}$$

in the volume and the boundary condition (4.10b) on the walls. Represent the fields $\vec{E}$ and $\vec{H}$ in the form of series

$$\vec{E} = \sum_N A_N \vec{E}^N - \mathrm{grad}\, \varphi, \tag{4.36a}$$

$$\vec{H} = \sum_N B_N \vec{H}^N. \tag{4.36b}$$

The gradient term should be introduced into (4.36a), since, according to (4.16), the electric field of any eigenoscillation is subjected to the condition $\mathrm{div}(\varepsilon \vec{E}^N) = 0$ and the quantity $\mathrm{div}(\varepsilon \vec{E})$, proportional to $\mathrm{div}\,\vec{j}^{\mathrm{ext}}$, is, in general, not zero for the field created by the extrinsic currents. If there exists a charge $\rho^{\mathrm{ext}} = i/\omega\,\mathrm{div}\,\vec{j}^{\mathrm{ext}}$, then $\mathrm{div}(\varepsilon \vec{E}) = -4\pi\rho^{\mathrm{ext}}$, and the function $\varphi(\vec{r})$ solves the equation

$$\mathrm{div}(\varepsilon\,\mathrm{grad}\,\varphi) = 4\pi\rho^{\mathrm{ext}} \tag{4.37}$$

with the boundary condition $(\nabla\varphi)_{\tan} = 0$ on the walls The function $\varphi(\vec{r})$ is a potential of the electrostatic field excited by the charge introduced in the resonator. This field does not have a resonant nature and almost always is small in comparison with the part of the field $\vec{E}$ represented by the series by fields $\vec{E}^N$. The vector $\nabla\varphi$ is orthogonal to all the fields $\vec{E}^N$, that is, the equality

$$\int_V \varepsilon\vec{E}^N \nabla\varphi\,dV = 0 \tag{4.38}$$

holds for any N. This equality can be easily obtained by integrating the expression $\mathrm{div}(\vec{H}^N \times \nabla\varphi)$ over V. According to the boundary condition for $\nabla\varphi$, this integral equals zero. Substituting $\mathrm{rot}\,\vec{H}^N$ from (4.16) into the obtained expression, we get (4.38).

After substituting the series (4.36) into (4.35), applying the "rot" operation termwise and using (4.16), we obtain the two equalities

$$\sum_N (kA_N - k_N B_N)\,\varepsilon\vec{E}^N - k\varepsilon\nabla\varphi = \frac{4\pi i}{c}\vec{j}^{\mathrm{ext}}, \tag{4.39}$$

$$\sum_N (k_N A_N - kB_N)\,\mu\vec{H}^N = 0.$$

We multiply these equalities by $\vec{E}^M$ and $\vec{H}^M$, respectively, and integrate over V. Taking into account the orthogonality conditions (4.17) and (4.38), we obtain the system of two algebraic equations for A_N and B_N for each N, from which the sought coefficients are calculated as

$$A_N = -\frac{4\pi i}{c}\frac{k}{k^2 - k_N^2}\frac{\int_V \vec{j}^{\mathrm{ext}}\vec{E}^N\,dV}{\int_V \varepsilon\left(\vec{E}^N\right)^2 dV},$$

$$B_N = -\frac{4\pi i}{c}\frac{k_N}{k^2 - k_N^2}\frac{\int_V \vec{j}^{\mathrm{ext}}\vec{E}^N\,dV}{\int_V \varepsilon\left(\vec{E}^N\right)^2 dV}. \tag{4.40}$$

At the frequencies close to eigenfrequencies, the frequency dependence is mainly determined by the denominator of (4.40). The amplitudes A_N and B_N

are inversely proportional to the difference of the frequencies, that is, they depend on k as $1/(k - k_N)$.

Note that in the case of excitation of the waveguide, the frequency dependence is weaker than that for the resonator, and the amplitude of the eigenmode with number N is proportional to $(k - k_N^{\text{cut}})^{1/2}$, where k_N^{cut} is the cutoff frequency of the mode. Weakening of the resonant property of the waveguide in comparison with that of the resonator can be explained by the fact that the waveguide is nonclosed in one of the directions, and the resonance takes place only in the cross-section.

The denominator in the amplitudes of eigenoscillation fields has the same form as that in expression (3.77) for the eigenmodes in the problem of excitation of the waveguide. To excite certain eigenoscillation, the extrinsic current should be directed along the electric field of this oscillation.

If the frequency k coincides with one of the eigenfrequencies, then, according to (4.40), the amplitude of the corresponding eigenoscillation is infinite. There is no solution to the nonhomogeneous system (4.35) with arbitrary right-hand side if $k = k_N$, that is, when the homogeneous system (4.16) is solvable. The solution exists only in the case when the right-hand side of the nonhomogeneous system is orthogonal to the solution of the homogeneous system in the following sense:

$$\int_V \vec{j}^{\text{ext}} \vec{E}^N \, dV = 0. \tag{4.41}$$

We have already had such a demand in Section 3.2 when finding the correction to the wave number of the eigenmode in the waveguide, caused by replacing the ideal-conducting walls with the impedance ones (see the text below (3.55)).

If conditions (4.41) are fulfilled, then a solution to system (4.35) exists, but it is nonunique. The fields of the Nth eigenoscillation may be added to this solution, with any coefficient. As we have already noted, this nonphysical result (the solution either does not exist or is nonunique) is caused by the contradictory idealizations accepted.

The resonator can be excited not only by the current introduced inside it, but also by the electric field on a hole in the wall, that is, by the field created by external sources. We denote this two-dimensional vector field by $\vec{e}$ and the hole area by Σ. We find the fields $\vec{E}$, $\vec{H}$ satisfying equations (4.41) with $\vec{j}^{\text{ext}} = 0$ and the boundary condition (4.10b) on the whole wall except for the hole on which

$$\vec{E}_{\text{tan}}\Big|_{\Sigma} = \vec{e}. \tag{4.42}$$

We again express the sought fields in the form of series (4.36), dropping only the gradient term in the series for $\vec{E}$. Expansion (4.36a) for $\vec{E}$ is not valid on the whole surface, since $E_{\tan}^N = 0$ for all N on the whole surface, but $E_{\tan}$ is not zero on its part. In order to avoid termwise application of the "rot" operation to the series (4.36), we use an auxiliary technique which neither requires the termwise differentiation of the nonuniformly convergent series, nor uses the value of the series for $\vec{E}$ on the wall. Strictly speaking, this technique should have been used when deriving formulas (4.40), since for the problem on excitation of the field by the metallic wire with the current, the series (4.36b) is not applicable on the wire surface. The field $\vec{H}$ has a jump on this surface, but the fields $\vec{H}^N$ are continuous. However, the above-used technique of the immediate substitution of the series into equation (4.35) leads to the proper results and it is logically more natural.

Calculate the following quantities:

$$\mathrm{div}\left(\vec{E}^N \times \vec{H}\right) = -ik_N \mu \vec{H}\vec{H}^N - ik\varepsilon\vec{E}\vec{E}^N, \tag{4.43a}$$

$$\mathrm{div}\left(\vec{E} \times \vec{H}^N\right) = -ik\mu\vec{H}\vec{H}^N - ik_N\varepsilon\vec{E}\vec{E}^N \tag{4.43b}$$

using only identity (1.55) and equations (4.35), (4.16). Integrate both the identities over the volume. The integral of the left-hand side of identity (4.43a) is $\int_S(\vec{E} \times^N \vec{H})_N \, dS$, that is, it equals zero, since $E_{\tan}^N = 0$ on the whole surface. The integral of the left-hand side of (4.43b) equals $\int_S(\vec{E} \times \vec{H}^N)_N \, dS$, which is $\int_\Sigma(\vec{e} \times \vec{H}^N) \, dS$. The series (4.36) can be used in the volume integrals of the right-hand sides of (4.43). Taking into account also the identity which in natural way generalizes relations (4.12) for the case $\varepsilon \neq 1$, $\mu \neq 1$, we obtain the system of two equations for A_N and B_N:

$$kA_N - k_N B_N = 0, \tag{4.44a}$$

$$k_N A_N - kB_N = \frac{i \int_\Sigma \left(\vec{e} \times \vec{H}^N\right)_\nu \, dS}{\int_V \varepsilon \left(\vec{E}^N\right)^2 \, dV}. \tag{4.44b}$$

Here index ν indicates the normal component to S. As a result, the sought coefficients of the series (4.36) are

$$A_N = -\frac{ik_N}{k^2 - k_N^2} \frac{i \int_\Sigma \left(\vec{e} \times \vec{H}^N\right)_\nu \, dS}{\int_V \left(\varepsilon\vec{E}^N\right)^2 \, dV}, \tag{4.45a}$$

$$B_N = \frac{ik}{k^2 - k_N^2} \frac{i \int_\Sigma \left(\vec{e} \times \vec{H}^N\right)_\nu \, dS}{\int_V \varepsilon \left(\vec{E}^N\right)^2 \, dV}. \tag{4.45b}$$

The frequency dependence of the amplitudes of the eigenoscillations is the same as in (4.40), that is, in the case when the resonator is excited by the given current. The numerators in (4.45) are the same as in formulas (3.79) for amplitudes of the eigenmodes in the problem on the waveguide excited by the given field on the hole in its wall. If the value of $\vec{H}_{\text{tan}}^N$ on the wall is expressed, according to (1.80), by the density of the surface current $\vec{I}^N$ flowing on the solid wall in the eigenoscillation, then the component $(\vec{e} \times \vec{H}^N)_N$ is proportional to the scalar product $\vec{e}\,\vec{I}^N$ of two two-dimensional vectors.

Similarly as in the case when the resonator is excited by the current at the frequency k_N of its eigenoscillation, the solution to the Maxwell equations (nonuniquely) exists only if the electric field $\vec{e}$ given on the hole is orthogonal to the surface current with density $\vec{I}^N$, flowing on the part of the wall which becomes the hole. The orthogonality condition, similar to (4.41), has the form

$$\int_{\Sigma} \vec{e}\,\vec{I}^N \, dS = 0. \qquad (4.46)$$

4.2
Resonators with impedance walls

4.2.1
Complex eigenfrequencies

The main distinction between the resonator with losses and a nonfilled (or containing the objects without losses) resonator having ideal-conducting walls, lies in the complexity of the eigenfrequencies: $k_N = k'_N + ik''_N$. The time dependence of the complex amplitudes of the eigenoscillation is described by the multiplier $\exp(i\omega_N t)$, where $\omega_N = ck_N$. The physical quantities vary with time as $\exp(-ck''_N t)\cos(ck'_N t)$. If the field energy outgoes into the walls or into the lossy material, then the eigenoscillations should damp with time, that is,

$$k''_N > 0. \qquad (4.47)$$

In the resonator with impedance walls, the field must satisfy equations (4.35) and the boundary conditions (1.67). Correspondingly, the eigenoscillations must satisfy equations (4.16) with the same boundary conditions.

The parameters ε, μ and impedance w should be the same in both the problems: nonhomogeneous and homogeneous. Otherwise, the series (4.36) by the fields of the homogeneous system do not satisfy the nonhomogeneous one (4.35). However, in general, these parameters depend on the frequency. Therefore, in equations (4.16), ε, μ and w should be treated not as the values of the parameters at the eigenfrequency, but as the values which these parameters take at the frequency of the nonhomogeneous problem; for solving this problem the fields of eigenoscillations are used. In this sense, as already explained in the text below formula (4.16), when finding the eigenfrequencies, the resonator parameters should be treated as those independent of frequency.

We derive the expression for the eigenfrequency of the resonator with lossy walls. If the thickness d of the skin layer (1.77) of the impedance wall is small in comparison with all linear sizes of the problem, then this wall can be replaced by the ideal-conducting wall covered with a thin film having ε with large imaginary part. The thickness of such film should be several times larger than d. After such a replacement the field in the resonator does not change. The replacement is equivalent to the way of treating the resonator boundary not as a surface of the impedance wall but as a parallel to it surface with $E_{\tan} = 0$ located inside the impedance wall. It allows us to use the formulas applicable for the resonator with ideal-conducting walls, inside which, however, objects with any $\varepsilon(\vec{r})$ and $\mu(\vec{r})$ are placed.

We find the expression for k_N similar to (4.24), which is valid at $\varepsilon \neq 1$, $\mu \neq 1$. Eliminating the field $\vec{E}^N$ from the expression for $\mathrm{div}(\vec{E}^N \times \vec{H}^{N*})$ with the usage of formula (1.55) yields

$$\mathrm{div}\left(\vec{E}^N \times \vec{H}^{N*}\right) = -i\mu k_N \left|\vec{H}^N\right|^2 + \frac{i}{\varepsilon k_N}\left|\mathrm{rot}\,\vec{H}^N\right|^2. \tag{4.48}$$

The volume integral of the left-hand side is zero, since the fields satisfy the condition $E_{\tan}^N = 0$ on the chosen boundary.

Consequently,

$$k_N^2 = \frac{\int_V \varepsilon^{-1}\left|\mathrm{rot}\,\vec{H}^N\right|^2 dV}{\int_V \mu \left|\vec{H}^N\right|^2 dV}. \tag{4.49}$$

Formula (4.24) is a partial case of this.

For simplicity, we assume that $\mu'' = 0$, that is, the losses are caused only by the nonzero ε''. Then

$$\mathrm{Im}\,k_N^2 = -\frac{\int_V \varepsilon'' |\varepsilon|^{-2}\left|\mathrm{rot}\,\vec{H}^N\right|^2 dV}{\int_V \mu \left|\vec{H}^N\right|^2 dV}. \tag{4.50}$$

Replace $\mathrm{rot}\,\vec{H}^N$ under the integral in the numerator by its expression from (4.16):

$$\frac{\mathrm{Im}\,k_N^2}{|k_N|^2} = \frac{\int_V (-\varepsilon'') \left|\vec{E}^N\right|^2 dV}{\int_V \mu \left|\vec{H}^N\right|^2 dV}. \tag{4.51}$$

In order to find the contribution into $\operatorname{Im} k_N^2$ caused by the losses in the walls, we temporarily introduce the z-coordinate directed inside the wall; $z = 0$ on the wall surface. According to (1.76), $|\vec{E}^N(z)|^2 = |\vec{E}^N(0)|^2 \exp(-2z/d)$ in the skin layer. The integral over the wall thickness equals $-\varepsilon''|\vec{E}^N(0)|^2 \cdot d/2$. According to (1.75), (1.67), and (1.74), $-\varepsilon'' d/2 = w'/(k_N|w|^2)$. Substituting this equality together with $|\vec{E}_{\tan}^N(0)| = |w||\vec{H}_{\tan}^N(0)|$ into (4.51), we obtain the sought expression

$$\frac{\operatorname{Im}\left(k_N^2\right)}{|k_N|} = \frac{\int_S w' \left|\vec{H}_{\tan}^N\right|^2 dS}{\int_V \mu \left|\vec{H}^N\right|^2 dV}. \tag{4.52}$$

Of course, this formula conforms with the fact that the energy flux outgoing into the impedance wall per unit area is $c/4\pi \cdot |\vec{H}_{\tan}|^2 w'$ (see Subsection 1.3.2); it equals the energy absorbed by the wall material. The imaginary part of k_N is proportional to the ratio of the energy absorbed by the wall to the entire energy in the resonator volume.

The frequency dependence of the eigenoscillation amplitude is mainly determined by the multiplier $1/(k^2 - k_N^2)$ (4.40), as in the lossless resonator. Similarly as in oscillation theory, we introduce the notion of the dimensionless resonator quality Q (*Q-factor*) by the formula

$$Q = \frac{|k_N|}{2k_N''}. \tag{4.53}$$

Then the time dependence $\exp(-k_N'' ct)$ of the field amplitudes is written as $\exp(-\omega/(2Q) \cdot t)$. The relative energy decrease over the period is $2\pi/Q$.

According to (4.52), (4.53), the Q-factor equals

$$Q = \frac{2}{d} \frac{\int_V \mu \left|\vec{H}^N\right|^2 dV}{\int_S \left|\vec{H}^N\right|^2 dS}. \tag{4.54}$$

It is proportional to the ratio of the resonator volume to its surface area S and has the order of $V/S \cdot d$. Since d is proportional to $\lambda^{1/2}$ (if the dependence of the conductivity σ on the frequency is not considered) and V/S is proportional to λ, then Q is proportional to $\lambda^{1/2}$. For the wavelength shorter than several millimeters, the resonator must be very small and its Q-factor cannot be large.

According to (4.53), the frequency dependence may be written in the same form as for the LCR contour; it is determined by the multiplier proportional to the quantity

$$\frac{1}{k^2 - \left(k_N'\right)^2 - i\left(k_N'\right)^2 \Big/ Q}. \tag{4.55}$$

At the frequencies such that $|k^2 - (k'_N)^2| \sim (k'_N)^2/Q$, this multiplier is large, of the order $Q/(k'_N)^2$, that is, approximately Q times larger than that outside this interval. The width of this interval ("width of resonant curve") is k'_N/Q. The resonator may be used as a single-frequency contour if $1/Q < |k_N - k_{N+1}|/k_N$, that is, if the relative distance between two neighbor eigenfrequencies is larger than the inverse Q-factor.

The Q-factor of the circular waveguide resonator for the TM_{010}-mode is

$$Q = \frac{1}{d}\frac{aL}{a+L}. \tag{4.56a}$$

For the TE_{011}-mode the Q-factor is larger; it is

$$Q = \frac{1}{d}\frac{aL}{a\,\pi^2 / (k^2 L^2) + L \cdot (3.83)^2 / (k^2 a^2)}. \tag{4.56b}$$

Recall that in the circular waveguide, the TE_{01}-mode damps slower than TM_{01} (see Section 3.2).

4.2.2
Displacement of eigenfrequencies

Replacement of the ideal-conducting walls of the resonator by the impedance ones not only supplies the imaginary part, proportional to w', to the eigenfrequencies, but also leads to the displacement of their real parts proportional to w''. This effect is interesting mainly owing to the fact that the displacement is different for different eigenoscillations. This difference can be increased by making the impedance different in different parts of the wall. In particular, w'' can be made positive in one part of the wall and negative in another. Creating an impedance with the imaginary part $w''(S)$ (S is a point on the wall) of an appropriate form, we can, for instance, lower the eigenfrequency of the main eigenoscillation, leaving the eigenfrequency of the first overtone almost unchanged (or even increasing it) and widening, in this way, the frequency range in which the resonator (at its main oscillation) is equivalent to the single-frequency oscillation contour.

Assume, for simplicity, that $\mu = 1$ in the resonator and there are no losses inside it, that is, $w' = 0$, $\varepsilon'' = 0$, so that the eigenfrequencies are real. Then the expression

$$k_N^2 = (k_N^0)^2 - \frac{k_N \int_S w'' \left|H_{\tan}^N\right|^2 dS}{\int_V \left|\vec{H}^N\right|^2 dV} \tag{4.57}$$

can be easily obtained, where k_N^0 is the eigenfrequency of the resonator of the same shape with ideal-conducting walls; the expression for k_N^0 is given by

(4.24). The integral in the numerator of (4.57) is taken over the actual surface
of the resonator, not over the "equivalent" one (deepen into the wall), as it was
described in the text before formula (4.48).

At not large values of $|w|$, the field $\vec{H}^N$ can be replaced by the field in the
resonator with the ideal-conducting walls; H_{tan}^N is proportional to the current
flowing on the wall of such a resonator at the Nth eigenoscillation. For decreasing k_N, an area with a large positive value of w'' should be created in the
part of the walls where this current is large. If $w'' < 0$ in this part, then k_N
is larger than k_N^0. The mentioned effect of the one-frequency range widening
is based on the fact that the current distribution on the walls is different for
different eigenoscillations.

4.2.3
Impedance as a spectral parameter

In Section 4.1 the spectral method was used for solving the nonhomogeneous
Maxwell equations. The system of the eigenoscillation fields was developed,
which satisfied the corresponding homogenous equations in which the frequency was replaced by the spectral parameter. At certain values of this parameter different from the actual frequency value, the homogeneous equations are solvable. In the homogeneous problem, the impedance boundary
condition (1.67) is taken the same as in the nonhomogeneous one, that is, with
the same value of the impedance w.

The spectral method can be generalized. Namely, not only the frequency
but also any other parameter participating in the problem may play a role of
the spectral parameter. In these variants the frequency in the homogeneous
problem has the same value as that in the nonhomogeneous one. Another parameter, which plays a role of a spectral one, has a different value than in the
nonhomogeneous problem. In this subsection, such a variant of the generalized method is described. Solutions to this homogeneous problem make up a
system of eigenoscillation fields. The impedance w is chosen as a spectral one
in it.

The problem is to find the fields $\vec{E}$, $\vec{H}$ from the nonhomogeneous equations
(4.35) satisfying the conditions

$$E_{\text{tan}} = \pm w H_{\text{tan}} \tag{4.58}$$

on the resonator surface S (see (1.67)). The two signs relate to two tangential components. For simplicity, we assume that w is constant on the whole
surface. The homogeneous equations generating the system of the eigenoscillation fields have the form

$$\text{rot}\,\vec{H}^N - ik\varepsilon\vec{E}^N = 0, \quad \text{rot}\,\vec{E}^N + ik\mu\vec{H}^N = 0. \tag{4.59}$$

The spectral parameter w_N is introduced into the boundary conditions for the fields $\vec{E}^N$, $\vec{H}^N$, as follows:

$$E_{\tan}^N = \pm w_N H_{\tan}^N. \tag{4.60}$$

Similar to the actual impedance w in (4.58), the eigenimpedances w_N are constant on the whole boundary. In contrast to (4.16), the frequency k in the homogeneous equations (4.59), is the same as that in the nonhomogeneous problem. The eigenimpedances w_N depend on k, similarly as the eigenfrequencies k_N depend on w. In this variant of the spectral method, the eigenoscillation fields do not coincide with those of the variant described in Section 4.1.

We find the expression of w_N by the eigenoscillation fields, similar to (4.49). Assume that $\mu = 1$. Multiply the wave equation $\operatorname{rot} \operatorname{rot} \vec{E}^N - k^2 \varepsilon \vec{E}^N$ by $\vec{E}^{N*}$ and transform the obtained product by (4.27). Then we integrate the obtained equality

$$\operatorname{div}\left(\vec{E}^{N*} \times \vec{H}^N\right) = \frac{i}{k}\left(\left|\operatorname{rot} \vec{E}^N\right|^2 - k^2 \varepsilon \left|\vec{E}^N\right|^2\right) \tag{4.61}$$

over the volume V. Substituting $E_{\tan}^{N*}$ by (4.60) yields

$$w_N^* = \frac{i}{k} \frac{\int_V \left(\left|\operatorname{rot} \vec{E}^N\right|^2 - k^2 \varepsilon \left|\vec{E}^N\right|^2\right) dV}{\int_S \left|H_{\tan}^N\right|^2 dS}. \tag{4.62}$$

The eigenimpedance w_N does not depend on the impedance w of the actual system. If in the actual system the losses exist only in the walls, that is, $w' > 0$, but $\varepsilon'' = 0$, then, according to (4.62), $w_N' = 0$, that is, the auxiliary system described by equations (4.59), (4.60) has no losses.

We can obtain another expression

$$w_N^* = \frac{-ik \int_S \left|E_{\tan}^N\right|^2 dS}{\int_V \left(\left|\operatorname{rot} \vec{E}^N\right|^2 - k^2 \varepsilon \left|\vec{E}^N\right|^2\right) dV} \tag{4.63}$$

from formula (4.62), which possesses the stationarity property similarly as for expression (4.24). We consider the right-hand side of this formula as a functional $\mathcal{L}(\vec{E})$ which possesses the property

$$\mathcal{L}\left(\vec{E}^N + v\vec{\varphi}\right) = w_N^* + O\left(v^2\right) \tag{4.64}$$

(see (4.22)), that is, $\mathcal{L}$ is stationary at $\vec{E} = \vec{E}^N$. We apply the variational technique described in Subsection 4.1.4 to this functional.

The auxiliary problem (4.59), (4.60) has its own physical meaning. It describes the resonator having the wall impedance for which the undamping oscillation of the frequency k exists. It follows from (4.62) that if the energy is not absorbed in the resonator, that is, if the dielectric contained inside it has no losses ($\varepsilon'' = 0$), then $w_N' = 0$. If the dielectric with losses is placed into the real resonator, that is, $\varepsilon'' < 0$, then, according to (4.62), $w_N' < 0$. This means that there exists the energy flux in the auxiliary problem which enter into the volume from the wall. The auxiliary problem describes the field in the resonator having such a wall material which radiates the energy under the influence of the field in which it is contained, that is, the wall material has the negative absorption. In this case the auxiliary problem describes a certain although exotic but, in principle, possible physical system, as well.

The fields of two eigenoscillations corresponding to different eigen-impedances are orthogonal in the following sense:

$$\int\limits_{S} \left(\vec{E}^N \times \vec{E}^M \right)_t dS = 0, \quad \int\limits_{S} \left(\vec{H}^N \times \vec{H}^M \right)_t dS = 0, \quad w_N \neq w_M. \quad (4.65)$$

Here $(\vec{A}\vec{B})_t$ implies $A_{t_1} B_{t_1} + A_{t_2} B_{t_2}$, where $\vec{t}_1$, $\vec{t}_2$ are two mutually perpendicular unit vectors tangential to the surface. These formulas can be easily obtained by integrating the expression $\mathrm{div}(\vec{E}^N \times \vec{H}^M - \vec{E}^M \times \vec{H}^N)$ over the volume. Since all the fields satisfy the same equation system (4.59), this expression equals zero. Consequently, the surface integral of the normal component of the vector under the divergence, is zero, either. This fact together with conditions (4.60) leads to conditions (4.65).

When solving the problem (4.35), (4.58), it is more convenient to express not the complete fields $\vec{E}$, $\vec{H}$, but the differences $\vec{E} - \vec{E}^0$, $\vec{H} - \vec{H}^0$ in the form of the series by $\vec{E}^N$, $\vec{H}^N$; here $\vec{E}^0$, $\vec{H}^0$ are the fields created by the sources $\vec{j}^{\mathrm{ext}}$ in vacuum.

We represent the fields $\vec{E}$, $\vec{H}$ in the form

$$\vec{E} = \vec{E}^0 + \sum A_N \vec{E}^N, \qquad \vec{H} = \vec{H}^0 + \sum A_N \vec{H}^N. \quad (4.66)$$

The gradient term presented in (4.36a) does not appear in (4.66), because it is contained in $\vec{E}^0$. The series (4.66) satisfy equations (4.35) termwise. The coefficients A_N, the same in both the series, are found from condition (4.58). It must be fulfilled on the surface, where the eigenoscillation fields are orthogonal. Therefore, the coefficients A_N are easily found. They are

$$A_N = \frac{1}{w - w_N(k)} \cdot \frac{\int \left[\left(\vec{E}^0 \times \vec{H}^N \right)_N - w \left(\vec{H}^0 \times \vec{H}^N \right)_t \right] dS}{\int |H_{\mathrm{tan}}^N|^2 \, dS}. \quad (4.67)$$

The values of the tangential components of the fields $\vec{E}^0$, $\vec{H}^0$ on the resonator surface play a role of the excitation current $\vec{j}^{\mathrm{ext}}$ explicitly involved into the analogous formula (4.40). Formula (4.67) is valid, in particular, at $w = 0$, that is, for the resonator with the ideal-conducting walls.

The frequency dependence of the amplitudes is mainly determined by the multiplier $1/[w - w_N(k)]$, that is, by the dependence of w_N on k. Since $w' \geq 0$, $w'_N \leq 0$, this multiplier may become infinite only if there are no losses in the resonator both in its walls ($w' = 0$) and inside it ($\varepsilon'' = 0$, hence $w'_N = 0$). The "infinite resonance" implies that the problem either has no solution at the frequency for which $w_N(k) = w$, or the solution is not unique if the numerator in (4.67) is zero.

In contrast to the *method* of *eigenfrequencies*, the *eigenimpedances method*, as well as some other variants of the *generalized eigenoscillations method*, is easily transferred to the case of the open resonators and, in general, to the diffraction problems in the nonclosed domains (see Subsection 7.1.6).

5
Open Lines

5.1
Dielectric waveguides

5.1.1
Circular dielectric waveguides

Eigenmodes may propagate along the dielectric cylinder with arbitrary cross-section. They are the fields with all the components dependent on z as $\exp(-ihz)$; z is the coordinate directed along the cylinder axis. The fields are mainly concentrated in the dielectric or in its neighborhood. For the circular cylinders the fields of these modes are expressible by the special functions so that finding these fields and propagation constants h consists in solving simple equations.

In this subsection, we present the theory of eigenmodes of the circular homogeneous dielectric waveguides, that is, the transmitting lines with $\varepsilon = const > 1$ at $r < a$ and $\varepsilon = 1$ at $r > a$, where a is a radius of the circular cylinder. We assume that $\mu = 1$. The theory of dielectric waveguides is based on the same formulas (3.1), (3.3), (3.4), (3.5) as the theory of closed ones. The fields are expressed by the two scalar functions $\chi(r, \varphi)$, $\psi(r, \varphi)$ satisfying the wave equations (3.5). However, in the dielectric waveguides there is, in general, no division into the modes with $E_z \equiv 0$ and $H_z \equiv 0$. The functions χ and ψ are not the solutions to the two independent problems of the types (3.11) and (3.13).

The exceptions are the symmetric modes ($\partial/\partial\varphi \equiv 0$), for which the division into the TE- and TM-modes is kept. For instance, for the symmetric TM-modes the fields are expressed by formulas (3.3), (3.4) in which $\psi \equiv 0$. Inside the waveguide, the function χ satisfies the same equation (3.5a) having the form

$$\frac{d^2\chi}{dr^2} + \frac{1}{r}\frac{d\chi}{dr} + \left(k^2\varepsilon - h^2\right)\chi = 0, \quad r \leq a \tag{5.1a}$$

in cylindrical coordinates. In the waveguide exterior, the equation for $\chi(r)$ is

High-frequency Electrodynamics. Boris Z. Katsenelenbaum
Copyright © 2006 WILEY-VCH Verlag GmbH & Co. KGaA, Weinheim
ISBN: 3-527-40529-1

of the form

$$\frac{d^2\chi}{dr^2} + \frac{1}{r}\frac{d\chi}{dr} + \left(k^2 - h^2\right)\chi = 0, \quad r \geq a. \tag{5.1b}$$

Solutions to these equations are the cylindrical functions of the zeroth order

$$\chi(r) = AJ_0(\tau r), \quad \tau^2 = k^2\varepsilon - h^2, \qquad r \leq a, \tag{5.2a}$$

$$\chi(r) = BH_0^{(2)}(\alpha r), \quad \alpha^2 = k^2 - h^2, \quad \text{Im}\,\alpha \leq 0, \quad r \geq a. \tag{5.2b}$$

The cylindrical function J_0 appears in (5.2a) since it is the only cylindrical function of the zeroth order which has no singularity at the point $r = 0$ lying inside the domain $r \leq a$. A function decreasing at $r \to \infty$ must participate in (5.2b). At the chosen sign of $\text{Im}\,\alpha$ only the function $H_0^{(2)}(\alpha r)$ possesses this property; it has the asymptotic $\sqrt{2i/\pi} \cdot \exp(-ik\alpha r)/\sqrt{\alpha r}$ at $r \to \infty$.

Three components E_z, H_φ, E_r differ from zero. The first two must be continuous on the cylinder surface (at $r = a$) where the dielectric constant is discontinuous. This demand leads to the two relations

$$A\tau^2 J_0(\tau a) - B\alpha^2 H_0^{(2)}(\alpha a) = 0, \quad \varepsilon\tau J_1(\tau a) - B\alpha H_1^{(2)}(\alpha a) = 0. \tag{5.3}$$

Eliminating A and B from them results in the following dispersion equation for h:

$$\frac{\alpha a H_0^{(2)}(\alpha a)}{H_1^{(2)}(\alpha a)} = \frac{\tau a J_0(\tau a)}{\varepsilon J_1(\tau a)}. \tag{5.4}$$

After the propagation constant h is found from (5.4) the ratio A/B (which describes the field structure in the whole space), can be determined from (5.3).

At the real $\varepsilon > 1$, equation (5.4) has only the real roots h with $\text{Im}\,\alpha < 0$, all lying in the two symmetrical intervals

$$-k\sqrt{\varepsilon} < h \leq -k, \qquad k \leq h < k\sqrt{\varepsilon}. \tag{5.5}$$

The phase velocity $v = k/h$ of all the eigenmodes lies in the interval

$$c \geq v > \frac{c}{\sqrt{\varepsilon}}, \tag{5.6}$$

that is, v is larger than the velocity of the plane wave in the infinite dielectric and smaller than the wave velocity in vacuum.

According to (5.5), $\tau^2 > 0$, $\alpha^2 < 0$. The field decreases as $\exp(-|\alpha|r)/\sqrt{r}$ at $r \to \infty$.

Equation (5.4) has also the complex roots for which $\text{Im}\,\alpha > 0$. The fields with such α grow with r increasing. Such waves are not the eigenmodes of the waveguide. However, at certain frequencies it is expedient to introduce them for describing the field in the bounded domains. We will return to this case in Subsection 5.1.3.

The eigenmode corresponding to the qth root h of equation (5.4), is called the TM_{0q}-mode. For each such mode, h varies from the value $h \approx k\sqrt{\varepsilon}$ at high frequencies ($k \gg 1/a(\sqrt{\varepsilon - 1})$) to the value $h = k$ at a certain frequency denoted by k_{0q}^{cut} (cutoff frequency). At the frequency lower than k_{0q}^{cut}, the TM_{0q}-mode does not exist. When k decreases and becomes lower than k_{0q}^{cut} for any q, then the number of eigenmodes becomes smaller by one. At the cutoff frequency, the parameter α is zero for the given mode. Since the left-hand side of (5.4) is zero at $\alpha = 0$, then the quantity $k_{0q}^{cut}a\sqrt{\varepsilon - 1}$ is equal to the qth zero of the function J_0.

At high frequencies the field is fast decreasing at $r \to \infty$; it is almost wholly concentrated inside the dielectric and v is close to its minimal value $c/\sqrt{\varepsilon}$. At frequencies only a little larger than the cutoff one of the eigenmode of the given type, the field is slowly decreasing as r increases and almost wholly lies outside the dielectric. The velocity v is close to its maximal value c. The eigenmode ceases to exist at its cutoff frequency; it would transfer the infinite energy at $\alpha = 0$.

The theory of the TE_{0q}-modes differs from that of the TM_{0q}-modes only by the fact that the other three components, namely, H_z, E_φ, H_r are present in the field. They are expressed by the function $\psi(r)$ which satisfies the same equations (5.1) and has the same form (5.2). The boundary conditions for the components H_z and E_φ (continuity conditions at $r = a$) lead to the equation which differs from (5.4) only by the absence of the multiplier $1/\varepsilon$ in the right-hand side. The roots h of this equation lie in the same intervals (5.5) and possess the same properties as the roots of (5.4). In particular, at $ka\sqrt{\varepsilon - 1} < 2.40$ there are no modes of the types TE_{0q} and TM_{0q}. The cutoff frequencies of these modes coincide.

The theory of the nonsymmetrical ("hybrid") modes, the fields of which depend on φ, is constructed by the same scheme. The fields contain all the six components; they are expressed by both the functions $\chi(r.\varphi)$ and $\psi(r, \varphi)$. These functions are proportional to the functions $\cos(m\varphi)$, $\sin(m\varphi)$, $m \geq 1$, and $J_m(\tau r)$, $H_m^{(2)}(\alpha r)$. The boundary conditions at $r = a$ lead to an equation similar to (5.4), but more cumbersome. The roots of this equation possess the same properties as the propagation constants of the symmetric modes. There is only one distinction: the cutoff frequency of the mode with indices $m = 1$, $q = 1$ is zero. This eigenmode (a *dipole mode*) exists at any frequency, in particular, at the arbitrarily low one.

Near the cutoff frequency of a certain mode, that is, at small positive value of $k - k_{mq}^{cut}$, the difference $h - k$ is also small. At these frequencies the equations for h, similar to (5.4), can be simplified by using the asymptotic of the cylindrical functions at small value of $|\alpha a|$. The obtained approximate dispersion equations make possible the qualitative investigation of the field properties, as well as of the behavior of the phase and group velocities of the eigenmodes when the frequency approaches the cutoff one.

Such approximate equations for the dipole, symmetric and hybrid (but not dipole) modes are

$$\ln\left(-\alpha^2\right) = -\frac{C}{k^2}, \qquad -\alpha^2 = h^2 - k^2 > 0, \tag{5.7a}$$

$$-\alpha^2 \ln\left(-\alpha^2\right) = -C\left[k^2 - \left(k^{\text{cut}}\right)^2\right], \tag{5.7b}$$

$$-\alpha^2 = C\left[k^2 - \left(k^{\text{cut}}\right)^2\right], \tag{5.7c}$$

respectively. In these three dispersion equations the positive coefficients C are different for the modes of different numbers.

It follows from (5.7a) that the dipole mode is connected with a cylinder weaker than others. The quantity $-\alpha^2$ is zero at the zero frequency; it remains small for a long time when the frequency increases. The phase velocity $v = k/h$ and the group velocity $v_{\text{gr}} = (dh/dk)^{-1}$ remain close to c at relatively large k.

According to (5.7b), the symmetric modes are stronger connected with a cylinder. With increasing the frequency (i. e., with moving it from the cutoff one), the difference $h - k$ increases faster than for the dipole mode. The velocities v and v_{gr}, equal to c at the cutoff frequency, are varying also faster than for the dipole mode. The hybrid modes (see (5.7c)), except for the dipole one, are connected with the dielectric cylinder stronger than others. The group velocity v_{gr} of these modes remains smaller than c even at the cutoff frequency. At $k = k^{\text{cut}} + 0$ the Poynting vector flux of these modes through the dielectric remains to be a finite part of the flux through the whole plane $z = const$. For other modes, the entire flux goes outside the cylinder at $k = k^{\text{cut}} + 0$.

5.1.2
Eigenmodes of dielectric waveguides with arbitrary cross-section

The eigenmodes of dielectric waveguides with arbitrary cross-section possess the same main properties as those of the circular waveguides. However, there is no mathematical technique for the arbitrary case, as simple as for the circular waveguides. There are neither the explicit expression (5.2) for the potential functions nor a simple dispersion equation similar to (5.4) for the propagation constant h. For such waveguides it can be shown that the propagation constant h and the phase velocity v lie in the same intervals (5.5) and (5.6), respectively; the number of the eigenmodes is finite and grows as $ka\sqrt{\varepsilon - 1}$ increase, where a is a linear scale of the cross-section.

All the modes except one exist only at the frequencies larger than their cut-off modes. The existence of the specific mode, not having the cutoff frequency (a dipole mode), may be simply explained. Let a dielectric cylinder be inserted into the field of the linearly polarized plane wave, such that the cylinder director coincides with the wave propagation direction (Fig. 5.1). The field partially penetrates into the cylinder and the phase velocity decreases. The field decreases in the radial direction keeping its structure at infinity. The eigenmode arises. Such a mode exists at any frequency. It corresponds to the indices $m = 1, q = 1$ for the circular waveguide.

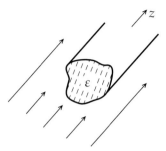

Fig. 5.1 Dielectric waveguide in the field of plane wave

It can be shown that near the cutoff frequencies the dispersion equations for all other modes of any waveguide have either the form (5.7b) or (5.7c). Equations (5.7) are the universal notation of the dispersion equations for the eigenmodes of any dielectric waveguide at frequencies a little larger than the cutoff ones. Each mode belongs to one of the three types described in the text below formula (5.7). These three types differ in degree of the field-to-dielectric connection and, in particular, in the frequency dependencies of the phase and group velocities.

Propagation of the eigenmode along the dielectric waveguide may be treated as a multiple complete inner reflection of the plane waves falling from the waveguide onto its boundary. Consider the aggregate of the plane waves propagating in the dielectric, normals to the equiphase surfaces of which make up a circular cone surface having the z-axis as its director. Denote the apical angle of this surface by 2γ. The fields of such waves depend on z by means of the multiplier $\exp(-ik\sqrt{\varepsilon}\cos\gamma \cdot z)$. If the angle γ is defined by the equation $\cos\gamma = h/k\sqrt{\varepsilon}$, where h is the propagation constant of the eigenmode, then the dependence on z of the waves aggregate is determined by the multiplier $\exp(-ihz)$, that is, the same as that in the eigenmode. When interfering, the fields of these plane waves create the field of the eigenmode. The similar derivation was carried out in Section 3.1 where the eigenmodes of the closed rectangular waveguide were considered (see formulas (3.34), (3.35)). Since h lies in the interval (5.5), then $\cos\gamma < 1$, that is, the angel γ is real. However, it is not large, $\cos \gamma > 1/\sqrt{\varepsilon}$. The waves fall onto the boundary at the glancing angle γ. According to (2.86) (where $\beta = \pi/2 - \gamma$), the complete inner reflection occurs when the waves fall at such small angle of glancing from the medium with $\varepsilon > 1$ onto the interface with the medium in which $\varepsilon = 1$. The energy does not outgo into this medium, the field decreases exponentially when moving away from the boundary. This corresponds to the structure of the eigenmode field.

Two properties of the eigenmodes, namely, "their velocity is smaller than c" and "their fields concentrate near the waveguide" are mutually connected. As we will find out later, the fields of the slow modes of other open lines decrease exponentially as $r \to \infty$, as well. This fact follows from equation (5.1b). The function $\nabla^2 \chi$ is proportional to χ outside the line; the proportionality factor is $h^2 - k^2$. For the slow modes, $h > k$, that is, this factor is positive. Consequently, χ depends on r with the real exponent, approximately as $\exp(-\sqrt{h^2 - k^2})$, that is, it decreases with r increasing.

The fields of the eigenmodes are orthogonal for any dielectric waveguide. They are related as (3.17) or, in general, as (3.72). The derivation of these relations which was carried out in Section 3.2 for the modes in the closed waveguide filled by the material with ε being an arbitrary function of the transverse coordinates, remains valid also for the open waveguides. These orthogonality conditions are fulfilled if the fields are decreasing at the radial infinity, that is, for any two slow modes.

At high frequencies, we have $h \approx k\sqrt{\varepsilon}$ for any mode, that is, τa is small. In the zeroth order of the expansion of all fields by k^{-1} as the small parameter, the only nonzero components are transverse ones H_x, H_y inside the waveguide. As the Cartesian components, they satisfy the two-dimensional homogeneous wave equation

$$\nabla^2 H_{x,y} + \tau^2 H_{x,y} = 0 \tag{5.8}$$

in the cross-section. Owing to the continuity conditions for the z- and s-components of the electric field on the boundary, the above components fulfill the approximate boundary conditions

$$H_{x,y}\big|_S = 0. \tag{5.9}$$

Consequently, both these components are solutions to the same homogeneous boundary value problem. In absence of the degeneration, they are proportional to each other so that the mode polarization is linear.

The usage of the transverse components H_x, H_y as potential functions, is easy when investigating the dielectric waveguides (instead of E_z, H_z as in the metallic ones) of arbitrary cross-section not only at high frequencies but also in the general case. Such an approach simplifies the numerical technique for solving the homogeneous problems, in particular, the variational technique.

5.1.3
Excitation of dielectric waveguides

We begin with solving a model problem. An elementary electric dipole is located on the axis of the circular waveguide and oriented along it, that is, the current $j_z^{\text{ext}} = \delta(|\vec{r}|)$. We repeat with nonessential modifications, the first stage of the derivations carried out in Subsection 3.3.4, for the similar but more simple problem about the field created by the same current in the closed waveguide. The field consists of the *incident* field and the *diffracted* field. The incident field is the field created by the same dipole in the infinite homogeneous dielectric. This field is described by the electric Hertz vector having only the z-component. According to (3.85), it equals

$$\Pi_z^{\text{inc}} = -\frac{i}{\omega} \frac{\exp\left(-ik\sqrt{\varepsilon}R\right)}{R}, \tag{5.10}$$

where $R = \sqrt{r^2 + z^2}$ is a radius of the spherical coordinate system (R, ϑ, φ). We use this expression only inside the waveguide, at $r < a$. Outside the cylinder we put $\Pi_z^{\text{inc}} = 0$. Since the boundary conditions must be fulfilled on the cylindrical surface $r = a$, this function should be represented as a superposition of the product of functions of z and r, that is, similarly to (3.87),

$$\Pi_z^{\text{inc}} = \frac{1}{2\omega} \int_{-\infty}^{\infty} H_0^{(2)}(\tau r) \exp\left(-ihz\right) dh, \quad \tau^2 = k^2\varepsilon - h^2, \tag{5.11}$$

where $\operatorname{Re}\tau > 0$ at $\tau^2 > 0$, $\operatorname{Im}\tau < 0$ at $\tau^2 < 0$.

The diffracted field must be expressed by the function $J_0(\tau r)$ (having no singularity in this domain) at $r < a$, and by $H_0^{(2)}(\alpha r)$ at $r > a$. As $r \to \infty$, this function describes an outgoing wave at $\alpha^2 > 0$, and a damping wave at $\alpha^2 < 0$. We represent the component Π_z^{diff} in the form of the Fourier integral similar to (5.11), that is, as the superposition of functions, similar to (5.2),

$$\Pi_z^{\text{diff}} = \int_{-\infty}^{\infty} A(h) J_0(\tau r) \exp\left(-ihz\right) dh, \quad r < a \tag{5.12a}$$

$$\Pi_z^{\text{diff}} = \int_{-\infty}^{\infty} B(h) H_0^{(2)}(\alpha r) \exp\left(-ihz\right) dh,$$

$$\alpha^2 = k^2 - h^2, r > a, \tag{5.12b}$$

where $\operatorname{Re}\alpha > 0$ at $\alpha^2 > 0$, $\operatorname{Im}\alpha < 0$ at $\alpha^2 < 0$.

Integrals (5.12) satisfy the wave equations for Π_z^{diff} with different values of ε at $r < a$ and $r > a$, at any functions $A(h)$, $B(h)$ for which the integrals converge. These two functions are found from the boundary conditions on the cylinder surface $r = a$. The components E_z and H_φ, tangential to this surface must be continuous on it. These components are expressible by the total Hertz vector $\vec{\Pi} = \vec{\Pi}^{inc} + \Pi^{diff}$ by formulas

$$E_z = k^2 \varepsilon \Pi_z + \frac{\partial^2 \Pi_z}{\partial z^2}, \qquad H_\varphi = -ik\varepsilon \Pi_z \frac{\partial \Pi_z}{\partial z}, \tag{5.13}$$

which follow from (3.25a). The boundary conditions lead to the equality of the two Fourier integrals. Since these equalities are valid at any z, $-\infty < z < \infty$, then the coefficients at $\exp(-ihz)$ are equal under the integrals. Differentiating with respect to z and r under the integrals yields two nonhomogeneous linear equations for $A(h)$ and $B(h)$. The left-hand sides of these equations coincide with those of equations (5.3) for the coefficients A and B in the fields of the symmetric eigen TM-modes. The functions $H_0^{(2)}(\tau a)$ and $J_0(\tau a)$ and their derivatives participate in the right-hand side of the equations for $A(h)$ and $B(h)$.

The solutions to the nonhomogeneous algebraic equation system become infinite if the corresponding homogeneous system has a solution. The first solution contains the denominator vanishing at the values of h at which solutions to the homogeneous system (5.3) exist, that is, at such h which are the roots of equation (5.4). Recall that the similar situation occurs in the problem about the closed waveguide excitation. The coefficient at $\exp(-ihz)$ under the Fourier integral (3.88) contains the function $J_0(a\sqrt{k^2 - h^2})$ in the denominator, which equals zero at the values of h corresponding to the eigenmodes.

However, the structures of fields arising when exciting the closed and open waveguides are different. This corresponds to the difference of analytical properties of the functions under the integrals in (3.88) and (5.12). Continuation of these functions from the real axis of the complex plane h ($h = h' + ih''$) onto the whole plane, has different singularities.

The coefficient at $\exp(-ihz)$ under the integral (3.88) has an infinite number of poles. A finite number of them lies on the real axis, and the rest (infinite number) lies on the imaginary axis. In the integrals (5.12), the functions $A(h)$ and $B(h)$ have only a finite number of poles and all of them are located on the real axis (we will specify this statement below). The analytical continuation of the integrand in (3.88) is unique on the whole complex plane h. In the problem about the dielectric waveguide, the corresponding analytical continuation is not unique.

The integral representation (5.12) for the fields allows us to find the approximate asymptotic expressions. They are different in different domains of space.

We find the field inside the cylinder (at $r < a$) and in the adjacent domain in which $z > 0$ and

$$kr^2 \ll z, \qquad kz \gg 1. \tag{5.14}$$

Deform the integration contour in (5.12), so that the real axis $h'' = 0$ is deformed into an infinitely remote half-circle lying in the lower part of the complex plane, that is, at $h'' \to -\infty$. This procedure and further applying of the residue theorem (3.89) can be carried out only if the function under the integral is single valued. However, the functions $A(h)$ and $B(h)$ are ambiguous since they contain the ambiguous functions $H_0^{(2)}(\tau a)$ and $H_0^{(2)}(\alpha a)$ having the singularities $h = \pm k\sqrt{\varepsilon}$ and $h = \pm k$. As follows from the explicit expressions for $A(h)$ and $B(h)$ which are not shown here because they are too cumbersome, the ambiguity of the Hankel functions at the points $h = \pm k\sqrt{\varepsilon}$ does not cause the functions $A(h)$ and $B(h)$ to be also ambiguous. However, when passing around the points $h = k$ and $h = -k$, the functions $A(h)$ and $B(h)$ change the sign. In order that they be single-valued one should make cuts from the points $h = \pm k$ (*branching points*) – dashed lines in Fig. 5.2.

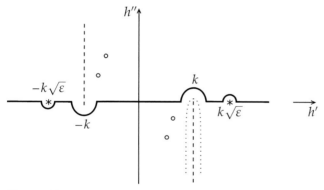

Fig. 5.2 The integration contour for the problem of dielectric waveguide excitation

In order to find the field at the points with $z > 0$, the integration contour should be deformed into the lower half-plane (see text below (3.88)). The initial integral taken over the real axis, equals the sum of residues which the contour sweeps during the displacement (according to the residue theorem (3.89)), and the integral over the loop on the cut (dotted line in Fig. 5.2), or (which is the same) over the cut when it is taken of the difference of integrand values on both sides of this cut. If the cut is made downward from the point $h = k$, so that $\text{Re}\, h = k$ on it and $\text{Im}\, h$ varies from zero to $-\infty$, then the multiplier $\exp(-ihz)$ is fast decreasing at $kz \gg 1$ and the integral over the cut may be asymptotically estimated. The field described by this integral (so-called *additional* field) has the phase factor $\exp(-ikz)$. At frequencies not close to the cutoff one of some eigenmode, it decreases inversely proportional to z^2. Separation of the domain (5.14) is caused by the fact that only in this domain the additional field is small and can be easily found, thus, the method of the integral estimation (5.12) consisting in the deformation of the integration contour is efficient. For finding the approximate expression of the fields in other domains, below we use another method of the asymptotic calculation of integrals (5.12).

The residues at poles, lying on the real axis in the interval $k < h < k\sqrt{\varepsilon}$, describe the eigenmodes excited by the dipole. Their fields have the multipliers $\exp(-ih_n z)$, where h_n are the roots of equation (5.4). The amplitudes of these modes are determined by (3.89). Without writing the corresponding cumbersome formulas similar to (3.91), we only outline that when the frequency decreases and approaches the cutoff frequency of this mode, its amplitude tends to zero.

The expression for the amplitudes of the eigenmodes excited by the given current in the dielectric waveguide, can be found without using the technique of integration in the complex plane. Similarly as in Section 3.3 when deriving the formula (3.77) for the closed waveguides, the relation (3.76) can be used here. The field of the back mode of the same number should be chosen as the Green function and the orthogonality condition should be used. Since the system of the eigenmodes is incomplete because the field also contains the additional one, the expression for the sought amplitude should also contain the integral of the Green function multiplied by this additional field. Since the amplitude does not depend on z, and the integral depends on it, then the integral equals zero. This method for finding field in the domain (5.14) will be specified at the end of the next subsection.

As it was already noted, equation (5.4) has not only the roots lying in the intervals (5.5) but also ones for which $\operatorname{Im} \alpha > 0$. The poles h_n corresponding to these roots in the integrals (5.12), lie in the complex plane h between the cut and imaginary axis, in the domain $0 < \operatorname{Re} h < k$, $\operatorname{Im} h < 0$ (small circles in Fig. 5.2). When the integration contour is deformed into the lower half-plane, residues at these poles are also introduced into the field expression. The fields corresponding to these residues depend on z and r as $\exp(-ih_n z)H_0^{(2)}(\alpha_n r)$. They decrease proportionally to $\exp(\operatorname{Im} h_n \cdot z)$, $\operatorname{Im} h_n < 0$ when moving away from the source along the cylinder and increase nearly as $\exp(\operatorname{Im} \alpha_n \cdot z)$, $\operatorname{Im} \alpha_n > 0$ when moving away from the cylinder. These waves are present in the field only in the domain (5.14). When there are no sources, these waves do not exist: they are not the eigenmodes. In fact, their existence is visible only if poles exist at which $-\operatorname{Im} h_n$ is small in comparison with k. Such poles appear only at the frequencies a little lower than the cutoff one of a certain eigenmode. These waves as if replace the eigenmodes which disappear when the frequency becomes slower than the cutoff one.

In the previous subsection, the propagation of the eigenmode was compared with the multiple complete inner reflection of the plane waves falling from the dielectric onto its boundary at the glancing angle γ, where $\cos \gamma = h_n/k\sqrt{\varepsilon}$. Such an illustration is also possible for the wave with $\operatorname{Im} h_n < 0$

considered in the preceding paragraph. However, since the wave is fast (Re $h_n < k$), the glancing angle γ determined from the condition $\cos \gamma = \text{Re}\, h_n / k\sqrt{\varepsilon}$ is larger than the angle at which the complete inner reflection occurs; the cosine of such an angle should be larger than $1/\sqrt{\varepsilon}$, but $\cos \gamma < 1/\sqrt{\varepsilon}$ for the considered wave. Each reflection is not complete, the field energy partially goes out of the cylinder; the field decreases with z increasing. As for any fast wave, the field grows when moving away from the cylinder. Therefore, in particular, all the eigenmodes are slow. The ray illustration of this wave propagation describes this "highlighting" effect as well. The further the given point, located outside the cylinder, is moved away from the boundary, the closer to the source is the place on the dielectric boundary, where the ray reaching this point left the cylinder, and, therefore, the larger the energy carried away by this ray is (Fig. 5.3). Such waves are called the *leaky waves*. The figure illustrates the ray model of creating such modes. Two rays are shown, which outgo from the source and do not reflect completely. The ray 1 transmits more energy, than the ray 2. In the domain where the field is expressed by these rays, it decreases with the distance from the surface decreasing.

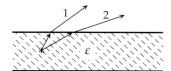

Fig. 5.3 Formation of the leaky wave

The main result consists in the fact that when the circular waveguide is excited by the axial dipole, then the total field in the domain consists of the fields of the eigenmodes and the leaky waves together with the additional field which decreases by the algebraic law with respect to the distance from the source. In this formulation, the result is valid for the waveguide of arbitrary cross-section at any local excitation.

According to the text at the end of Subsection 5.12, all eigenmodes of any waveguide are divided into three types. It can be shown that their dispersion equations have the universal form (5.7) at the frequencies a little larger than the cutoff one. Some properties of the field created by an extrinsic current, also depend only on the type of the arisen eigenmodes. All the properties of the fields arisen in the above problem are also inherent to the fields created at arbitrary waveguide excitation, if the modes of the second type arise. If the modes of the third type are arisen, then their amplitudes do not tend to zero when the frequency approaches the cutoff one. If the dipole mode (mode of the first type) arises, then the leaky waves do not appear; the additional field decreases slower than $1/z^2$, approximately as $1/z$.

We consider an approximate method for calculating the integral (5.12b), which is applicable in the domain

$$kr^2 \gg z, \qquad kR \gg 1, \tag{5.15}$$

"complementary" to the domain (5.14). This method does not require the analytical continuation of the integrand onto the complex plane h. For the points lying in domain (5.15), the main contribution into the integral value is made by a small subdomain of the integration. This subdomain lies in the interval $0 < h < k$, its location depends on the ϑ-coordinate of the point for which the integral is calculated. In the subdomain $\alpha^2 > 0$.

Assume that $\alpha r \gg 1$ in the subdomain (we will verify this assumption further). Then the function $H_0^{(2)}(\alpha r)$ can be replaced by its asymptotic, and the integral (5.12b) takes the form

$$\frac{1}{\sqrt{R}} \int_{-\infty}^{\infty} F(h) \exp\left[-ikR\psi(h)\right] dh, \tag{5.16}$$

where

$$\psi(h) = \frac{\alpha}{k} \sin \vartheta + \frac{h}{k} \cos \vartheta. \tag{5.17}$$

Here it is considered that $r = R \sin \vartheta$, $z = R \cos \vartheta$. Introduce the new integration variable ϑ_h instead of h by the equations $\cos \vartheta_h = h/k$, $\sin \vartheta_h = \alpha/k$. Then the angle ϑ_h is real in the interval $0 < h < k$, and $dh = -k \sin \vartheta_h \cdot d\vartheta_h$. The function $\psi(h)$ is equal to $\cos(\vartheta - \vartheta_h)$.

The method of the approximate calculation of the integral (5.16) is based on the fact that the phase of integrand involves the large factor kR in front of the function $\psi(\vartheta_h)$. Even if the integration variable is varied only a little, the phase changes by π. The function $F(h)$ almost does not change and the contributions of the neighbor areas to the integral is almost completely compensated. The larger the parameter kR, the more complete the compensation and the smaller the value of integral.

If the integration domain did not contain the point at which

$$\frac{d\psi}{dh} = 0, \tag{5.18}$$

then the integration by parts would be possible. The whole integral in (5.16) would be of the order $1/kR$ or smaller. However, equation (5.18) has the root $\vartheta_h = \vartheta$ lying in the integration interval. The compensation is weakened near this root, and the whole integral approximately equals the integral taken over the small segment near the root. This point, located on the integration contour, is called the *stationary phase point*. The considered method is called the *stationary phase method*.

Denote the values of h and ϑ_h at the stationary phase point by h_s and ϑ_s, respectively. Equation (5.18) becomes $\sin(\vartheta_h - \vartheta) = 0$, so that $\vartheta_s = \vartheta$ and $h_s = k\cos\vartheta$. Near this point, $F(h)$ can be replaced by $F(k\cos\vartheta)$ in (5.16) and then taken out of the integral sign. The phase can be replaced by two terms of its Taylor series, as follows:

$$\psi(\vartheta_h) \approx 1 - \frac{1}{2}(\vartheta_h - \vartheta)^2.$$ (5.19)

The assumed condition $\alpha r \gg 1$ must be valid also for $\alpha = \alpha_s = k\sin\vartheta$; it becomes of the form $kr\sin\vartheta \gg 1$. This condition holds in domain (5.15).

Substituting the approximate formula (5.19) into (5.16), we can calculate the integral and obtain the expression as

$$F(k\sin\vartheta) \cdot \sin\vartheta \cdot \frac{e^{-ikR}}{kR},$$ (5.20)

where the constant factor is omitted. Formula (5.20) gives the first term of the expansion of integral (5.12b) into the asymptotic series by the powers of small parameter $(kR)^{-1/2}$. The next term has the order $(kR)^{-3/2}$.

In domain (5.15), the Hertz potential and, therefore, the fields, are the spherical wave outgoing from the source. The presence of the dielectric cylinder leads not to forming the cylindrical waves in this domain, as it does in domain (5.14), but to altering the spherical wave pattern, that is, the field dependence on the angle ϑ. Note that any straight ray beginning at the origin comes to the domain (5.15) at the sufficiently large R.

The paraboloid of revolution $kr^2 = z$ can be treated as a conventional boundary separating the domains of the cylindrical and spherical waves. The field determination in the transition domain adjacent to this boundary is the problem solvable only by the numerical methods.

5.1.4
Nonregular dielectric waveguides

The theory of nonregular waveguides, that is, the waveguides with the parameters (size and shape of the cross-section, axis direction, and dielectric permittivity) different in different domains, is constructed by the same principle scheme as the theory of closed nonregular waveguides (see Section 3.4). At any cross-section $z = const$, a certain regular waveguide (*reference waveguide*) is associated with the given nonregular one. The reference waveguide has the same parameters through its length, as the nonregular waveguide has in the given cross-section. The field at any cross-section is represented as a linear combination of the fields of eigenmodes of the reference waveguide corresponding to this cross-section. The coefficients of this combination are the functions of z. They satisfy a certain linear equation system. The three-dimensional vector problem is thereby reduced to the two-dimensional vector problem on finding the fields of the eigenmodes of regular waveguides and to an one-dimensional scalar equation system.

The application of this method to the closed nonregular waveguides was based on the fact that the system of fields of the eigenmodes of regular waveguides is complete (with some reservations; see text around formulas (3.75)). For the open waveguides, the system of fields of the eigenmodes is incomplete. It consists of the finite number of modes; in some cases there are no eigenmodes at all. In order to use this method for the dielectric waveguides, it is necessary to enlarge the class of functions which can play a role of the fields of the eigenmodes and by which the field of the nonregular waveguide can be expressed. For this end, the demand imposed on the eigenmodes should be weakened.

In the problem about the symmetric eigenmodes of the circular waveguide, the function $H_0^{(2)}(\alpha r)$ standing in formula (5.2b) for the potential function in the cylinder exterior, was chosen among all the cylindrical functions because only this function satisfies the imposed condition

$$\lim_{r \to \infty} \sqrt{r} \chi(r) = 0. \tag{5.21}$$

Equation (5.4) for the eigenvalues, eigenfunctions of which satisfy the above demand, has only few roots h_n.

A demand which weakens (5.21) in such a way that the enlarged class of the fields of the eigenmodes becomes complete, only requires that the product $\sqrt{r} \chi(r)$ does not turn to be infinite at $r \to \infty$, that is,

$$\sqrt{r} \chi(r) < C \quad \text{at} \quad r \to \infty. \tag{5.22}$$

In this way, the notion of the *generalized eigenmode* is introduced. At $r > a$ the function $\chi(r)$ of such a mode may not be proportional to $H_0^{(2)}(\alpha r)$ at $\alpha^2 < 0$, but can be any linear combination of the functions $J_0(\alpha r)$ and $N_0(\alpha r)$ at $\alpha^2 > 0$. Formula (5.2b) should be replaced by

$$\chi(r) = B^{(1)} J_0(\alpha r) + B^{(2)} N_0(\alpha r), \quad \alpha^2 > 0. \tag{5.23}$$

At any coefficients $B^{(1)}$ and $B^{(2)}$, this function satisfies the wave equation (5.1b) and condition (5.22).

At $r = a$, the boundary conditions for E_z and H_φ lead to two equations for the three coefficients A, $B^{(1)}$, and $B^{(2)}$. At any value of α, there exist the values of ratios $B^{(1)}/A$ and $B^{(2)}/A$ at which these conditions are fulfilled. All real values of α (i.e., such that $\alpha^2 \geq 0$) are the eigenvalues of the problem (5.1), (5.22) and the eigenmode of a certain structure corresponds to each of them.

These modes of the continuous spectrum are "numbered" by the number α analogous to the integer n for the discrete spectrum. If α lies in the segment $-k < \alpha < k$, then $h^2 > 0$ and these modes propagate without damping. It can be accepted that all $\alpha < 0$ correspond to the direct modes, whereas $\alpha > 0$ correspond to the back modes.

The fields of different modes are functionally orthogonal. The orthogonality conditions are similar to (3.71) with the integral taken over the infinite plane $z = const$. Condition (3.71) holds for two different modes of the discrete spectrum. Each mode of the discrete spectrum is orthogonal to each mode of the continuous one, that is, (3.71) with the index m replaced by α, holds. If both modes belong to the continuous spectrum, then the Dirac function $\delta(\alpha_1 - \alpha_2)$ occurs on the right-hand side of (3.71), multiplied by the norm which can be chosen arbitrary.

The application of the cross-section method to the problem about the non-regular dielectric waveguide is based on the fields representation

$$\vec{E}(x,y;z) = \sum_{n=-N}^{N} A_n(z)\, \vec{E}^n(x,y;z) + \int_{-\infty}^{\infty} A(\alpha,z)\, \vec{E}(\alpha;x,y;z)\, d\alpha,$$

$$\vec{H}(x,y;z) = \sum_{n=-N}^{N} A_n(z)\, \vec{H}^n(x,y;z) + \int_{-\infty}^{\infty} A(\alpha,z)\, \vec{H}(\alpha;x,y;z)\, d\alpha,$$

(5.24)

analogous to (3.92). There are two distinctions in the formulas. Firstly, in (5.24) the sums related to the discrete spectrum contain the finite number of terms; N is the number of such modes of one direction. Secondly, the integrals of the discrete spectrum modes are presented in this formula; taken over all real values of α.

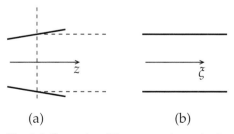

(a)   (b)

Fig. 5.4 Geometry of the nonregular and reference waveguides

The vectors $\vec{E}^n(x,y;z)$, $\vec{H}^n(x,y;z)$ are the coefficients at the multipliers $\exp(-ih_n\xi)$ in the fields of eigenmodes of the reference waveguide (Fig. 5.4(b)), where ξ is the longitudinal coordinate, and z is a parameter. The fields of these modes have the form $\vec{E}^n(x,y;z)\exp(-ih_n\xi)$. In the regular waveguide $A_n(z) = \exp(-ih_nz)$. The functions $\vec{E}(\alpha;x,y;z)$, $\vec{H}(\alpha;x,y;z)$, and $A(\alpha,z)$ have similar meaning.

Substituting (5.24) into the Maxwell equations and using the orthogonality of the fields of eigenmodes, we obtain the system of integro-differential equations for $A_n(z)$ and $A(\alpha, z)$, similar to the system of differential equations (3.93). The left-hand side of the equations consists of the expressions $dA_n(z)/dz + ih_n A_n(z)$ or $dA(\alpha, z)/dz + ih(\alpha, z) A(\alpha, z)$. The right-hand side involves the finite sum of the modes of discrete spectrum and the integral of those of the continuous spectrum. The coupling coefficients participate in these sums and integrals. The corresponding formulas for them are very cumbersome and we do not present them.

The same methods of the approximate solving are applicable to this system as those considered, in detail, in Section 3.4 when applying them to the system (3.93). If the waveguide parameters are slowly varying, then it is possible to obtain the explicit (approximate) expressions for $A_n(z)$ and $A(h, z)$. When a discrete spectrum wave (that is, a slow one) falls onto the nonhomogeneous waveguide segment, then not only other slow waves, but also a spherical one represented as the integral over the continuous spectrum, appears.

The existence of the complete orthonormal system of fields of the generalized eigenmodes allows us to solve the problem about the open waveguide excitation without using the methods of integration in the complex plane. The Green function method may be used as well. Its detailed description is given in Section 3.3 when solving the problem about the closed waveguide excitation. The method allows to find the amplitudes of all generalized eigenmodes arisen when the field is excited by any given source. This technique is applicable not only to the dielectric waveguides but also to other open lines.

5.1.5
The optical fiber

A long glass or quartz fiber in which the signals are transmitted in the visible optical or infrared band, is called an *optical fiber*. The recently developed technologies allow to manufacture the fibers having very small optical losses. The losses caused by the arisen divergent spherical waves are also small since the wavelength is small in comparison with the nonregularity sizes, for instance, the bending radii. The losses in the optical fiber are so small that the information can be transmitted in tens of kilometers without its intermediate recovery.

In the optical fibers, the information is transmitted in the form of a sequence of the optical impulses. In this case the question about the impulse deformation when passing the long distances is essential. If the impulse widening becomes of the same order as the distance between two neighbor impulses, then the transmitted information is distorted.

   The impulse widening in the regular multimode optical fiber is caused by the simultaneous excitation of several eigenmodes. They propagate with different phase velocities and, therefore, with different group velocities. The impulses transmitted by different modes come to the receiving device at different time, thus, the total received impulse is wider than the sent one.

   Before describing the method which allows us to avoid or to lessen the impulse widening caused by the multimodeness of the line, we mention another cause of widening. The frequency spectrum of an impulse of the finite length contains not only the carrier frequency but also a band of those close to it. The width of this band has the order of the inverse impulse duration. The wave velocity as well as the dielectric permittivity of the material depend on the frequency. Therefore, nonmonochromaticness of the signal also leads to the distortion of the impulse shape, that is, to its widening. Since the optical carrier frequency is very high, the width of the frequency band is relatively small. The impulse widening caused by the nonmonochromaticness should be taken into account only if the multimodeness, as a principal cause of the widening, is eliminated or compensated. Further this effect is not mentioned.

   There are two ways of avoiding the impulse widening caused by the multimodeness of the optical fiber. The first one suggests to use optical fibers of so small sizes that only the dipole mode propagates in it. For this end the parameter $ka\sqrt{\varepsilon - 1}$ must be about several units. The second way consists in making the optical fiber to be nonhomogeneous over the cross-section. The dielectric permittivity must be large near the fiber axis decreasing toward the periphery. By choosing its decreasing rate, it is possible to achieve the equality of the velocities of all or at least of the majority of the eigenmodes.

   Below the basis of the theory of dielectric multimode circular waveguide is given for the case when its dielectric permittivity is a function of radius $\varepsilon = \varepsilon(r)$. The construction of this theory is easier without using the notions of the wave theory, but with considering the ray propagation in the nonhomogeneous medium in geometrical terms.

   In Subsection 5.1.2, the possibility to connect any symmetric eigenmode with a certain ray cone, was shown. The rays undergo multiple sequential complete inner reflections on the waveguide boundary. Such a ray cone can be associated with the nonsymmetrical waves. In Subsection 6.2.6, the correspondence between rays and waves will be considered in more general case. The rays coincide with the normals to the equiphase surface of the field. In the homogeneous medium, the rays are the straight lines; in the nonhomogeneous one where ε is not constant, the rays are the curvilinear. On the surfaces where ε is discontinuous, the rays are reflected and refracted.

   First, we qualitatively describe the rays in the dielectric waveguide in which ε is a function of r monotonously decreasing toward the periphery. In the ray interpretation, the waveguide excitation is described as appearance of the ray cone outgoing from a certain point of the waveguide. We assume that this point is located on the waveguide axis, so that the system of rays does not depend on the angle φ.

The larger the mode number is, the faster this field varies with r varying and, therefore, the larger the angle which the rays associated with this mode make with the z-axis. The modes with the large numbers transmit only a little part of the total energy. In geometrical terms, this means that the impulse widening mainly depends on the behavior of rays making a small angle with the axis.

The rays are bending when propagating in the nonhomogeneous medium. This process is called the *refraction*. In some distance from the axis, the rays are directed parallel to it. They continue bending and then intersect the axis again (see Fig. 5.5). The ray trajectory depends on the angle γ_0 made up by the ray with the axis at its exit point. The larger the angle, the larger the distance by which the ray moves away from the axis. However, if ε varies with r as

$$\varepsilon(r) = \varepsilon_0 - Cr^2, \tag{5.25}$$

then (as it will be shown below) the rays for which $\gamma_0 \ll 1$ intersect the axis nearly at the same point. Then this process is repeated.

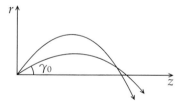

Fig. 5.5 Rays in nonhomogeneous medium

This property of the nonhomogeneous medium with ε varying as (5.25) is similar to that of the thin lens. In the so-called *paraxial approximation* coincident with the condition $\gamma_0 \ll 1$, after passing the lens, all the rays outgoing from the same point meet together again at another point. In the lens theory, it is explained by the fact that the rays outgoing at the larger angle fall onto the lens surface further from its center where the surface slope is larger and, therefore, the rays are refracted by the larger angle. In the optical fiber, the change of the ray trajectory is not stepwise but continuous. The refraction becomes stronger with increasing gradient of ε. According to (5.25), the refraction grows with r increasing. The more steeper the rays are, the larger refraction they have.

In the medium with ε dependent only on one coordinate, the ray trajectory can be found without applying the general optics theory in the nonhomogeneous media, but by considering the medium with $\varepsilon = \varepsilon(r)$ as the limiting case of the layered nonhomogeneous one. The thickness of each layer tends to zero at the limit. The ray is refracted at each interface between media (Fig. 5.6). According to (2.82), the glancing angle $\gamma_n = \pi/2 - \beta_n$, where β_n is the angle of incidence connected with the dielectric permittivities ε_n and ε_{n+1} of the nth and $(n+1)$th layers, as

$$\cos\gamma_{n+1} \cdot \sqrt{\varepsilon_{n+1}} = \cos\gamma_n \cdot \sqrt{\varepsilon_n}. \tag{5.26}$$

At the limit the broken ray turns to be a bent one for which the product $\cos\gamma \cdot \sqrt{\varepsilon}$ is constant along the ray. At the initial point $r = 0$, $\gamma = \gamma_0$, $\varepsilon = \varepsilon_0$; equality (5.26) implies

$$\cos\gamma(r) \cdot \sqrt{\varepsilon(r)} = \cos\gamma_0 \cdot \sqrt{\varepsilon_0}. \tag{5.27}$$

The equation of the ray trajectory, that is, the function $r(z)$ is easily found from this relation. Since $\tan\gamma = dr/dz$, then expressing $\tan\gamma$ by $\cos\gamma$ in (5.27), we obtain

$$\frac{dr}{dz} = \sqrt{\frac{\varepsilon(r)}{\varepsilon_0 \cos^2\gamma_0} - 1}. \tag{5.28}$$

This equation describes the ray from the initial point ($r = 0$, $z = 0$) up to the returning one where $r(z)$ reaches its maximum value and $dr/dz = 0$. We denote the value of r at this point as r_{max}; $\varepsilon(r_{max}) = \varepsilon_0 \cos^2\gamma_0$. The larger the angle γ_0, the larger the value r_{max}. If $\varepsilon(r)$ varies as (5.25), then

$$r_{max} = \sqrt{\frac{\varepsilon_0}{C}} \sin\gamma_0. \tag{5.29}$$

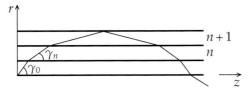

Fig. 5.6 Ray in plano-layered medium

Integrating equation (5.28), we obtain the equation for the ray $r = r(z)$,

$$r(z) = r_{max} \cdot \sin\left(\frac{z}{\cos\gamma_0 \cdot \sqrt{\varepsilon_0/C}}\right). \tag{5.30}$$

The second half of the ray trajectory, from $r = r_{max}$ to $r = 0$, is the mirror reflection of the first one described by (5.30). The value z_{max} at which the ray intersects the axis again, equals the doubled value of z at which the curvature of the ray trajectory changes the sign, so that

$$z_{max} = \pi\sqrt{\frac{\varepsilon_0}{C}} \cdot \cos\gamma_0. \tag{5.31}$$

With accuracy to γ_0^2, that is, in the paraxial approximation, all the rays outgoing from the same point on the axis at different angles, mutually intersect at another point of this axis.

The length of rays measured from the one gathering point to the next one, depends on γ_0. The more steeper rays have the larger length. The shortest is the ray with $\gamma_0 = 0$, that is, propagating along the axis. However, the optical length, that is, the integral of the phase incursion over the ray

$$\int \sqrt{\varepsilon}\,ds, \tag{5.32}$$

taken along the ray, is the same for each ray trajectory; it does not depend on γ_0. The steep rays pass the waveguide part with the smaller ε, the flat rays pass a domain with the larger ε.

The equality of the optical lengths of all the rays outgoing from the same point and meeting at another follows from the *Fermat principle*. According to this principle, the value of the integral (5.32) taken over the ray connecting any two points is minimal (more precisely, extremal) among values of the integral taken over neighbor lines connecting these points. If there are many rays connecting two points, then there are other rays among the lines close to any of them. Their optical length is also minimal. It is possible only if the optical lengths of the rays coincide.

Equality of the values of integral (5.32) implies that the field phase changes by the same quantity for any ray. Time needed for the impulse passing between these two points is the same for any ray. Consequently, according to the geometro-optical theory , in the nonhomogeneous optical fibers with the quadratic dependence of ε on r, the impulse (5.25) transmitted by the optical carrier frequency is not widening in the paraxial approximation. The same conclusion can be drawn by investigating the propagation of different eigenmodes in the waveguide with such dependence of ε on r. Of course, it is clear that the structure of the fields of eigenmodes is more complicated in the waveguide with nonhomogeneous filling, than at $\varepsilon = const$.

5.2
The lines with surface wave

5.2.1
Nonideal metallic cylinder

Similarly to the dielectric cylinder, the nonideal metallic cylinder can serve as a guideline for the surface wave. The phase velocity of such a wave is smaller than c and its field exponentially decreases at radial infinity. The wave slowing is explained by the fact that the wave partially propagates in the skin layer where $\mathrm{Re}\,\varepsilon^{1/2} \gg 1$. Since the layer is very thin, the major part of the energy flux leaks out of the cylinder, and the slowing is small. Therefore, the concentration of the field in the cylinder neighborhood is small as well. The domain, in which the field has the same order as on the cylinder surface, is large in comparison with that occupied by the cylinder. The wave propagates with damping, because $\mathrm{Im}\,\varepsilon \neq 0$ in the skin layer, and a part of the energy is absorbed by it; being transformed into heat.

We begin with the theory of surface wave propagation along the circular cylinder of radius a much larger than the skin layer thickness d (see (1.77)). In this theory, the field at $r < a$ may not be considered. The tangential components of the fields in the exterior domain $r \geq a$ are connected by the impedance relations (1.67), (1.78) on the cylinder surface. Consider the symmetrical *TM*-mode. Its field does not depend on the azimuthal coordinate φ, it has only the components E_z, H_φ, E_r. According to formulas (3.3), (3.7), they are expressed by the potential function $\chi(r)$, as follows:

$$E_z = \alpha^2 \chi, \qquad H_\varphi = -ik\varepsilon \frac{\partial \chi}{\partial r}, \tag{5.33}$$

where the factor $\exp(-ihz)$ is omitted. According to (5.2b),

$$\chi(r) = H_0^{(2)}(\alpha r), \qquad \alpha^2 = k^2 - h^2, \quad \mathrm{Im}\,\alpha < 0. \tag{5.34}$$

On the cylinder surface, the condition

$$E_z = -w H_\varphi, \tag{5.35a}$$

$$w = \frac{1+i}{2}\mu k d, \tag{5.35b}$$

is imposed (see (3.48), (1.78)). The sign in (5.35a) is opposite to that in (3.48), because the normal in (3.48) is assumed to be directed into metal, oppositely to the r-coordinate.

The equation for the propagation constant of this mode is

$$-i\frac{\alpha}{k}\frac{H_0^{(2)}(\alpha a)}{H_1^{(2)}(\alpha a)} = w. \tag{5.36}$$

It can be simplified if using the smallness of the parameter w: $|w| \ll 1$. The argument αa of the cylindrical functions in (5.36) is also small; it has the order $|\alpha a| \approx |w|^{1/2}|ka|^{1/2}$. Substituting the values of these functions by their asymptotics at small arguments, we have

$$\left(\frac{\alpha}{k}\right)^2 \ln(\alpha a) = \frac{d}{a} \cdot \frac{1-i}{2}. \tag{5.37}$$

In a very rough approximation, putting $\ln(\alpha a) = -1$, we obtain (in the value order)

$$\frac{h^2}{k^2} - 1 \approx \frac{d}{a}(1 - i). \tag{5.38}$$

Both the slowing of the wave and its damping, characterized by the relations $(\operatorname{Re} h - k)/k$ and $\operatorname{Im} h/k$, respectively, have the order d/a. The radius of the field concentration domain, that is, the value $1/|\alpha|$ has the order $k^{-1}(a/d)^{1/2}$. It is much larger than the cylinder radius a and the wave length $2\pi/k$. Only the currents occupying the domain with cross-section area not smaller than $1/|\alpha|^2$ can effectively excite the surface wave. Otherwise, the major part of the energy will be radiated in the form of the spherical wave.

The larger the conductivity σ of metal is (i. e., the thinner the skin layer is), the smaller the slowing and the field concentration are. This also relates to the modes of the other types, which are even less concentrated than the TM-mode. The surface modes cannot propagate along the ideal metallic cylinder.

If the skin layer depth calculated by (1.77), is not small in comparison with the cylinder radius, then the structure of the field in the cylinder differs from that in the skin layer near the plane surface. The field does not decrease exponentially when moving away from the surface, that is, it cannot be described by formula (1.76) (where the z-coordinate is directed into the metal). In this case the surface wave theory must be constructed in the same general way, as the theory of the modes in the dielectric waveguide. The propagation constant h is determined from equation (5.4). Since the cylinder consists of material with $|\varepsilon| \gg 1$, the parameter $\tau = (k^2\varepsilon - h^2)^{1/2}$ can be taken approximately equal to $k\varepsilon^{1/2}$ and its dependence on h can be neglected. Then inside the cylinder, the field E_z is proportional to $J_0(k\varepsilon^{1/2}r)$, and the equation for h takes the form

$$\frac{\alpha a H_0^{(2)}(\alpha a)}{H_1^{(2)}(\alpha a)} = \frac{k a J_0(k a \sqrt{\varepsilon})}{\sqrt{\varepsilon} J_1(k a \sqrt{\varepsilon})}, \tag{5.39}$$

where $\alpha^2 = k^2 - h^2$, $\operatorname{Im} \alpha \leq 0$.

Expressing $\varepsilon^{1/2}$ by w and d according to (1.78) and (5.35), and putting $\mu = 1$, we reduce equation (5.39) to the form

$$\frac{\alpha}{k} \frac{H_0^{(2)}(\alpha a)}{H_1^{(2)}(\alpha a)} = w \frac{J_0((1 - i)a/d)}{J_1((1 - i)a/d)}. \tag{5.40}$$

In contrast to (5.36), this equation is valid for any value of a/d. If this value is large, then the ratio of the Bessel functions in (5.40) equals $-i$, and this equation transforms into (5.36).

If the cylinder radius a has the order not larger than d or smaller, then the field is almost homogeneous inside the cylinder. At $a \ll d$, the right-hand side of (5.40) has the order kd^2/a. Then $|\alpha|/k$ has the order of unity, that is, the domain occupied by the field, has the order of the wavelength. Recall that at $d \ll a$ the size of this domain is much larger. The thinner the nonideal metallic cylinder and the smaller its conductivity are, the more concentrated the field of the surface wave around the cylinder, and, consequently, the easier it can be excited.

5.2.2
The ideal metallic cylinder covered by a thin dielectric layer

The ideal metallic cylinder, covered by a thin dielectric layer, can also support the surface wave. In the problem considered in the preceding subsection, the role of such a layer was played by the skin layer on the surface of nonideal metal. The difference between these two cases lies in the fact that the permittivity of the layer material can be real and not necessarily large, and the layer thickness is not connected with the value of $|\varepsilon|$.

The rigorous theory of the surface waves in such a line is constructed in the same way, as the theory of waves in the dielectric waveguide. However, in the layer (at $b < r < a$, where b and a are the radii of the metallic cylinder and of the line, respectively) the potential function $\chi(r)$ for the symmetrical mode does not have the form (5.2a), but also involves the Neumann function, as follows:

$$\chi(r) = A^{(1)} J_0(\tau r) + A^{(2)} N_0(\tau r), \qquad \tau^2 = k^2\varepsilon - h^2. \tag{5.41}$$

The introduction of the function $N_0(\tau r)$ is permitted (and necessary), since the point $r = 0$, at which it has a singularity, lies outside the domain where $\chi(r)$ is expressed. The presence of the two unknown coefficients in (5.41) is sufficient for the boundary conditions to be fulfilled both on the metallic surface (at $r = a$) where $E_z = 0$, and on the exterior boundary of the dielectric layer (at $r = b$) where the components E_z, H_φ are continuous. The conditions lead to a homogeneous system of three linear algebraic equations for $A^{(1)}$, $A^{(2)}$ and B, where B is the coefficient at the function $H_0^{(2)}(\alpha r)$ in the potential function $\chi(r)$ outside the line (as in (5.2b)). The existence condition for nontrivial solutions to this system gives a transcendental equation for h, involving the Bessel and Neumann functions of the arguments τa and τb, respectively, and the Hankel function of αa. The equation is similar to (5.4), but has a more cumbersome form. It transforms into (5.4) at $b = 0$. The equation also has a finite number of the roots; all they are real with $h^2 > k^2$, that is, $\alpha^2 < 0$.

If the thickness $\delta = a - b$ of the dielectric layer is small in comparison with both the line radius ($\delta \ll a$) and the wavelength in the material

($\delta \ll 2\pi/(k\varepsilon^{1/2})$), then simple approximate equations for α can be obtained. At small δ, the slowing is small, h is close to k, and $|\alpha|^2/k^2 \ll 1$. The expression $\tau = [k^2(\varepsilon - 1) + \alpha^2]^{1/2}$ can be replaced by the first terms of its expansion in the series by the small parameter $|\alpha|/(k\varepsilon^{1/2})$. After performing not complicated but cumbersome derivations, the equation becomes

$$\frac{\alpha}{k}\frac{H_0^{(2)}(\alpha a)}{H_1^{(2)}(\alpha a)} = -k\delta\left(1 - \frac{1}{\varepsilon}\right). \tag{5.42}$$

The terms of the order $k\delta(|\alpha|a)^2$ are omitted in the right-hand side of this expression. They should be considered only at very small values of $\varepsilon - 1$. Further we will not mention this clarification; the obtained formulas are valid at $\varepsilon \neq 1$.

Equation (5.42) is analogous to (5.36), where the impedance w of a nonideal metallic surface, which is proportional (according to (1.78)) to the skin layer thickness d, is substituted by the value $ik\delta(1 - \varepsilon^{-1})$. It follows from (5.42) that $|\alpha|a \ll 1$. Substituting the cylindrical functions by their values at small arguments, we obtain the simpler equation

$$\frac{\alpha^2}{k^2}\ln(\alpha a) = \frac{\delta}{a}\left(1 - \frac{1}{\varepsilon}\right). \tag{5.43}$$

In a very rough approximation, similar to that used when passing from (5.37) to (5.38), we obtain the following explicit formula for h

$$\frac{h^2}{k^2} - 1 \approx \frac{\delta}{\alpha}\frac{\varepsilon - 1}{\varepsilon}, \qquad \varepsilon \neq 1. \tag{5.44}$$

The slowing, that is, the ratio $(h - k)/k$ has the order δ/a. The wave is not damping at real ε. The size of the domain occupied by the field has the order $(a/\delta)^{1/2}/k$. It is smaller than that in the case of the similar wave propagating along the nonideal metallic cylinder (since, usually, $\delta \gg d$) and hence the wave along the cylinder with a dielectric layer can be excited much easier.

The surface of the thin dielectric layer lying on the metal cannot be characterized by an impedance. The field structure in the dielectric and the ratio of the tangential components of $\vec{E}$ and $\vec{H}$ on the surface depend on the field structure outside the dielectric. Only at $\varepsilon \gg 1$, the field in the layer remains to be a standing wave regardless of the exterior field structure, in which $E_{\text{tan}} = 0$ on the metal. In this case, the ratio of the tangential components does not depend on the exterior field; it equals $ik\delta$.

For a finite ε and arbitrary field structure, the field components are connected as

$$E_z = ik\delta H_\varphi + \frac{\delta}{\varepsilon}\frac{\partial E_r}{\partial z} \tag{5.45}$$

on the surface. This relation is not an impedance one, because it involves a derivative. To obtain (5.45), we should use the equation $(\mathrm{rot}\,\vec{E})_\varphi = -ikH_\varphi$ following from (1.31). Integrate this equation over the rectangle of length l in the layer ($l \gg \delta$, $kl \ll 1$) (Fig. 5.7) and use the Stokes theorem, according to which the surface integral of the flux of the vector rotor equals the contour integral of the tangential component of the vector. On the metal $E_z = 0$. The components E_z, H_φ are continuous on the exterior boundary of the layer; in the layer they are approximately the same as on the surface. Inside the dielectric, the component E_r is ε times smaller than outside. Canceling l out of both the sides of the obtained integral expression for H_φ, we get (5.45).

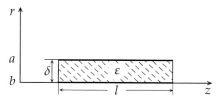

Fig. 5.7 Integration domain for obtaining (5.45)

The application of formula (5.45) simplifies solving the boundary value problems about the field in a domain bounded by an ideal metallic surface covered with a thin dielectric layer, since it allow us not to consider the field inside the layer.

5.2.3
The short-periodical surface

In certain cases, the surface wave can propagate along the ideal metallic rod if its radius is a periodical function of the longitudinal coordinate z. For simplicity, we assume that inside each periodical cell the rod surface is symmetrical with respect to the plane $z = const$ passing through the center of the cell (Fig. 5.8). Then the rod radius can be expressed as the series

$$r(z) = r_0 + \sum_{n=1}^{\infty} A_n \cos \frac{2\pi n z}{L}, \qquad (5.46)$$

where L is the period (cell length). The rod is fully contained inside the virtual cylindrical surface $r = a = r_0 + \sum |A_n|$. On this surface, all the field components are periodical (with the same period L) functions of z multiplied by $\exp(-ihz)$ with certain value of h. Since the space outside the surface is homogeneous with respect to z, then all the components of the surface wave depend on z by the same rule there.

We confine ourselves to the case of symmetrical modes which propagate along the rod being a body of revolution. The division onto the TE- and TM-

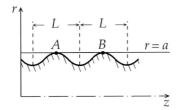

Fig. 5.8 Periodical surface

modes remains possible for this case, even if the rod radius depends on z. Further we consider the TM-modes.

Outside the cylinder $r = a$, the component E_z can be expressed as

$$E_z(r,z) = \exp(-ihz)\left[B_0 H_0^{(2)}(\alpha_0 r) + \sum_{n\neq 0} B_n H_0^{(2)}(\alpha_n r)\exp\left(-\frac{2\pi i n z}{L}\right)\right].$$

(5.47)

The multiplier $\exp(-ihz)$ describes the alteration of both the phase and amplitude (if $\mathrm{Im}\, h \neq 0$) when passing from one cell to the next one. The quantity h depends on the cell shape. For a simple shape it will be found below. The second multiplier in (5.47) describes the field alteration inside the cell. It follows from the wave equation for E_z, that

$$\alpha_n^2 = k^2 - \left(h + \frac{2\pi n}{L}\right)^2.$$

(5.48)

If α_n^2 is positive (i.e., α_n is real) for at least one n, $-\infty < n < \infty$, then the field is slowly (only as $r^{-1/2}$) decreasing at $r \to \infty$ (unless $B_n = 0$ for this n). The field contains the term, which describes the wave outgoing into the radial infinity and transmitting the energy out of the rod. In this case the surface wave cannot exist. The existence condition for this wave is

$$\left(h + \frac{2\pi n}{L}\right)^2 > k^2$$

(5.49)

which should be fulfilled for all n.

We give a geometrical interpretation of this condition. The external field can be treated as a sum of the fields created by currents flowing on the surface of the line. Similarly as the field (5.47), the current consists of the terms depending on z as $\exp(-ih_n z)$, where $h_n = (h + 2\pi n/L)$. The current of each cell generates a spherical wave, diverging with the velocity c. If such waves, arisen in different cells, are not in-phase, then they mutually cancel almost completely. The energy does not outgo from the line and the surface wave exists. If there exists a direction in which the waves created by the adjacent cells are parallel and in-phase, then the waves of all the cells are in-phase in this direction. The energy outgoes away off the line and the surface wave is destroyed. This fact is inherent also to the homogeneous lines. An example is the leaky wave of the dielectric waveguide: it is fast, destroyed and not the eigenwave.

Consider the surface waves together with the spherical waves, radiated by the cell centered at point A (Fig. 5.9). Each surface wave propagates with the velocity ck/h_n and reaches the next cell (centered at B) after time $(L/c)(h_n/k)$. During this time the spherical wave covers the distance Lh_n/k. If the surface wave is fast ($h_n < k$), then this distance is smaller than L (Fig. 5.9(a)). On the front of the spherical wave, the phase is the same as at point B. The spherical wave radiated by the point B is in-phase with the wave passing through the point C, and these waves are in-phase with all the waves radiated by other cells. In the direction making the angle γ ($\cos\gamma = h_n/k$) with the z-axis, the cone wave is formed. This is the scheme of the fast surface wave destroying. If condition (5.49) holds, that is, $h_n > k$, then the spherical waves outgoing from the cell B is delayed in the phase from the wave generated in the cell A (Fig. 5.9(b)) and the cone wave does not appear. The slow waves are not destroyed.

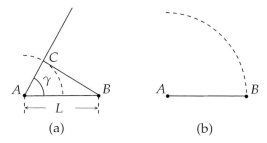

Fig. 5.9 Illustration of the fast wave destroying

The field of the surface wave over the short-periodical surface possesses certain properties, which will be considered in the example of the TM-mode propagating over the rod of variable radius. In expression (14.15) for $E_z(r,z)$, all the terms under the sum are fast-decreasing when moving away from the surface $r = a$. According to (5.48), (5.49), this decrease is approximately described by the factor $\exp[-2\pi|n|(r-a)/L]$. The sum in (5.47) describes small-scale "ripple" on the field near the rod surface. The ripple disappears at distances of the order L, and the field is smoothed at $r - a \gg L$. It is described only by the first term in (5.47) there.

The same term also describes the period-average field $\bar{E}_z(r,z)$ calculated as

$$\bar{E}_z(r,z) = \exp(-ihz)\frac{1}{L}\int_0^L \left[B_0 H_0^{(2)}(\alpha_0 r) \right.$$

$$\left. + \sum B_n H_0^{(2)}(\alpha_n r) \exp\left(-\frac{2\pi i n z}{L}\right) \right] dz. \quad (5.50)$$

When averaging, the periodical terms cancel, only the term with $H_0^{(2)}(\alpha_0 r)$ remains. This field coincides with the asymptotic of the field at $z - a \gg L$:

$$\exp(-ihz)\frac{1}{L}\int_0^L E_z(r,\varsigma)\,d\varsigma = \lim_{r/a\to\infty} E_z(r,z). \quad (5.51)$$

In accordance with the above, the surface nonsmoothness can be neglected when calculating the field outside the surface: it is possible to impose the condition on a smooth surface, close to the short-periodical one, that the field is equal to its average (per the period) value. This average value equals the value of the field at the distance from the surface, large in comparison with the period L, but small in comparison with the distance scale, where the exterior field is noticeably varied.

5.2.4
The corrugated surface

According to the above assertion, in the problem about the rod with the short-periodical surface, we can put

$$E_z(a,z) = B_0 H_0^{(2)}(\alpha a) \exp(-ihz) \quad (5.52)$$

on the smooth virtual surface $r = a$, where the index 0 at α is omitted. According to (3.3), at $r > a$, the potential function $\chi(r)$ equals

$$\chi(r) = \frac{1}{\alpha^2} B_0 H_0^{(2)}(\alpha r). \quad (5.53)$$

Then (3.4) yields

$$H_\varphi(a) = \frac{ik}{\alpha} B_0 H_1^{(2)}(\alpha a) \exp(-ihz). \quad (5.54)$$

In order to obtain the dispersion equation for h, we should determine the field structure in the domain between the surface $r = a$ and the metallic rod surface, and then write the continuity condition for the field.

This technique can be easily realized for the periodical structure in the form of a corrugated surface. Each cell of the surface is a rectangular canal (Fig. 5.10). Denote its depth by δ. The metallic bottom of the canal is located at $r = a - \delta$. The neighboring cells are separated by partitions. The partition thickness is denoted by t, so that the canal width is $L - t$. The component E_z equals zero on the cylindrical surface $r = a$ at $L - t < z < L$ (here z is the local longitudinal coordinate, equal zero on the left edge of the cell).

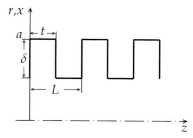

Fig. 5.10 Corrugated surface

The field in the canal consists of the wave, field of which does not depend on z, and of infinite number of higher waves, fields of which depend on z. These fields exist at $r \leq a$, they make up the small-scale ripple at $r > a$. They cancel when averaging over the period, only the main wave remains, for which $E_z(r, z)$ does not depend on z. In this wave, only the components E_z, H_φ differ from zero. According to (1.29), $H_\varphi = -(i/k)dE_z/dr$ in the field independent of z.

We confine ourselves to the case, when the canal depth is small in comparison with the cylinder radius, $\delta \ll a$. Then the curvature of the surfaces $r = a$ and $r = a - \delta$ may be ignored and the field in the canals can be expressed as a combination of not cylindrical, but trigonometric functions. The argument of these functions contains the value $x = r - a$ which can be treated as a Cartesian coordinate. Since $E_z = 0$ on the canal bottom (at $x = -\delta$), then

$$E_z = C \sin[k(x + \delta)], \qquad H_\varphi = -iC \cos[k(x + \delta)]. \tag{5.55a}$$

On the cylindrical surface $r = a$ we have $x = 0$, so that in the canal orifice the field components are

$$E_z = C \sin(k\delta), \qquad H_\varphi = -iC \cos(k\delta). \tag{5.55b}$$

On the surface $r = a$ outside the canal (i. e., on metal), $E_z = 0$. On the partition top (at $r = a$), the component H_φ has approximately the same value as on its side walls, because the current continuously passes from the partition wall onto its top. The average value of $E_z(a, z)$ is $\bar{E}_z = Cq \sin(k\delta)$, where $q = (L - t)/L$, that is, it equals the ratio of the canal width to the cell length.

At $r = a - 0$, the ratio of the average fields is

$$\frac{\bar{E}_z}{\bar{H}_\varphi} = iq \tan(k\delta). \tag{5.56}$$

Strictly speaking, the corrugated surface cannot be characterized by the impedance, that is, by a value which does not depend on the exterior field structure. The relation (5.56) between the period-average values of the field components is valid only if the exterior field, and hence the field in the canals, does not depend on φ.

The impedance of the corrugated surface is anisotropic. Formula (5.56) relates only to the component of the electric field, perpendicular to canals. The waves with the only tangential component, parallel to canals, cannot propagate in them. This component must be zero on both the canal walls, that is, it should be fast varying in the z-direction at the distance of the order L. Consequently, it must decrease in the x-direction, because it follows from the wave equation that if $\partial^2/\partial z^2 \gg k^2$, then $\partial^2/\partial x^2 \gg k^2$ as well. For the component E_φ, the corrugated surface is equivalent to the metallic one located a little lower the surface $x = 0$, at the distance of the order L from it. If the exterior field does not involve the component E_z, then the corrugated surface cannot support the surface wave.

Equating the ratio E_z/H_φ for the symmetrical TE-mode propagating along the rod with the corrugated surface (see (5.52)), to the impedance (5.56), we obtain the following equation:

$$-\frac{\alpha}{k} \frac{H_0^{(2)}(\alpha a)}{H_1^{(2)}(\alpha a)} = q \tan(k\delta) \tag{5.57a}$$

for the wave number h of this mode. It is analogous to equation (5.42) for the wave propagating along the cylinder covered by a thin dielectric layer. At $\text{Im}\,\alpha < 0$, that is, at $\alpha a = -i|\alpha a|$, the Hankel functions $H_m^{(2)}(\alpha a)$ can be expressed by the modified Hankel functions $K_m(|\alpha a|)$, which are real and positive at all the real values of their argument. Then equation (5.57a) becomes

$$\frac{|\alpha a|}{ka} \frac{K_0(|\alpha a|)}{K_1(|\alpha a|)} = q \tan(k\delta). \tag{5.57b}$$

Equation (5.57) can also be replaced by an approximate one of the type (5.43). In a much more rough approximation, we can obtain the explicit expression for h^2, similar to (5.44), as follows:

$$\frac{h^2}{k^2} - 1 \approx \frac{1}{ka} q \tan(k\delta). \tag{5.58}$$

A simpler expression is obtained if the canal depth is small in comparison with the wavelength in vacuum, that is, at $k\delta \ll 1$. In this case

$$\frac{h^2}{k^2} - 1 \approx q\frac{\delta}{a}. \tag{5.59}$$

According to (5.57), (5.58), in order to provide $h > k$, necessary for the slow wave existence, the inequality $\tan(k\delta) > 0$ should be fulfilled. The surface wave can propagate along the corrugated rod with the canal depth δ such that $\tan(k\delta) > 0$. Hence, δ should be smaller than the quarter of the wavelength in vacuum ($k\delta < \pi/2$) or lie in the intervals ($n\pi < k\delta < (2n+1)\pi/2$). At $\tan(k\delta) < 0$, the corrugated surface does not support the surface wave.

Formulas (5.57)-(5.59) are not valid at $k\delta \approx \pi/2$ (or $k\delta \approx (2n+1)\pi/2$). It follows from (5.57) that $|\alpha a| \gg 1$ (i.e., $h/k \ll 1$) at $\tan(k\delta) \gg 1$. The length $2\pi/h$ of the surface wave becomes too small. Then the condition $hL \ll 1$ (i.e., the exterior field is weakly varied over the period), used when deriving equations (5.57), is violated. Formula (5.50) for the average field turns to be wrong. The higher waves in canals participate in the formation of the exterior field. The corrugated surface with the canal depth close to $\lambda/4, 3\lambda/4, \ldots$, cannot be characterized by the surface impedance.

5.2.5
Slow waves in closed waveguides

The structures considered above (dielectric layer on metal, corrugated surface), along which slow waves can propagate, can also be used for construction of the closed metallic waveguides in which slow waves can exist. Such waves are used in devices, in which the electromagnetic field interacts with the axial electron flux, for instance, in the accelerators and travelling wave tubes. The interaction is effective if the phase velocity v of the wave is close to the electron velocity (which, of course, is smaller than c). In the nonfilled metallic waveguides $v > c$ ($h < k$). For decreasing the phase velocity (i.e., to slow down the wave), we should either fill a part of the waveguide with the dielectric, or construct the waveguide walls on which the field satisfies certain impedance condition.

Only the longitudinal component E_z of the electric field effectively interacts with the electron flux. On the axis of circular waveguide this component differs from zero only for the symmetrical TM-mode. In this and the next subsections we consider three devices, providing the existence of the slow symmetric TM-mode in the circular waveguide.

Let the waveguide filled with the dielectric have a canal along the axis (Fig. 5.11). The radii of the waveguide and canal are a and b, respectively ($0 \leq b \leq a$); $\varepsilon = 1$ at $r < b$, and $\varepsilon > 1$ at $b < r < a$. In dielectric, the potential function $\chi(r)$ has the form (5.41). This field satisfies the following boundary conditions: $E_z = 0$ at $r = a$; E_z and H_φ are continuous at $r = b$. The conditions lead to a system of three homogeneous linear equations for the coefficients $A^{(1)}$, $A^{(2)}$, C. The existence condition for solutions to this system gives the sought equation for α.

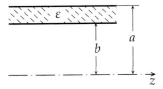

Fig. 5.11 Closed waveguide with internal dielectric coating

It is easier not to apply this general technique, but to write the function $\chi(r)$ in the dielectric in the form

$$\chi(r) = A\left[J_0(\tau r)N_0(\tau a) - N_0(\tau r)J_0(\tau a)\right], \tag{5.60}$$

providing the immediate fulfillment of the condition at $r = a$ for E_z. According to (3.3), $E_z = \tau^2\chi$, $H_\varphi = -ik\varepsilon d\chi/dr$ in the dielectric. In the axial canal, $\chi(r) = CJ_0(\alpha r)$, and $E_z = \alpha^2\chi$, $H_\varphi = -ikd\chi/dr$. Equating the ratio E_z/H_φ at $r = b + 0$ (in the dielectric) and $r = b - 0$ (in the canal) leads to the following equation for α:

$$\frac{\alpha J_0(\alpha b)}{J_1(\alpha b)} = \varepsilon\tau\frac{J_0(\tau b)N_0(\tau a) - N_0(\tau b)J_0(\tau a)}{J_1(\tau b)N_0(\tau a) - N_1(\tau b)J_0(\tau a)}. \tag{5.61}$$

At $b = a$, that is, in the absence of the dielectric layer, the equation becomes $J_0(\alpha a) = 0$, that is, the same as in the nonfilled waveguide (see Subsection 3.1.6). Its first root is $\alpha = \nu_{01}/a$, where $\nu_{01} = 2.40$. In the nonfilled waveguide, $\alpha^2 > 0$, $h < k$, and all the modes are fast. At $b = 0$, that is, when the whole waveguide is filled with the dielectric, equation (5.61) transforms into $J_0(\tau a) = 0$, and its first root is $\tau = \nu_{01}/a$. Then $\alpha^2 = (\nu_{01}/a)^2 - k^2(\varepsilon - 1)$. Introduce the parameter

$$\widehat{\varepsilon} = 1 + \left(\frac{\nu_{01}}{ka}\right)^2, \tag{5.62}$$

so that $\alpha^2 = k^2(\widehat{\varepsilon} - \varepsilon)$. Then $\alpha = 0$ at $\varepsilon = \widehat{\varepsilon}$, and $\alpha^2 > 0$ at $\varepsilon < \widehat{\varepsilon}$. Even the complete filling of the waveguide with the material having a not large ε does not cause the first mode to be slow. If $\varepsilon > \widehat{\varepsilon}$, then the first mode slows down so much that its phase velocity is smaller than c.

The case $b > 0$ (i.e., the canal radius if finite) is qualitatively the same as at $b = 0$, but the value $\widehat{\varepsilon}$ at which the phase velocity equals c, is larger than that obtained by (5.62). In this case $\widehat{\varepsilon}$ should be calculated from equation (5.61) at

$\alpha = 0$, $\tau = k(\widehat{\varepsilon} - 1)^{1/2}$, $\varepsilon = \widehat{\varepsilon}$. At $a - b \ll a$, that is, when the dielectric forms a thin film on the metallic surface, the equation gives

$$\widehat{\varepsilon} = \frac{1}{ka}\sqrt{\frac{2a}{a-b}}, \tag{5.63}$$

that is, $\widehat{\varepsilon} \gg 1$. Only the layer with a very large ε, $\varepsilon > \widehat{\varepsilon}$, coated on the waveguide walls, can cause the slowing of the waveguide mode such that its phase velocity is smaller than c.

The second example of the slowing system is a corrugated surface coated on the inner wall of the circular waveguide. Denote the waveguide radius and radius of the axial canal free of the partitions between the cells, by a and b, respectively (Fig. 5.12). In the axial canals, the potential function is $\chi(r) = A J_0(\alpha r)$. In the radial canals, we take into consideration only the term in the component E_z, independent of z and equal to $C \sin(kx)$, where $x = r - a$. In this form, $E_z(r) = 0$ at $r = a$ (not at $r = b$, as in (5.55a)). For the field with $\partial/\partial z \equiv 0$, we have $H_\varphi = -(i/k)\partial E_z/\partial r = -iC\cos(kx)$. On the corrugated surface $r = b$, the field components are $E_z = -C\sin(k\delta)$, $H_\varphi = -iC\cos(k\delta)$, where $\delta = a - b$ is the canal depth. The impedance of the corrugated surface is

$$\frac{\bar{E}_z}{H_\varphi} = -iq\tan(k\delta). \tag{5.64}$$

It differs from that given by (5.56) only in sign, which is explained by the fact that (5.56) relates to the case when not the interior, but the exterior surface of the cylinder is corrugated.

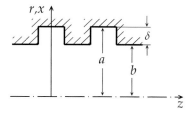

Fig. 5.12 Closed waveguide with corrugated surface

On the periphery of the axial canal (at $r = b - 0$), $E_z/H_\varphi = -i(\alpha/k)J_0(\alpha b)/J_1(\alpha b)$. The equation for h has the form

$$\frac{\alpha}{k}\frac{J_0(\alpha b)}{J_1(\alpha b)} = q\tan(k\delta). \tag{5.65}$$

Of course at $\delta = 0$ it follows from (5.65) that $\alpha = v_{01}/b$, that is, α is real. With δ growing, α decreases and vanishes at

$$\tan(k\delta) = \frac{2}{qkb}. \tag{5.66}$$

For the larger δ, it follows from (5.65) that $\alpha^2 < 0$, $\alpha = -i|\alpha|$, and the wave becomes slow. Equation (5.65) takes the form, similar to (5.57b)

$$\frac{|\alpha|}{k} \frac{I_0(|\alpha|b)}{I_1(|\alpha|b)} = q \tan(k\delta). \tag{5.67}$$

Here I_0, I_1 are the modified Bessel functions; at real values of the argument they are real and positive.

In order to decrease the phase velocity of a waveguide mode to the value smaller than c, the canal depth should be larger than that obtained by (5.66). The note given at the end of Subsection 5.2.4 concerning nonapplicability of the obtained results for $\delta \approx \lambda/4,\ldots$, is valid for the waveguide considered here, either.

5.2.6
The helix line; the array

The metallic wire making up a helix line (Fig. 5.13(a)), is also a slowing system. In a very rough approximation we may assume that a current wave propagates along the wire, the wave velocity is c and it depends neither on the wire shape nor on the current in other segments. Then the field created by the current is a wave; its velocity v along the axis equals $c \cos \gamma$, where γ is the angle made by a wire turn with the axis, so that $v < c$.

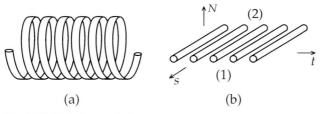

(a)   (b)

Fig. 5.13 Helix line and wire array

In this subsection we outline the foundation of the helix line theory, constructed not on the above concept of the current wave propagating along the wire, but on the electromagnetic consideration accounting the mutual influence of different segments of the line. The theory is based on the local relations between different components of the fields near short-periodical array made up by the metallic wires (Fig. 5.13(b)). The helix line is considered as an array convolved into the cylinder. The theory deals with the fields of the wave propagating along the cylinder.

The main property of the array is its anisotropy, which makes it similar to the corrugated surface. In contrast to the case of such a surface, the field exists on both sides of the array. The boundary conditions determine the relations between the field components on both of its sides.

We consider a plane array of straightline wires and assume that the conditions imposed on it are local, that is, they are the same as on the curvilinear one making up the cylindrical surface. Denote the direction along the wires by s, and the perpendicular direction lying in the array plane by t. In the array neighborhood, the field can be divided into two parts linearly polarized along the directions s and t, respectively. The first part has the components E_s, H_t, the second one has E_t, H_s. The boundary conditions for the fields of the polarizations are different and independent of each other.

The conditions for the fields of the first polarization are

$$E_s^{(1)} = 0, \qquad E_s^{(2)} = 0, \tag{5.68a}$$

where indices (1), (2) refer to the opposite sides of the array. For the field of this polarization, the array is equivalent to the ideal-metallic surface. The wave falling onto the array completely reflects. The current flows along the s-direction, the component H_t is discontinuous. The current can be determined after solving the problem and finding the fields E_s, H_t in the whole space.

For the fields of the second polarization the local boundary conditions are

$$E_t^{(1)} = E_t^{(2)}, \qquad H_s^{(1)} = H_s^{(2)}. \tag{5.68b}$$

The fields of this polarization are not disturbed by the array, they remain continuous. The wave falling onto the array completely passes through it. The currents do not flow in the t-direction, that is, perpendicularly to the wires, the component H_s is continuous.

Conditions (5.68) describe the array as a smooth surface having an infinite anisotropic conductivity. The conditions are valid if the array period L is small. In a layer near the array, the thickness of which has the order L, the fields are fast varying. Outside this layer, at the distance from the array, much larger than L, these fast variations ("ripple") disappear. The fields participating in (5.68) are the limiting values of the fields at the large distance from the array. The boundary conditions (5.68) have a sense if the exterior fields vary a little over the distance of the order L. Under this condition, formulas (5.68) are boundary conditions for these far fields. The two-sided conditions (5.68) have the same sense as the one-sided impedance one (5.56) on the corrugated surface.

Conditions (5.68) are approximate. They involve neither the array density, nor the wire shape. Obviously, the waves of the first polarization ($E_t = 0$) partially penetrate through the wide-spaced array, as well as those of the second polarization ($E_s = 0$) partially reflect from the dense one. Both these effects contradict conditions (5.68). The effects are essential in the problems connected with radiation from resonators, walls of which contain the arrays. The more exact theory of arrays will be explained in Subsection 6.3.4 where such a problem is considered.

In the theory of the slow wave lines, considered here, all main results can be obtained under assumption that the averaged boundary conditions (5.68) are valid on the cylindrical surface made up by the helix line. It may also be assumed that the wires are located not inside a certain waveguide with the radius larger than the wavelength, but in free space. Accounting the conditions on the waveguide walls does not change anything in principle.

Although the fields of the symmetrical waves do not depend on φ, such waves are hybrid, because the conditions imposed on the cylindrical surface, depend on φ. Both the axial components E_z, H_z differ from zero in these waves, and the fields are expressed by both the potential functions χ and ψ. They are proportional to the Bessel function of αr inside the cylinder and to the Hankel function of the same argument outside it, as follows:

$$\chi(r) = A J_0(\alpha r), \quad \psi(r) = B J_0(\alpha r), \qquad r < a, \tag{5.69a}$$

$$\chi(r) = C H_0^{(2)}(\alpha r), \quad \psi(r) = D H_0^{(2)}(\alpha r), \qquad r > a. \tag{5.69b}$$

The four tangential components E_z, E_φ, H_z, H_φ should be calculated by formulas (3.3), (3.4), and then the components participating in (5.68) should be expressed as

$$A_s = A_z \cos \gamma + A_\varphi \sin \gamma, \qquad A_t = -A_z \sin \gamma + A_\varphi \cos \gamma. \tag{5.70}$$

valid for any vector $\vec{A}$. Substituting these expressions into (5.68), we obtain a system of four linear homogeneous equations for the coefficients in (5.69). The existence condition for its nontrivial solutions gives the sought equation for α. Since the waves in the line are slow, the values α obtained from this equation are imaginary: $\alpha = -i|\alpha|$. Introducing the modified Bessel functions I_n, K_n instead of J_n, $H_n^{(2)}$ by the well-known formulas, we write the equation in the form

$$\alpha^2 = -k^2 \frac{K_1(|\alpha|a) I_1(|\alpha|a)}{K_0(|\alpha|a) I_0(|\alpha|a)} \tan^2 \gamma. \tag{5.71}$$

In this formula, the fraction is always positive so that the assumption $\alpha^2 < 0$ is fulfilled.

If $|\alpha|a \gg 1$ is assumed, then the above fraction is approximately equal to unity, and $\alpha = -ik \tan \gamma$. The assumption is valid if $ka \tan \gamma \gg 1$. Then $h = k/\cos \gamma$, that is, the phase velocity along the z-axis is $c \cos \gamma$. This result was obtained above under the assumption that a current wave propagates along the wires with velocity c. The theory based on the boundary conditions (5.68), confirms this assumption for the case $ka \tan \gamma \gg 1$.

The condition that the phase velocity does not depend on the frequency, under which $v = c \cos \gamma$, is fulfilled in a wide frequency range. The range is limited not only from below (ka should be sufficiently large) but also from above (ka should be sufficiently small). The demand $hL \ll 1$, necessary for condition (5.68) (and, therefore, for equation (5.71)) to be valid, is violated at very high frequencies. For the one-thread helix (one wire is coiled in such a way that it makes up the cylindrical surface), the period L equals $2a\cos\gamma$, so that if $h = k/\cos\gamma$, then $hL = 2ka$. Consequently, this equation is valid only at $ka \ll 1$. Therefore, the simple result $v = c\cos\gamma$ is valid in the range

$$\cot\gamma \ll ka \ll 1. \tag{5.72}$$

At $\pi/2 - \gamma \ll 1$ this range is very wide. At the lower frequencies, $|\alpha|$ tends to zero and the phase velocity increases tending to c.

5.2.7
The microstrip line

The microstrip line consists of a wide metallic strip, a dielectric plate lying on it, and a narrow metallic strip lying on the dielectric (Fig. 5.14). The system is a double-conducting line, with the conductors being the metallic strips. The field of the wave propagating along this line, is more complicated than the field of the TEM-mode, due to the presence of the dielectric layer. The field has six components, $E_z \neq 0$, $H_z \neq 0$ in it. The theory of such a line is more complicated than the theory of the co-axial one. Since the metallic surfaces do not coincide with the coordinate ones, the fields cannot be presented as a product of two functions, each of which is a function of one transverse coordinate. The variable separation method, used above when investigating the cylindrical lines, cannot be applied in this case. The detail theory of such lines is very cumbersome. We confine ourselves to its qualitative description.

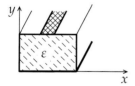

Fig. 5.14 Geometry of microstrip line

First, consider a conventional line which differs from the microstrip line in the absence of the dielectric. As in any two-conductor line, in such a line with arbitrary shaped conductors, there exists a wave with the propagation constant $h = k$. The phase velocity equals c; it does not depend on the frequency. The field of the wave is expressible by the potential function $\chi(x, y)$, satisfying the Laplace equation $\nabla^2\chi = 0$. The function χ has different constant values on the conductors. With accuracy to a nonessential factor, the field components are expressed by (3.39). The same opposite-directed currents flow along the conductors.

If the width of the lower strip is much larger than that of the upper one and than the distance between the strips, then the field exists only over the plane of the lower strip. The field is the same as that between two narrow strips located at the distance twice as large as that between the wide and narrow strips. In this case, not the wide strip, but the reflection of the narrow strip in the wide one, being like a mirror, may be treated as the second conductor.

This simple consideration allows us to estimate the rate of the field decreasing in the radial direction for the line without the dielectric. It is convenient to treat the field as one radiated by the currents on the conductors. The current flowing along each conductor creates the field decreasing inversely proportional to the distance from the conductor. This assertion is exact for the conductor in the form of a circular cylinder. In the case of the arbitrary-shaped conductors, it is approximately valid for the fields at the large distance from them. Two conductors – the narrow strip and its reflection – with the same currents, different only in sign, create the field decreasing inversely proportional to the squared distance from the line.

If the distance between the strips is small in comparison with the width of the narrow one, then the field is mainly concentrated between the strips, under the narrow one. The components E_y and H_x ($E_y \approx -H_x$) are larger than others and weekly dependent on the coordinates. In this domain χ almost linearly depends on the coordinate y, normal to the strips. As it has been shown in Subsection 3.1.7, the electric field of the *TEM*-mode of the two-conductor line coincides with the electrostatic field of the long condenser. The field outside the domain between the narrow strip and its reflection, coincides with the "scattering field" of the long condenser consisting of two parallel narrow metallic strips.

In the domain between the strips, the potential function $\chi(x, y)$ depends on both the coordinates x, y and decreases when moving away from the line. This function can be found by using the conformal mapping technique. The obtained formulas are very cumbersome. Recall that all these considerations relate to a simplified, conventional model, namely, the line without the dielectric.

Introduction of the dielectric plate into the space between the strips results in a field disturbance, in particular, in appearance of the components E_z, H_z. Then the potential functions χ and ψ satisfy not the Laplace equation but the wave equations different in the dielectric and outside it. The most noticeable effect is the wave slowing. It is caused by the fact that the energy flux partially flaws in the dielectric where the field is mainly concentrated. In the very rough approximation, the propagation constant equals $k\varepsilon^{1/2}$, as in unbounded dielectric.

At the radial infinity, the field of the slow wave decreases exponentially, approximately as $\exp(-|\alpha|r)$, where $\alpha^2 = k^2 - h^2$. If we assume that $h = k\varepsilon^{1/2}$, then $|\alpha| = k(\varepsilon - 1)^{1/2}$. Consequently, the field of the wave in the microstrip line decreases when moving away from the line, approximately as $\exp(-kr(\varepsilon - 1)^{1/2})/r^2$, and propagates with the phase velocity $v = c/\varepsilon^{1/2}$. Both these estimations are very rough. They give only the order of the linear size of the domain occupied by the field, and the order of the phase velocity.

5.3
The wave beam

5.3.1
The wave beam

The energy can be transmitted not only by an electromagnetic wave propagating along a certain material guide, but also by a wave making up a long narrow beam in the free space. The beam can be formed by a set of radiators located in the plane domain $z = 0$, all linear sizes of which are much larger than the wavelength. An appropriate choice of the amplitudes and phases of the radiators permits to form the beam due to the interference of the fields created by them. The total field is mostly located inside the beam, and it fast decreases when moving away from the beam. The distance at which the beam widens insignificantly, has the order of ratio of the area of the domain, where radiators are located, to the wavelength, that is, it is much larger than the size of radiator set. The beam falls onto a set of receivers, lying in the plane $z = d$ parallel to the radiator plane. Such a line can transmit the energy for a very large distance with losses of a few per cent. The sets of radiators (transmitting antenna) and receivers (receiving antenna) are called the *antenna* and *rectenna*, respectively.

The theory of such a line involves the methods of solving the direct and inverse problems. The direct problem consists in finding the field in the whole space (in particular, on the rectenna and in its plane) by the given field on the antenna. The inverse problem consists in finding the field on the antenna such that certain functional of the field in the space (in particular, in the rectenna plane) is extreme (minimal or maximal, depending on the functional type). Further, we consider three functionals, describing the energy and ecology demands. Three corresponding formulations of the inverse problem are considered and the field structures on the rectenna, providing the minimum of the functionals, are found. Another optimization problem is considered, consisting in finding the optimal shapes of the antenna and rectenna with the given fixed areas.

The theory of long beam can be constructed in the scalar approximation. A linear polarization created on the antenna remains in the entire beam with almost no distortion. Neglecting the depolarization, we describe the field by only one scalar function $U(X, Y, Z)$, where X, Y, Z are the Cartesian coordinates. The function U is any of the Cartesian components of the fields $\vec{E}$ or $\vec{H}$, parallel to the antenna. The function satisfies the wave equation

$$\frac{\partial^2 U}{\partial X^2} + \frac{\partial^2 U}{\partial Y^2} + \frac{\partial^2 U}{\partial Z^2} + k^2 U = 0. \tag{5.73}$$

Let the coordinate origin be located on the antenna and the Z-axis be direct toward the rectenna; then the axes X, Y lie in the antenna plane (see Fig. 5.15). Denote the domain, occupied by the antenna, by D. The field in the antenna plane, that is, the function $U(X, Y, 0)$, differs from zero only at points X, Y lying inside D or on its boundary.

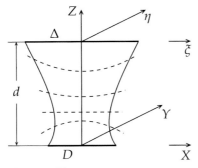

Fig. 5.15 The antenna-rectenna transmitting line

The direct problem consists in expressing $U(X, Y, Z)$ at $Z > 0$ by $U(X, Y, 0)$. Besides equation (5.73), the field $U(X, Y, Z)$ should satisfy the radiation condition at $Z \to \infty$ (see (1.33)). Since the value of the sought function is given on the very simple surface (the plane), the problem can be solved without using the general technique of the Green function, but directly continuing the function given at $Z = 0$ into the domain $Z > 0$.

Expand the function $U(X, Y, 0)$ into the double Fourier integral

$$U(X, Y, 0) = \int\!\!\!\int_{-\infty}^{\infty} f(\beta, \gamma) \exp(-i\beta X - i\gamma Y) \, d\beta d\gamma. \tag{5.74}$$

The function $f(\beta, \gamma)$ equals

$$f(\beta, \gamma) = \int\!\!\!\int_{D} U(X, Y, 0) \exp(i\beta X + i\gamma Y) \, dX dY, \tag{5.75}$$

with accuracy to a nonessential numeral factor, omitted in the next formulas as well. At $Z > 0$, the function $U(X, Y, Z)$ is obtained from (5.74) by replacing the factor $\exp(-i\beta X - i\gamma Y)$ with $\exp[-i\beta X - i\gamma Y - i(k^2 - \beta^2 - \gamma^2)^{1/2}Z]$, where $(k^2 - \beta^2 - \gamma^2)^{1/2} = -i(\beta^2 + \gamma^2 - k^2)^{1/2}$ at $\beta^2 + \gamma^2 > k^2$. The function

$$U(X,Y,Z) = \iint\limits_{-\infty}^{\infty} f(\beta,\gamma) \exp\left(-i\beta X - i\gamma Y - i\sqrt{k^2 - \beta^2 - \gamma^2}Z\right) d\beta d\gamma$$

(5.76)

satisfies equation (5.73) at $Z > 0$ and the radiation condition at $Z \to \infty$; it becomes $U(X,Y,0)$ at $Z = 0$. Substituting (5.75) into (5.76) and changing the integration order, we obtain a formal solution to the above boundary value problem, as follows:

$$U(X,Y,Z) = \iint\limits_{D} U(\widehat{X},\widehat{Y},0) \iint\limits_{-\infty}^{\infty} \exp(-i\beta(X - \widehat{X}) - i\gamma(Y - \widehat{Y})-$$

$$i\sqrt{k^2 - \beta^2 - \gamma^2}Z)\, d\beta d\gamma d\widehat{X}\, d\widehat{Y}. \quad (5.77)$$

We assume that the field $U(X,Y,0)$ is smooth, that is, it is slightly varying over the distance of the order of the wavelength. Then the function $f(\beta,\gamma)$ is "concentrated" at the values $|\beta|, |\gamma|$, small in comparison with k. At larger values of these variables the function is small. Therefore, we can put

$$\sqrt{k^2 - (\beta^2 + \gamma^2)} = k - \frac{\beta^2 + \gamma^2}{2k} \quad (5.78)$$

in (5.76). Substituting this approximate expression into (5.77), we can keep the integration with respect to β, γ over the infinite limits. Although the expression (5.78) is not valid at large β^2, γ^2, the error in the value of the field is small, because the integrand is small at such β^2, γ^2. After substituting (5.78) into (5.77), the factor $\exp(-ikZ)$ is taken out of the integral, and the inner integrals (with respect to β and γ) are calculated in the explicit form. The expression (5.77) becomes

$$U(X,Y,Z) = \exp(-ikZ)\frac{k}{Z} \iint\limits_{D} U(\widehat{X},\widehat{Y},0)$$

$$\times \exp\left\{-\frac{ik}{2Z}\left[(X - \widehat{X})^2 + (Y - \widehat{Y})^2\right]\right\} d\widehat{X}\, d\widehat{Y}. \quad (5.79)$$

Here a constant factor is omitted again.

Introduce the function

$$\Psi_Z(X,Y) = \frac{k}{2Z}(X^2 + Y^2). \quad (5.80)$$

Then the formula (5.79) can be written as

$$U(X,Y,Z) = \exp(-ikZ)\exp\left[-i\Psi_Z(X,Y)\right]\frac{k}{Z}\iint\limits_{D} U(\widehat{X},\widehat{Y},0)$$

$$\times \exp\left[-i\Psi_Z(\widehat{X},\widehat{Y})\right]\exp\left[\frac{ik}{Z}(X\widehat{X}+Y\widehat{Y})\right]d\widehat{X}\,d\widehat{Y}. \quad (5.81)$$

This formula determines the field at any Z and, in particular, in the rectenna plane, that is, at $Z = d$, where d is the distance between the antenna and rectenna.

Further we seek for the field $U(X,Y,0)$ on the antenna from the demand for the certain functionals of $U(X,Y,d)$ to be extreme. It turns out that in many cases the field $U(X,Y,0)$ should be such that $U(X,Y,0)\exp[i\Psi_d(X,Y)]$ is real. The modulus of this product is different for different functionals, but, as a rule, its phase is zero. Hence, the function $U(X,Y,0)$ should have the phase $\Psi_d(X,Y)$. In the antenna neighborhood, the total field phase is $\Psi_d(X,Y) - kZ$. This result has a simple geometrical interpretation. The equation of the equiphase surface

$$\frac{X^2 + Y^2}{2d} - Z = const \qquad (5.82a)$$

describes a paraboloid, which approximates a part of the sphere centered at $X = 0$, $Y = 0$, $Z = d$, that is, at the rectenna center. The normals to this surface, that is, the sphere radii, intersect in its center. Consequently, the field focused on the rectenna in the geometro-optical sense should be created on the antenna.

At certain symmetry conditions, which are usually fulfilled by the field on the antenna, the reality of $U(X,Y,0)\exp[i\Psi_d(X,Y)]$ causes the reality of the integral in (5.81) at $Z = d$. In this case the field on the rectenna has the phase $-\Psi_d(X,Y) - kZ$. The equiphase surface described by the equation

$$\frac{X^2 + Y^2}{2d} + Z = const, \qquad (5.82b)$$

is a part of the sphere with the curvature, which differs from that of the surface (5.82a) only in sign. Equations (5.82a), (5.82b) describe the equiphase surfaces of the convergent and divergent waves, respectively. The curvature of the equiphase surface of the field propagating along the beam (in the Z-direction) firstly decreases to zero, then changes its sign and linearly increases in modulus. The wave transforms from convergent to divergent one (see the dotted lines in Fig. 5.15).

The effective width of the beam is changing as well. It decreases when the wave goes out from the antenna (while the equiphase surface curvature is positive), then reaches its minimal value (the curvature is zero) and then grows (the curvature is negative). The process of the effective width changing depends on the distribution of the field modulus on the antenna, its physical parameters (i. e., the size and shape of the antenna), and the beam length.

5.3.2
Parabolic equation

The field in the beam is a wave propagating along the Z-axis; the structure of the field is slow varying, that is, it varies a little at the distance of the wavelength. The approximate formula (5.79) consists of two multipliers, one of them is the fast-varying exponent $\exp(-ikZ)$, and another is the slow-varying integral divided by Z. The slow-varying multiplier depends on all three coordinates; we denote it by $W(X, Y, Z)$.

The field has a similar structure in many other cases when the plane wave propagates with a slow deformation. The deformation can be caused either by the presence of certain external objects, or (as for the beam in free space) by diffusion laws which will be specified below. These laws lead to the field smoothening, that is, to elimination of its sharp jumps. In solutions to the wave equation, this process is "concealed" by the more intensive one describing the wave propagation, that is, by the fast varying of the field phase. The processes can be investigated by the artificial technique given below.

Express the sought solution to the wave equation as a product

$$U(X, Y.Z) = \exp(-ikZ)W(X, Y, Z). \tag{5.83}$$

Substituting (5.83) into (5.73) yields the equation

$$\frac{\partial^2 W}{\partial X^2} + \frac{\partial^2 W}{\partial Y^2} + \frac{\partial^2 W}{\partial Z^2} - 2ik\frac{\partial W}{\partial Z} = 0 \tag{5.84}$$

for $W(X, Y, Z)$. Introduction of the new unknown function W instead of U does not change the type of the equation – it remains to be elliptic (as (5.73)).

Denote the spatial scale of the function W varying with respect to the transverse coordinates X, Y by $L_\perp$, and to the longitudinal one Z by $L_\parallel$. In the dimensionless coordinates $x = X/L_\perp, y = Y/L_\perp, z = Z/L_\parallel$ equation (5.84), is rewritten as

$$\frac{\partial^2 W}{\partial x^2} + \frac{\partial^2 W}{\partial y^2} + \left(\frac{L_\perp}{L_\parallel}\right)^2 \frac{\partial^2 W}{\partial z^2} - 2ik\frac{L_\perp^2}{L_\parallel}\frac{\partial W}{\partial z} = 0. \tag{5.85}$$

All the derivatives with respect to the dimensionless coordinates have the same order. Assume, that

$$L_\parallel \gg L_\perp \gg \frac{2\pi}{k}, \tag{5.86}$$

that is, the function W is very slow varying (in the scale of the wavelength) in the transverse direction and much slower varying in the longitudinal one. Assume that the three scales participating in (5.86) are related as

$$kL_\perp^2 = L_\|.\tag{5.87}$$

Then the third term in (5.85) contains the small factor $(L_\perp/L_\|)^2 = (kL_\|)^{-1}$. Dropping this term reduces (5.85) to the form

$$\frac{\partial^2 W}{\partial x^2} + \frac{\partial^2 W}{\partial y^2} - 2i\frac{\partial W}{\partial z} = 0.\tag{5.88}$$

Of course, after solving (5.88) we should verify that the assumptions (5.87), (5.86) are fulfilled.

In contrast to the exact equation (5.85), the approximate one (5.88) belongs to the parabolic type. Equation (5.85) is an elliptic one having a small coefficient at one of the higher derivatives. Dropping this derivative changes the equation type. This dropping is permissible in the domain, in which the obtained equation has the property (5.86), assumed during its derivation. The appropriate multiplier in the solution (5.79), that is, the factor at $\exp(-ikZ)$ satisfies this demand. Function (5.79) is a solution to the parabolic equation (5.88). The approximation (5.78) used when deriving the solution (5.79) is equivalent to the dropping of the small term in the differential equation (5.85).

We obtain the solution to the parabolic equation (5.88) immediately using the technique of the Green function. The Green function of this equation has the form

$$G(X, Y, Z; X_0, Y_0, Z_0)$$
$$= \frac{1}{Z - Z_0} \exp\left\{\frac{-ik}{2(Z - Z_0)}\left[(X - X_0)^2 + (Y - Y_0)^2\right]\right\}.\tag{5.89}$$

It satisfies the nonhomogeneous equation corresponding to (5.88), with the product of the three δ-functions, of $X - X_0$, $Y - Y_0$, and $Z - Z_0$, respectively, on the right-hand side. Function (5.89) describes the field created by the point source located at the point (X_0, Y_0, Z_0). The integral

$$\iint\limits_D G|_{Z_0=0} \cdot U(X_0, Y_0, Z_0)\, dX_0\, dY_0\tag{5.90}$$

is the field created in the domain $Z > 0$ by sources located on the plane $Z = 0$ with the density $U(X_0, Y_0, 0)$, that is, the solution to the boundary value problem for equation (5.88) with the condition $W(X, Y, 0) = U(X, Y, 0)$, where $U(X, Y, 0)$ is a given function. Multiplying function (5.90) by $\exp(-ikZ)$, we obtain the sought approximate solution to equation (5.73) as the product (5.81) of the fast-varying exponent and the slow-varying function W.

Note that the Green function (5.89) of the parabolic equation (5.88) can be obtained from the Green function

$$\frac{\exp(-ikR)}{R}, \qquad R = \sqrt{(X - X_0)^2 + (Y - Y_0)^2 + (Z - Z_0)^2}, \qquad (5.91)$$

of the elliptic one, after using the approximation

$$R \approx Z - Z_0 + \frac{(X - X_0)^2 + (Y - Y_0)^2}{2(Z - Z_0)}, \qquad (5.92)$$

valid at $|Z - Z_0| \gg |X - X_0|$, $|Z - Z_0| \gg |Y - Y_0|$. This approximation is valid at the distance from the antenna much larger than its size, that is, almost in the whole domain $Z > 0$. According to (5.78), the narrow beam is a composition of the plane waves propagating in the directions making the small angles with the Z-axis. According to (5.92), the narrow beam is a field concentrated in the neighborhood of the Z-axis. In fact, these two definitions are equivalent.

The parabolic equation describes the processes of the diffusion type, such as the heat conduction or the diffusion of dissolved substances in fluid. These processes are time dependent. They are intensive in the domains where the gradients of the field (e. g., of the heat) are large. In our problems, the Z-coordinate plays a role of time, whereas the heat is represented by the amplitude of the field W. The amplitude diffusion proceeds across the beam. In the antenna neighborhood, the influence of the geometrical optics laws is stronger, a "focusing" takes place and the beam first is compressed. Then the diffusion processes, namely, repulsing (as the rays are thick elastic fibers) and sprawling manifest themselves. The beam stops compressing before it comes to the focus (i. e., to the rectenna), it starts extending. This conflict of the two tendencies (the ray and wave optics) manifests itself not only in the case of the wave beam propagation in free space, but also in many other cases when the field can be described by the product of the fast- and slow-varying factors.

5.3.3
The energy transmission coefficient

In this and the next subsections, the fields $U(X, Y, 0)$ will be found, which should be created on the antenna for the field in the rectenna plane to be optimal (in a certain sense). The consideration is based on formula (5.81).

We introduce the notations simplifying the derivations. Denote a characteristic linear scale of the antenna by a, and introduce the dimensionless coordinates $x = X/a$, $y = Y/b$ in its plane. Similarly, denote the rectenna scale by α, and introduce the coordinates $\xi = X/\alpha$, $\eta = Y/\alpha$ in its plane. Denote the product of the field on the antenna and its phase, taken with the opposite sign, by $u(x, y)$:

$$u(x, y) = U(X, Y, 0) \exp\left[-i\Psi_d(X, Y)\right]. \qquad (5.93)$$

Similarly, for the rectenna,

$$v(\xi, \eta) = U(X, Y, d) \exp\left[i\Psi_d(X, Y)\right]. \tag{5.94}$$

The signs in the phase factors correspond to the curvature signs of the equiphase surfaces near the antenna and rectenna (see Fig. 5.15). For fulfilling the demands which will be imposed on $v(\xi, \eta)$ further in this subsection, the function $u(x, y)$ should be real. Then the function $v(\xi, \eta)$ will be real, either. In fact, the functions $u(x, y)$ and $v(\xi, \eta)$ are the amplitudes of the fields in the antenna ($Z = 0$) and rectenna ($Z = d$) planes, respectively.

The main formula expressing $v(\xi, \eta)$ by $u(x, y)$ has the form

$$v(\xi, \eta) = \frac{c}{2\pi}\frac{a}{\alpha} \iint_D u(x, y) \exp[ic(x\xi + y\eta)]\, dxdy, \tag{5.95}$$

where

$$c = \frac{ka\alpha}{d} \tag{5.96}$$

is the main characteristics of the beam, D is the antenna domain. The nonessential factor $i\exp(-ikd)$ is omitted in (5.95). According to (5.95), the field in the rectenna plane is the double Fourier transformation of the field on the antenna. Strictly speaking, the functions u and v are not fields: they differ from those by phase factors.

In the neighborhood of the antenna and rectenna, the field structure is close to the plane wave, and therefore, the flux of energy radiated by the antenna and falling onto the rectenna is equal to the integrals of $|u|^2$ and $|v|^2$, respectively. The *energy transmission coefficient* is the ratio of the energy falling onto the rectenna to the energy radiated by the antenna:

$$l = \frac{\alpha^2}{a^2} \frac{\iint_\Delta |v(\xi, \eta)|^2\, d\xi d\eta}{\iint_D |u(x, y)|^2\, dxdy}, \tag{5.97}$$

where Δ is the rectenna domain. The factor l is smaller than unity. The value $1 - l$ is the relative part of the energy missing the rectenna, so-called the *diffraction losses*.

To simplify the formula, we introduce the vectors $\vec{x} = (x, y)$, $\vec{\xi} = (\xi, \eta)$. Then $d\vec{x} = dx \cdot dy$, $d\vec{\xi} = d\xi \cdot d\eta$, $\vec{x}\vec{\xi} = x\xi + y\eta$. The main formula (5.95) becomes

$$v(\vec{\xi}) = \frac{a}{\alpha} \int_D u(\vec{x}) K(\vec{x}, \vec{\xi})\, d\vec{x}, \tag{5.98}$$

where

$$K(\vec{x}, \vec{\xi}) = \frac{c}{2\pi} \exp(ic\vec{x}\vec{\xi}). \tag{5.99}$$

Substituting (5.99) into (5.98) and interchanging the integration order, we obtain the expression

$$l[u(\vec{x})] = \frac{\int_D u(\vec{x})u^*(\vec{x}')K_1(\vec{x},\vec{x}')\,d\vec{x}\,d\vec{x}'}{\int_D u(\vec{x})u^*(\vec{x})\,d\vec{x}} \tag{5.100}$$

for the factor l treated as the functional of $u(\vec{x})$; the iterated kernel K_1 in (5.100) is

$$K_1(\vec{x},\vec{x}') = \int_\Lambda K(\vec{x},\vec{\xi})K^*(\vec{x}',\vec{\xi})\,d\vec{\xi}. \tag{5.101}$$

The kernel is Hermitian, that is, $K_1(\vec{x},\vec{x}') = K_1^*(\vec{x}',\vec{x})$.

Denote the function providing the maximum of the functional $l[u]$ by u_l. Substituting the function $u(\vec{x}) = u_l(\vec{x}) + v\varphi(\vec{x})$ with arbitrary $\varphi(\vec{x})$ and small real v, into (5.100) and dropping the terms of the order v^2, we reduce the functional $l[u]$ to the form $l[u] = (M + Nv)/(m + nv)$. The function u_l provides the maximum to $l[u]$ if the linear terms are not present in the expansion of $l[u]$ by the powers of v, that is,

$$l[u_l(\vec{x}) + v\varphi(\vec{x})] = l[u_l(\vec{x})] + O(v^2). \tag{5.102}$$

This property of $u_l(\vec{x})$ takes place if $M/m = N/n$. Denote this ratio by Λ. Then $\Lambda = l[u_l]$ (with the same accuracy to v^2). It can be shown that the above condition holds if $u_l(\vec{x})$ satisfies the homogeneous equation

$$\int_D K_1(\vec{x},\vec{x}')u_l(\vec{x})\,d\vec{x} = \Lambda u_l(\vec{x}'). \tag{5.103}$$

The field on the antenna providing the maximum to the energy transmission coefficient (5.97) is an eigenfunction of equation (5.103). The value of this maximum equals the first (maximal) eigenvalue of the equation. We keep the notation Λ for this value. Since the kernel K_1 is Hermitian, Λ is real, as it must be according to (5.97). If the kernel $K_1(\vec{x},\vec{x}')$ is real (as it is in many symmetrical cases), then the field $u_l(\vec{x})$ is real, either.

The eigenvalues Λ depend on both the antenna and rectenna shapes, and on the parameter c. We give the appropriate formulas for the case when the antenna and rectenna have the circular shape. Denote their radii by a and α, respectively. The dimensionless radial coordinates on the antenna and rectenna are denoted by r, $0 \le r \le 1$, and ρ, $0 \le \rho \le 1$, respectively. Assume that the field on the antenna does not depend on the angular coordinate. Then the field on the rectenna is also symmetric, and formula (5.98) becomes

$$v(\rho) = c\frac{a}{\alpha}\int_0^1 u(r)J_0(cr\rho)r\,dr. \tag{5.104}$$

The homogeneous integral equation for the eigenfunction $u_l(r)$ providing the maximum of the energy transmission coefficient (5.97) is

$$c \int\limits_0^1 u_l(r') \frac{rJ_0(cr')J_1(cr) - r'J_0(cr)J_1(cr')}{r^2 - (r')^2} r' \, dr' = \Lambda u_l(r). \qquad (5.105)$$

In Figure 5.16 the first eigenfunction $u_l(r)$ of equation (5.105) is shown for several values of the parameter c. It is normalized by the condition[1] $\int_0^1 u_l^2(r)r\,dr = 1$. The larger c, the faster this optimal field decreases when moving toward the antenna edge, and the closer the function to the Gauss one $\exp(-\beta r^2)$ with some coefficient β. Simultaneously, the maximal transmission coefficient grows. The radiation losses $1 - \Lambda$ are presented in Fig. 5.17 as a function of c. Recall that, according to (5.96), the parameter c increases when the frequency is increasing or the distance between the antenna and rectenna is decreasing. For instance, the value $c = \pi$ is reached at the wavelength $\lambda = 0.1$ m, the distance between the antenna and rectenna $d = 1000$ km, the radius of the antenna $a = 100$ m and the rectenna $\alpha = 500$ m. In this case the diffraction losses equals about 9 per cent.

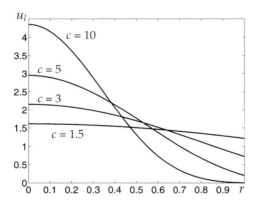

Fig. 5.16 Optimal field distribution for maximization of the transmission coefficient

5.3.4
Optimal antenna and rectenna shapes

The kernel of the transformation (5.95) involves the coordinates x, y and ξ, η in the combination $x\xi + y\eta$, invariant with respect to two simultaneous replace-

[1]In analogous Fig. 6.6 in book [18*] a different normalization was used.

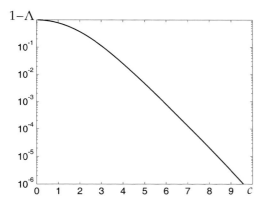

Fig. 5.17 Energy losses in the line with optimal transmission coefficient

ments of variables: x, y with $\widehat{x}, \widehat{y}$ and ξ, η with $\widehat{\xi}, \widehat{\eta}$, as follows:

$$\widehat{x} = tx, \qquad \widehat{y} = \frac{1}{t}y, \tag{5.106a}$$

$$\widehat{\xi} = \frac{1}{t}\xi, \qquad \widehat{\eta} = t\eta, \qquad t \neq 0. \tag{5.106b}$$

The invariance of the transformation (5.95) with respect to the replacements (5.106) implies that if the antenna is proportionally expanded in the x-direction and contracted it in the y-direction, then the field in the rectenna plane is contracted in the ξ-direction and expanded in the η-direction. An example of such a mapping is the known optical effect: if we contract the slot, through which a screen located in its Fresnel zone is illuminated, then the image of the slot on the screen is extended.

If the antenna and rectenna are replaced by those with the contours mapped by (5.106), then the new line is equivalent to the initial one. In particular, the areas of the antenna and rectenna do not change, because the Jacobian of both the transformations equals unity. If the field $\widehat{u}(\widehat{x}, \widehat{y})$ created on the antenna of the second line, is connected with the field on the first antenna by the formula $\widehat{u}(\widehat{x}, \widehat{y}) = u(\widehat{x}/t, t\widehat{y})$, then the fields on the rectennas are connected as $\widehat{v}(\widehat{\xi}, \widehat{\eta}) = v(t\widehat{\xi}, \widehat{\eta}/t)$. For instance, if the antenna and rectenna in the first line are circular (Fig. 5.18(a)), then they are ellipses (Fig. 5.18(b)) with the axes ratio t^2 in the second one. The large axes of the antenna and rectenna in second line are perpendicular. If the field on the first antenna is described by the function $u(x, y) = \cos(x^2 + y^2)$, then the field on the antenna in second line is $\widehat{u}(\widehat{x}, \widehat{y}) = \cos(\widehat{x}^2/t^2 + t^2\widehat{y}^2)$.

If the rectenna area is given, then the maximally possible energy transmission coefficient Λ depends on the rectenna shape. Its value is maximal if the field $v(\xi, \eta)$ is constant at all points of the rectenna contour. This fact can

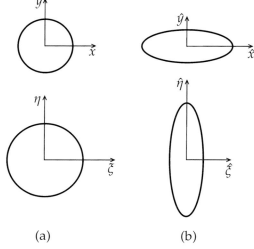

(a)   (b)

Fig. 5.18 Shapes of the radiating and receiving antennas in the equivalent transmitting lines

be proven by calculating the variation of the eigenvalue of equation (5.103); when slightly deforming the domain Δ with keeping its area. However, this result can be obtained by the following illustrative consideration. Let the field amplitude in a part of the rectenna contour be smaller than in a neighboring segment of the contour. This means that the density of energy flux, falling onto the rectenna domain neighboring to this contour segment, is smaller than that in the neighbor domain. Then the rectenna contour can be deformed with preserving its area, but contracting the domain with the small flux and enlarging that with the large one. The total flux falling onto the rectenna increases. Such a deformation is impossible only in the case when the energy flux is the same in the whole domain neighboring to the contour, and hence the field at all points of the rectenna contour is the same.

The same result also relates to the antenna. This fact can be proven in the above way applied to the "reciprocal" line, in which the antenna and rectenna are interchanged. In this line, the field must be constant on the antenna contour. The field on the antenna in the real line, must be equal to the field on the antenna in the reciprocal one. Obviously, the maximal energy transmission coefficient in both the lines is the same.

The condition for the fields on the antenna as well as for the field on the rectenna, hold in the line with circular antenna and rectenna and axially symmetrical fields on them.

Let the areas of antenna and rectenna be fixed; denote them by the same letters D and Δ as the corresponding domains were denoted above. Assume that the line described in the previous paragraph has the maximal value of Λ

among all lines having the same k, d, D, Λ. Of course, any line, antenna and rectenna of which are obtained from the circles by mapping (5.106) has the same value Λ.

The radii of the circles with areas D and Δ are equal to $a = (D/\pi)^{1/2}$, $\alpha = (\Delta/\pi)^{1/2}$, respectively. The parameter c is

$$c = \frac{k}{\pi d}\sqrt{D\Delta}. \tag{5.107}$$

If the above assumption holds, then the maximal value of the energy transmission coefficient l for a line with any shapes of the antenna and rectenna is equal to the value Λ for the circular antenna and rectenna with the parameter c given by (5.107).

5.3.5
Fields in the rectenna plane

The maximality of the energy transmission coefficient is not the only demand to the wave beam as an energy transmission line. In particular, the field distribution on the rectenna should be as uniform as possible, that is, the utilization factor of the rectenna should be maximally close to unity. Depending on the specific geographical conditions in the rectenna neighborhood and according to the ecological criteria, the field should either quickly decrease when moving away from the rectenna in its plane, or not to be concentrated near it.

We will seek for the field on the antenna, requiring that the field in the rectenna plane should be maximally close to a given field $v_0(\vec{\xi})$. In fact, only the modulus $|v_0(\vec{\xi})|$ of this function is of practical interest, its phase is nonessential. Free choice of the phase allows us to decrease the difference between the function $|v_0(\vec{\xi})|$ and modulus of the field which can be created on the rectenna. This possibility will be considered in the next subsection.

The demand for $v(\vec{\xi})$ to be equal to the given function $v_0(\vec{\xi})$ in the rectenna plane leads to the integral equation of the first kind

$$\frac{a}{\alpha}\int_D u(\vec{x})K(\vec{x}, \vec{\xi})\, d\vec{x} = v_0(\vec{\xi}), \qquad -\infty < |\vec{\xi}| < \infty, \tag{5.108}$$

for the field $u(\vec{x})$ on the antenna. This equation has a solution only for a very narrow class of functions $v_0(\vec{\xi})$: not any given field distribution in the rectenna plane can be created by the field $u(\vec{x})$ different from zero only on a finite part of the plane $Z = 0$. In any case, the solution does not exist if the function $v_0(\vec{\xi})$ is zero outside the rectenna. By this reason, we seek for the field $u(\vec{x})$ on the antenna, minimizing the functional

$$\sigma[u(\vec{x})] = \frac{\int_{-\infty}^{\infty}|v(\vec{\xi}) - v_0(\vec{\xi})|^2\, d\vec{\xi}}{\int_{-\infty}^{\infty}|v_0(\vec{\xi})|^2\, d\vec{\xi}}. \tag{5.109}$$

If equation (15.33) with given $v_0(\vec{\xi})$ has a solution, then it minimizes the functional σ and makes it zero. In other case, the minimal value of σ is positive. Denote the function, minimizing σ, by $u_\sigma(\vec{x})$, and the corresponding field in the rectenna plane, obtained from $u_\sigma(\vec{x})$ by formula (5.98), by $v_\sigma(\vec{\xi})$. The function $v_\sigma(\vec{\xi})$ is the *pseudo-solution* to equation (5.108).

To obtain the Lagrange-Euler equation for the functional σ, we substitute the field $u(\vec{x}) = u_\sigma(\vec{x}) + v\varphi(\vec{x})$ into (5.109); here $\varphi(\vec{x})$ is an arbitrary square-integrable complex function, v is a small real number. Equating the linear with respect to v term in the numerator of (5.109) to zero, we obtain

$$\nu \operatorname{Re} \int_D \varphi(\vec{x}) \int_{-\infty}^{\infty} \left[v_\sigma(\vec{\xi}) - v_0(\vec{\xi}) \right] K^*(\vec{x}, \vec{\xi}) \, d\vec{\xi} \, d\vec{x} = 0. \tag{5.110}$$

Choosing the arbitrary complex function $\varphi(\vec{x})$ as real and imaginary ones independently, we obtain that (5.110) holds only if the inner integral in it is zero. Substituting v_σ by (5.98), leads to the equation

$$\int_{-\infty}^{\infty} K^*(\vec{x}, \vec{\xi}) \int_D u_\sigma(\vec{x}') K(\vec{x}', \vec{\xi}) \, d\vec{x}' \, d\vec{\xi} = \int_{-\infty}^{\infty} v_0(\vec{\xi}) K^*(\vec{x}, \vec{\xi}) \, d\vec{\xi}. \tag{5.111}$$

In contrast to (5.108), equation (5.111) has a solution for any function $v_0(\vec{\xi})$. This solution can be easily found. It is easy to show that the integral

$$\int_{-\infty}^{\infty} K^*(\vec{x}, \vec{\xi}) K(\vec{x}', \vec{\xi}) \, d\vec{\xi} \tag{5.112}$$

is proportional to $\delta(\vec{x} - \vec{x}')$. Hence, the explicit solution to equation (5.111) is

$$u_\sigma(\vec{x}) = \int_\Delta v_0(\vec{\xi}) K^*(\vec{x}, \vec{\xi}) \, d\vec{\xi}; \tag{5.113}$$

here a constant factor is omitted .

We call the field $v_0(\vec{\xi})$ in the rectenna plane, constant on the rectenna and equal zero outside it, as the "ideal" one. Such field would provide the full utilization of the rectenna surface and the absence of the diffraction losses. Of course, it is impossible to create such a field by the antenna of the finite size. According to (5.113), the field on the antenna, creating the field closest (in the mean-square sense) to the ideal one in the rectenna plane, is

$$u_\sigma(\vec{x}) = \int_\Delta K^*(\vec{x}, \vec{\xi}) \, d\vec{\xi}. \tag{5.114}$$

For instance, for the line with circular antenna and rectenna, this field (with accuracy to a constant factor) is

$$u_\sigma(r) = \frac{J_1(cr)}{r}. \tag{5.115}$$

If the parameter c is smaller than the first zero of J_1, that is, $c < 3.83$, then the optimal field does not change its sign on the antenna. If $c > 3.83$, this field changes the sign. It turns out that in this case the field (5.115), creating the symmetrical (independent of the angular coordinate) field on the rectenna, is not optimal if the ideal field is given only by its modulus (see the next subsection). The larger the parameter c, the closest is the field in the rectenna plane to the ideal one. At $c \neq 3.83$, this field decreases when moving away from the rectenna edge (i.e., at $\rho > 1$) as $J_1(c\rho)/\rho$. At $c = 3.83$ when the field on the antenna is zero at the edge (at $r = 1$), the field in the rectenna plane decreases faster when moving away from the edge. In Chapter 6 it will be shown that this effect takes place for all large antennas - the smoother the field on the antenna decreases when approaching the edge, the stronger the far field concentrates around the main radiation direction.

Obviously, the field u_l on the antenna, providing the maximum to the energy transmission coefficient, gives the value σ larger than $\sigma[u_\sigma]$. Similarly, the function u_σ gives the value of functional l smaller than $l[u_l]$. The field, which should be created on the antenna, is determined by a compromise between the energy and ecology criteria.

5.3.6
Free phase of desired field on rectenna

The solution obtained in the preceding subsection can be improved in certain range of the parameter c, if only the modulus of the desired field v_0 on the rectenna is given, that is, if the functional to be minimized has the form

$$\sigma_1[u(\vec{x})] = \frac{\int_{-\infty}^{\infty} ||v(\vec{\xi})| - |v_0(\vec{\xi})||^2 \, d\vec{\xi}}{\int_{-\infty}^{\infty} |v_0(\vec{\xi})|^2 \, d\vec{\xi}}. \tag{5.116}$$

Applying the technique used when finding the function $u_\sigma(\vec{x})$ which minimizes the functional σ, we obtain the following nonlinear integral equation:

$$u_{\sigma_1}(\vec{x}) = \int_{-\infty}^{\infty} |v_0(\vec{\xi})| K^*(\vec{x}, \vec{\xi}) \exp\left\{i \arg\left[v_{\sigma_1}(\vec{\xi})\right]\right\} d\vec{\xi} \tag{5.117}$$

for the function $u_{\sigma_1}(\vec{x})$ minimizing the functional σ_1; here the function $v_{\sigma_1}(\vec{\xi})$ is connected with $u_{\sigma_1}(\vec{x})$ by (5.98). It is more convenient to use the nonlinear equation

$$v_{\sigma_1}(\vec{\xi}') = \int_\Delta |v_0(\vec{\xi})| K_2(\vec{\xi}, \vec{\xi}') \exp\left\{i\arg\left[v_{\sigma_1}(\vec{\xi})\right]\right\} d\vec{\xi}, \tag{5.118}$$

obtained by substituting (5.117) into the right-hand side of (5.98). Here

$$K_2(\vec{\xi}, \vec{\xi}') = \int_D K(\vec{x}, \vec{\xi}) K^*(\vec{x}, \vec{\xi}') d\vec{x} \tag{5.119}$$

is the iterated kernel, different from $K_1(\vec{x}, \vec{x})$ (5.101) in the integration domain. If equation (5.118) is solved, then the field on the antenna is calculated by (5.117). Note that these two equations are obtained from the necessary condition of the functional extremum, and they describe all its stationary points, not only minima.

We give the following simplest properties of equation (5.118):

(a) if $v_{\sigma_1}(\vec{\xi}')$ solves (5.118), then $\exp(i\alpha)v_{\sigma_1}(\vec{\xi}')$, also solves (5.118) with any real constant α;

(b) if $K_2(\vec{\xi}, \vec{\xi}')$ is real and $v_{\sigma_1}(\vec{\xi}')$ solves (5.118), then $v_{\sigma_1}^*(\vec{\xi}')$ also solves (5.118);

(c) if the function

$$v_{\sigma_1}^{(0)}(\vec{\xi}') = \int_\Delta |v_0(\vec{\xi})| K_2(\vec{\xi}, \vec{\xi}') d\vec{\xi} \tag{5.120}$$

is real and positive anywhere in Δ, then it solves (5.118). Obviously, $v_{\sigma_1}^{(0)}(\vec{\xi}')$ is the field in the rectenna plane, created by the field (5.114), if $v_0(\vec{\xi})$ is real and positive anywhere in Δ.

In the particular case, when the antenna and rectenna are circular, equation (5.118) becomes

$$v_{\sigma_1}(\rho, \psi) = c \int_0^{2\pi} \int_0^1 |v_0(\rho', \psi')| \frac{J_1(c|R - R'|)}{|R - R'|} \exp\left\{i\arg\left[v_{\sigma_1}(\rho', \psi')\right]\right\} \rho' \, d\rho' \, d\psi', \tag{5.121}$$

where $|R - R'| = [\rho^2 + (\rho')^2 - 2\rho\rho'\cos(\psi - \psi')]^{1/2}$ is the distance between two points on the rectenna. The field on the antenna, creating $v_{\sigma_1}(\rho, \psi)$ is (in accuracy to a nonessential factor)

$$u_{\sigma_1}(r, \varphi) = \int_0^{2\pi} \int_0^1 |v_0(\rho, \psi)| \exp[icr\rho\cos(\varphi - \psi)] \exp\left\{i\arg\left[v_{\sigma_1}(\rho, \psi)\right]\right\} \rho \, d\rho \, d\psi. \tag{5.122}$$

It turns out that at relatively small c, equation (5.121) have only solution (5.120) and it minimizes the functional (5.116). If $|v_0|$ does not depend on ψ, then the value c_0 such that (5.120) provides the minimum to (5.116) only at $c \leq c_0$, is the first root of the transcendental equation

$$\int_0^1 |v_0(\rho)| J_0(c_0\rho)\rho d\rho = 0. \tag{5.123}$$

In particular, if $|v_0(\rho)| = const$, then (5.123) has the form $J_1(c) = 0$ and its first positive zero is $c_0 = 3.83$. This value of c has been discussed in the preceding subsection as the value at which the optimal field is zero at the antenna edge. At $c > c_0$, function (5.115), optimal for (5.109), is not optimal for (5.116). In general, equation (5.121) should be solved by numerical methods there.

In certain range of the values $c > c_0$ the phase of the function $v_{\sigma_1}(\rho, \psi)$ is asymmetrical even if $|v_0|$ is independent of ψ: $\arg v_{\sigma_1}(\rho, -\psi) = -\arg v_{\sigma_1}(\rho, \psi)$. The corresponding field $u_{\sigma_1}(\rho, \psi)$ on the antenna is real, but nonsymmetrical. In the first approximation (valid for small $c - c_0$), $u_{\sigma_1}(\rho, \psi)$ is the field (5.113), shifted in the antenna plane.

If we are interesting only in the fields u_{σ_1} independent of φ at symmetric $|v_0(\rho)|$, then equation (5.121) becomes

$$v_{\sigma_1}(\rho) = c \int_0^1 |v_0(\rho')| \frac{\rho J_0(c\rho') J_1(c\rho) - \rho' J_0(c\rho) J_1(c\rho')}{\rho^2 - (\rho')^2} \exp\left\{i \arg\left[v_{\sigma_1}(\rho')\right]\right\} \rho' d\rho'. \tag{5.124}$$

This equation has analytical solutions having the property

$$\exp\left\{i \arg\left[v_{\sigma_1}(\rho)\right]\right\} = \frac{P_N(\rho^2)}{|P_N(\rho^2)|}, \tag{5.125}$$

where $P_N(\rho^2) = \prod_{n=1}^N (1 - \eta_{nN}\rho^{2n})$ are the polynomials of the variable ρ^2 of the degree N, having only the complex nonconjugated zeros η_{nN}^{-1}. The value of N, at which this solution is optimal in the chosen class, increases stepwise as c increases.

In Fig. 5.19, the optimal values σ_1 and σ are shown by the solid and dashed lines, respectively, for the circular antenna and rectenna at $|v_0(\rho)| = const$. In the investigated interval $c < 11$, the solutions with azimuthal symmetric u_{σ_1} are optimal for σ_1 only at $c_1 \leq c \leq c_2$ and $c \geq c_3$. In the intervals $c_0 < c < c_1$ and $c_2 < c < c_3$, the nonsymmetrical solutions (with field u_{σ_1} dependent on φ) are optimal. The values c_n are obtained from certain transcendental equations.

Comparing the curves $\sigma(c)$ and $\sigma_1(c)$ shows that the usage of the optimization criterion in the form (5.116) allows us to diminish the radius of

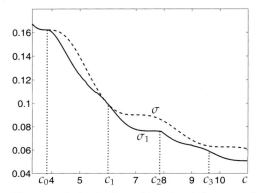

Fig. 5.19 Optimal values of functionals for two optimization criteria for the field in rectenna plane

the antenna and/or rectenna. More precisely, a value c_{opt} can be associated with any c, such that $\sigma_1(c_{opt}) = \sigma(c)$. We introduce the function $Q(c) = (c - c_{opt})/c$, as an efficiency of the free phase choice. It is seen from Fig. 5.20 that at certain values of c, the radius of the antenna or rectenna can be decreased up to 18 per cent, which corresponds to its area decreasing up to 40 per cent.

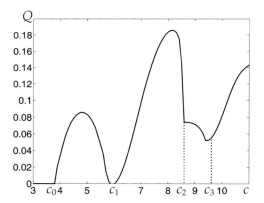

Fig. 5.20 Efficiency of the free phase choice

5.3.7
The lens line

In the first part of the wave beam, considered above, the *rays*, that is, the normals to the equiphase surface, make up a set of the convergent lines. In the second part of the beam, the curvature of this surface has the opposite sign, and the normals to it are the divergent lines. If a device is located at the beam

end, which transforms the divergent ray set into the convergent one, then the
beam can propagate forward, always constricting at first. In this way the energy can be transmitted by several subsequent beams repeating each other.

Such a *phase corrector* can be usual collecting (double convex) lens. The
lens line consists of the radiating system, a set of the lenses located in a line
(Fig. 5.21), and the receiving system. Such a line occupying relatively small
volume can transmit the energy for a long distance with small losses.

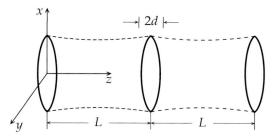

Fig. 5.21 Geometry of the lens line

In this system, the lens should be described not in terms of the geometrical optics, but in terms of the wave theory, similarly as the field in the
beam. Namely, the action of the lens should be described by a phase factor
$\exp[-i\varphi(X,Y)]$, connecting the field $U(X,Y,-d)$ behind the lens with that
$U(X,Y,d)$ in front of it, as follows:

$$U(X,Y,d) = U(X,Y,-d)\exp[-i\varphi(X,Y)]; \tag{5.126}$$

here $2d$ is the thickness of the lens. The amplitude of the field is the same on
both the sides of the lens. The function $\varphi(X,Y)$ equals the phase incursion
between the planes $Z = -d$ and $Z = +d$ along the line parallel to the Z-axis:

$$\varphi(X,Y) = k\int_{-d}^{d}\sqrt{\varepsilon(X,Y,Z)}\,dZ. \tag{5.127}$$

If the lens is double convex with the spherical surfaces of the radius b, then
$\varepsilon > 1$ at $|z| > d - (x^2 + y^2)/(2b)$ (see Fig. 5.22) and formula (5.127) gives

$$\varphi(X,Y) = -\frac{k(X^2 + Y^2)}{b}(\sqrt{\varepsilon} - 1), \tag{5.128}$$

where we drop a nonessential addend independent of the coordinates.

Formulas (5.126), (5.128) are equivalent to those of the lens theory, written
in the geometrical optics terms (*object* and its *image*). They are usually obtained from the law of the ray refraction on the lens surface. We deduce the
geometro-optical formulas from the above ones. Let a glowing point be located at the point $Z = -d_1$ of the Z-axis. The spherical wave centered at this

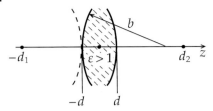

Fig. 5.22 The lens as a phase corrector

point falls onto the lens. In front of the lens (in the plane $Z = -d$) the phase of this wave equals $-k(X^2 + Y^2)/2d_1$. Behind the lens, the phase is

$$-\frac{k(X^2 + Y^2)}{2d_1} + \frac{k(X^2 + Y^2)}{b}(\sqrt{\varepsilon} - 1). \qquad (5.129)$$

The equiphase surface behind the lens is a part of the sphere centered at point $Z = d_2$ of the Z-axis, where

$$\frac{1}{d_2} = -\frac{1}{d_1} + \frac{2(\sqrt{\varepsilon} - 1)}{b}. \qquad (5.130a)$$

The normals to the equiphase surface intersect at the center of this sphere. Formula (5.130a) coincides with the formula

$$\frac{1}{d_2} + \frac{1}{d_1} = \frac{1}{F}, \qquad (5.130b)$$

describing the lens with the focus distance

$$F = \frac{b}{2(\sqrt{\varepsilon} - 1)}. \qquad (5.131)$$

This formula is obtained by consideration of the optical ray refraction with the parameters b and ε.

Therefore, the lens line can be described in terms of the wave theory. Propagating the wave beam between the lenses is described by formula (5.81), and its passing through the lens by (5.126).

If the line involves a lot of the lenses, it is expedient to introduce the notion of the eigenwaves of the lens line, analogous to those of a homogeneous line, for instance, of the dielectric waveguide. The eigenwave of the lens line is the field, the structure of which is repeated (with accuracy to a constant factor) after each lens. The dependence of the eigenwave of the lens line on the coordinates is a more complicated than in the homogeneous line. However, after passing the cell "beam+lens," the field is multiplied by a certain factor, similarly as in the homogeneous line. Denote this factor as $\exp(-ihL)$, where L is the period of the line. The "wave number" h introduced here, has roughly the same sense, as in the homogeneous line. In particular, if $h'' \neq 0$, then the energy losses per period are $\exp(-2h''L)$.

The condition of repeatability of the field of the eigenwave leads to the equation

$$U(X,Y,L) = U(X,Y,0) \exp[-i\varphi(X,Y)] \exp(-ihL), \qquad (5.132)$$

where $U(X,Y,L)$ is obtained from $U(X,Y,0)$ by formula (5.81). The equation is a homogeneous integral one, $U(X,Y,0)$ and $\exp(-ihL)$ are its eigenfunction and eigenvalue, respectively. They depend on the function $\varphi(X,Y)$, that is, on the phase correction made by the lenses, as well as on the distance L between them. In the transverse direction, the beam is limited by the lens size. The field is not completely intercepted by the lenses, and the eigenwaves propagate with losses. The eigenvalues are smaller than unity, the diffraction losses per cell are $1 - |\exp(-ihL)|^2$.

If the phase correction is large (more precisely, it lies in some range, which will be specified below) and the lens size is not small, then the fields of the several eigenwaves, in fact, do not outgo from the lens borders. The waves propagate without damping. Let the phase correction be a quadratic function of the coordinates:

$$\varphi(R) = -\frac{kR^2}{2F}, \qquad R^2 = X^2 + Y^2. \qquad (5.133)$$

Then the field of the main symmetric wave decreases in the transverse direction as $\exp[-R^2/(2R_s^2)]$, where the *effective beam radius* R_s is

$$R_s = \sqrt{\frac{L}{k}} \left(\frac{L}{F} - \frac{L^2}{4F^2} \right)^{-1/4}. \qquad (5.134)$$

It can be found from the integral equation mentioned above after formula (5.132). The ratio L/F should be large enough in order to have $R_s \ll a$, but smaller than 4 (i. e., $F > L/4$); for the wave beam can be restrained by the lens focussing. This means that the focussing should not be too strong and not too small.

The field is maximally pressed down to the line axis when $F = L/2$; then $R_s = (L/k)^{1/2}$. Such a lens line is called *confocal*. The focuses of the lenses are located at the half-distance between the lenses. Such line has minimal losses per period among the lines with the same value of ka^2/L, where a is the lens radius.

The phase correction can be proceeded not only by the lenses but also by a plano-parallel dielectric plate made of a nonhomogeneous dielectric with the permittivity depending on R and decreasing toward the periphery. Its phase correction can be calculated by formula (5.127). Similar action is proper for a periscope system consisting of two bent mirrors.

6
Backgrounds of Antenna Theory

6.1
Radiation of current set

6.1.1
The elementary electric dipole and multipole

A straight line segment with the length much smaller than the wavelength, having a current distributed along it, is the simplest radiator. It is called an *electric dipole*. The product of the segment length l_1 by the linear current density J is proportional to the quantity $\mathcal{P}$ called the *dipole momentum*; $l_1 J = i\omega\mathcal{P}$. The *elementary electric dipole* is a limiting case of the electric dipole at $l_1 \to 0$, $J \to \infty$, in which $\mathcal{P}$ remains to be constant. Field created by the elementary dipole, satisfies the Maxwell equation system (1.29) with the current having only one nonzero Cartesian coordinate

$$j_z^{\text{ext}} = i\omega\mathcal{P}\delta\left(|\vec{r} - \vec{r}_0|\right). \tag{6.1}$$

The z-axis is directed along the line on which the current flows, $\vec{r}_0$ describes the dipole location on this axis. The Dirac function δ is normalized by the condition $\int \delta dV = 1$, where dV is a volume element.

The field created by such current has already been introduced as the Green function of the Maxwell equations. In this subsection we consider it in detail. It is assumed that the dipole is located in free space and its field satisfies the radiation condition. Many properties of this field are inherent to fields radiated by an antenna with smooth current distribution, size of which is small in comparison with the wavelength.

In order to determine the field created by a certain current, first the electric Hertz vector $\vec{\Pi}^{(e)}(\vec{r})$ should be found from equation (1.42) with the radiation condition, and then the field components should be calculated by (1.43). Since the current (6.1) has the only Cartesian component j_z, the vector $\vec{\Pi}^{(e)}$ has also the only component $\Pi_z^{(e)}$. It satisfies the equation

$$\Delta\Pi_z^{(e)} + k^2\Pi_z^{(e)} = \frac{4\pi i}{\omega} j_z^{\text{ext}}. \tag{6.2}$$

High-frequency Electrodynamics. Boris Z. Katsenelenbaum
Copyright © 2006 WILEY-VCH Verlag GmbH & Co. KGaA, Weinheim
ISBN: 3-527-40529-1

We introduce the spherical coordinate system $\{R, \vartheta, \varphi\}$ having the axis coinciding with the z-axis of the Cartesian system, and originating at the point where the dipole is located. Since j_z does not depend on the angles ϑ, φ, the function $\Pi_z^{(e)}$ does not depend on them as well. Equation (6.2) becomes

$$\frac{1}{R^2}\frac{d}{dR}\left(R^2\frac{d\Pi_z^{(e)}}{dR}\right) + k^2\Pi_z^{(e)} = -4\pi\mathcal{P}\delta(R).$$ (6.3)

Its solution, satisfying the radiation condition, is

$$\Pi_z^{(e)}(R) = \mathcal{P}\frac{\exp(-ikR)}{R}.$$ (6.4)

Indeed, substituting this function into the left-hand side of (6.3) gives zero at $R \neq 0$. To calculate the coefficient in (6.4), equation (6.4) should be integrated over the volume of sphere with center at the origin. Using the definition of the δ-function, we found that the integral on the right-hand side equals $4\pi\mathcal{P}$. The first term on the left-hand side is div grad $\Pi_z^{(e)}$, and the volume integral taken of it over the sphere equals the integral of $d\Pi_z^{(e)}/dR$ over the sphere surface. At $ka \to 0$, where a is the sphere radius, this integral is $-4\pi\mathcal{P}$. The integral of the second term tends to zero at $ka \to 0$, and, thus, equation (6.3) is satisfied.

The vector $\vec{\Pi}^{(e)}$ has the two components $\Pi_R^{(e)} = \Pi_z^{(e)}\cos\vartheta$ and $\Pi_\vartheta^{(e)} = -\Pi_z^{(e)}\sin\vartheta$ in spherical coordinates. Three components E_φ, H_R, and H_ϑ of the field of the elementary electric dipole are zero, whereas the other three are

$$E_\vartheta = -\frac{k^3\mathcal{P}}{c}\left(1 + \frac{i}{kR} - \frac{1}{k^2R^2}\right)\sin\vartheta\frac{\exp(-ikR)}{kR},$$

$$E_R = \frac{ik^3\mathcal{P}}{c}\left(1 - \frac{i}{kR}\right)\cdot 2\cos\vartheta\frac{\exp(-ikR)}{kR},$$ (6.5)

$$H_\varphi = -\frac{k^3\mathcal{P}}{c}\left(1 - \frac{i}{kR}\right)\sin\vartheta\frac{\exp(-ikR)}{kR}.$$

In the near field zone, that is, at $kR \ll 1$, the fields in the higher order of $1/kR$ are

$$E_\vartheta = \mathcal{P}\frac{\sin\vartheta}{R^3}, \quad E_R = \mathcal{P}\frac{2\cos\vartheta}{R^3}, \quad H_\varphi = \mathcal{P}\frac{k\sin\vartheta}{R^2}.$$ (6.6)

In this zone, the magnetic component of the field is kR times smaller than the electric ones.

The electric field (6.6) coincides with the electrostatic one created by two electric charges equal in magnitude and opposite in sign, which are located on the z-axis close to each other. Denote the charge magnitude by Q and the distance between them by l_1. At the limit $Q \to \infty$, $l_1 \to 0$, such that the product Ql_1 is constant, the elementary electrostatic dipole appears. Each charge creates the Coulomb field $\vec{E}^{coul}$ having only the radial component $E_R^{coul} = Q/R^2$. The elementary electrostatic dipole creates the field

$$\vec{E}^{dip} = \frac{\partial \vec{E}^{coul}}{\partial \dot{z}_0} l_1, \tag{6.7}$$

where z_0 is the z-coordinate of point at which the charge is located. The co-ordinates R, ϑ of any point are changed by values $\partial R/\partial z_0 \cdot l_1$ and $\partial \vartheta/\partial z_0 \cdot l_1$, respectively, if the coordinate origin is shifted by l_1. They are expressed by z_0 as $R^2 = (z - z_0)^2 + x^2 + y^2$ and $\tan \vartheta = \sqrt{x^2 + y^2}/(z - z_0)$, respectively. Consequently,

$$\frac{\partial R}{\partial z_0} = -\cos \vartheta, \quad \frac{\partial \vartheta}{\partial z_0} = \frac{\sin \vartheta}{R}. \tag{6.8}$$

Applying (6.7) to the Cartesian coordinates yields

$$E_z^{dip} = \frac{\partial E_z^{coul}}{\partial z_0} l_1, \quad E_{x,y}^{dip} = \frac{\partial E_{x,y}^{coul}}{\partial z_0} l_1. \tag{6.9}$$

Expressing $\vec{E}^{coul}$ by E_R^{coul}, we obtain

$$E_R^{dip} = \frac{Ql_1}{R^3} \cdot 2 \cos \vartheta, \quad E_\vartheta^{dip} = \frac{Ql_1}{R^3} \sin \vartheta. \tag{6.10}$$

This field coincides with (6.6), since the current J in the elementary electric dipole and the charge Q in the elementary electrostatic one are connected by the relation $J = i\omega Q$ following from (1.18).

In the far field zone, at $kR \gg 1$, the fields are

$$E_\vartheta = -k^3 \mathcal{P} \frac{\exp(-ikR)}{kR} \sin \vartheta, \quad H_\varphi = -k^3 \mathcal{P} \frac{\exp(-ikR)}{kR} \sin \vartheta, \tag{6.11a}$$

$$E_R = ik^3 \mathcal{P} \frac{\exp(-ikR)}{(kR)^2} \cdot 2 \cos \vartheta. \tag{6.11b}$$

According to the radiation condition, the components E_ϑ and H_ϑ are the same and depend on R as $\exp(-ikR)/(kR)$. The component E_R decreases faster, namely, as $1/(kR)^2$, with R increasing (see (1.35)). The dipole does not radiate in the direction of its axis ($\vartheta = 0$, $\vartheta = \pi$), its maximal radiation is concentrated in the equatorial plane ($\vartheta = \pi/2$).

The energy carried out by the radiation is equal to the integral of the Poynting vector flux (1.37), taken over the surface surrounding the dipole. This integral does not depend on the surface shape. It is simply calculated as an integral of the flux over the sphere with radius so large that the field has the form (6.11) on its surface. The energy flux is

$$S^{\text{rad}} = \frac{c}{8\pi} k^2 \mathcal{P}^2 \int\limits_0^{2\pi} \int\limits_0^{\pi} \sin^3 \vartheta \, d\vartheta d\varphi. \tag{6.12}$$

Substituting $\mathcal{P} = -i l_1 J / \omega$, we obtain

$$S^{\text{rad}} = \frac{1}{c} \frac{k^2 l_1^2}{3} J^2. \tag{6.13}$$

For describing the radiation capacity of any configuration of the currents ("antenna"), the *radiation resistance* $\mathcal{R}$ is introduced, defined by the formula

$$S^{\text{rad}} = \frac{1}{2} \mathcal{R} J^2. \tag{6.14}$$

The larger the value $\mathcal{R}$, the more efficient the radiation.

For the elementary electric dipole, $\mathcal{R}^{\text{dip}} = 2(kl_1)^2/(3c)$. To express the resistance in ohms, one should replace the multiplier $1/c$ with the number 30 (see text below formula (3.40)). Hence,

$$\mathcal{R}^{\text{dip}} = 20 \, (kl_1)^2 \, \text{ohms}. \tag{6.15}$$

System of two close dipoles located on one line, being equal in magnitude but oppositely directed, makes up the so-called *electric quadrupole*. If the momenta of both the dipoles increase and the distance l_2 between them decreases such that the product $\mathcal{P}l_2$ remains finite and constant, then at the limit this radiator becomes the *elementary electric quadrupole*. Its field is expressible in terms of the dipole field by the following formula:

$$\vec{E}^{\text{quad}} = \frac{\partial \vec{E}^{\text{dip}}}{\partial z_0} l_2, \tag{6.16}$$

similar to (6.7).

At $kR \to 0$, that is, in the near field, the quadrupole field has a singularity of the higher order than the dipole one has.

We find the quadrupole field in the far field zone keeping only the term which decreases as $1/kR$. Only the first formula in (6.8) should be used. When differentiating the dipole field components given by (6.11) with respect to R in the far field zone, we should differentiate the exponent only. The dropped terms decrease as $1/R^2$ or faster. With accuracy to the terms of the order $1/R$, the quadrupole field in the far field zone is

$$E_\vartheta^{\text{quad}} = H_\varphi^{\text{quad}} = \frac{k^2 J l_1 l_2}{c} \sin \vartheta \cos \vartheta \frac{\exp(-ikR)}{R}. \tag{6.17}$$

The quadrupole radiation is zero both in the $\pm z$-directions ($\vartheta = 0, \pi$) and in the equatorial plane ($\vartheta = \pi/2$). It has the opposite signs on both the sides of this plane and reaches its maximum at $\vartheta = \pi/4$ and $\vartheta = 3\pi/4$.

The radiation resistance of the elementary electric quadrupole, defined by (6.14), is

$$\mathcal{R}^{\text{quad}} = 4 \left(k^2 l_1 l_2 \right)^2 \text{ohms}. \tag{6.18}$$

The quadrupole radiates less intensive than the dipole does: its radiation resistance is proportional to the fourth, not to the second power of the frequency.

Following the above scheme, other elementary sources with the same polarization, called the *electric multipoles of the higher order*, can be constructed. In the far field zone, their fields are proportional to the higher orders of the frequency.

6.1.2
Field of arbitrary currents

The field of arbitrary currents located in a finite domain is a sum or an integral of the fields created by the elementary electric dipoles. This fact follows from the linearity of equation (1.42) connecting $\vec{\Pi}^{(e)}(\vec{r})$ with $\vec{j}(\vec{r})$ and from the fact that any vector $\vec{j}(\vec{r})$ is expressible in the form of the integral $\int \vec{j}(\vec{r}')\delta(|\vec{r} - \vec{r}'|)\, dV_{\vec{r}'}$ taken over the domain in which $\vec{j}(\vec{r}) \neq 0$, that is, the integral of the elementary dipoles. According to (6.4), the electric Hertz vector of the fields created by the currents $\vec{j}(\vec{r})$ equals

$$\vec{\Pi}^{(e)}(\vec{r}) = -\frac{i}{\omega} \int \vec{j}(\vec{r}') \frac{\exp\left(-ik\,|\vec{r} - \vec{r}'|\right)}{|\vec{r} - \vec{r}'|} \, dV_{\vec{r}'}, \tag{6.19}$$

where the integral is taken over the whole domain in which $\vec{j}(\vec{r}) \neq 0$.

Consider the field in the far field zone. For the currents located in a finite domain, the definition $kR \gg 1$ is not sufficient; it will be specified further. Only the terms which decrease as $1/R$ with R increasing are kept in the fields whereas those decreasing faster are dropped. The kept terms make up the directivity pattern (1.32). It can be shown that the two functions introduced in (1.32) completely determine the field in the whole space, or, more precisely, in the domain which may be reached from the infinite point without intersecting the surface, on which the field is discontinuous. The one-to-one connection of the terms decreasing as $1/R$ with the field in the whole space, implies that if the field does not contain such terms, that is, it decreases faster, then it equals zero everywhere.

We explain the distance, involved into (6.19), between the point $\vec{r}$ at which the integral is calculated (observation point) and the integration point $\vec{r}'$. Introduce the spherical coordinate system originating in the domain occupied by the currents. Denote the coordinates of the observation and integration points as (R, ϑ, φ) and $(R', \vartheta', \varphi')$, respectively. In this notation

$$\left|\vec{r} - \vec{r}'\right| = R\left(1 - \frac{2R'}{R}\cos\gamma + \frac{(R')^2}{(R)^2}\right)^{1/2}. \tag{6.20}$$

Here γ is the angle between $\vec{r}$ and $\vec{r}'$,

$$\cos\gamma = \cos\vartheta\cos\vartheta' + \sin\vartheta\sin\vartheta'\cos\left(\varphi - \varphi'\right). \tag{6.21}$$

Denote the antenna size by D so that $R' \leq D$. If $R \gg D$, that is, the distance from the observation point to the antenna is large in comparison with the antenna size, then $\left|\vec{r} - \vec{r}'\right|$ can be replaced by R in the denominator of (6.19). In the exponent,

$$k\left|\vec{r} - \vec{r}'\right| = kR - kR'\cos\gamma + O\left[\frac{k(R')^2}{R}\right]. \tag{6.22}$$

The last term can be dropped at

$$R \gg kD^2. \tag{6.23}$$

This condition together with $R \gg D$ and $kR \gg 1$ defines the far field zone. If the antenna size is much larger than the wavelength ($kR \gg 1$), then according to (6.23), the far field zone begins in the distances much larger than D. In the far field zone, we approximately write

$$k\left|\vec{r} - \vec{r}'\right| = kR - kR'\cos\gamma. \tag{6.24}$$

Equation (6.19) becomes

$$\vec{\Pi} = -\frac{i}{\omega}\frac{\exp\left(-ikR\right)}{R}\vec{N}, \tag{6.25}$$

where the vector $\vec{N}$ is defined as

$$\vec{N}\left(\vartheta, \varphi\right) = \int \vec{j}\left(\vec{r}'\right)\exp\left(ikR'\cos\gamma\right)dV_{\vec{r}'}. \tag{6.26}$$

Here the integration domain is the same as in (6.19). Formula (6.24) has a simple geometric interpretation. The lines OA and $A'A$ (Fig. 6.1) connecting different points on the antenna with a point in the far field zone may be treated as parallel ones. According to (6.26), calculation of the field of a source system is reduced to the summation of the fields of elementary sources with accounting the current directions in them and the phase difference on the piece OC of the spherical waves coming from different sources. The phase incursions along the lines CA and $A'A$ are the same.

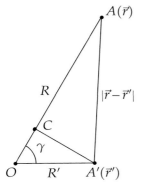

Fig. 6.1 The phase difference of the fields of two sources

Similarly as for the elementary dipole, only the terms of the order $1/R$ should be kept when calculating fields in the far field zone by formulas (1.43). In this case, only the exponent should be differentiated and the ϑ- and φ-derivatives may be dropped. The components E_R and H_R are small (proportional to $1/R^2$), the others are expressed by the components of the vector $\vec{N}$ as

$$E_\vartheta = -H_\varphi = \frac{ik}{c} N_\vartheta \frac{\exp(-ikR)}{R}, \tag{6.27a}$$

$$E_\varphi = H_\vartheta = -\frac{ik}{c} N_\varphi \frac{\exp(-ikR)}{R}. \tag{6.27b}$$

In the far field zone the field of the currents arbitrarily distributed in a certain domain is represented by a spherical wave. The directivity pattern is determined by the vector $\vec{N}$. Location of the coordinate origin influences the pattern by a simple phase multiplier: if in the two-dimensional problem the origin is displaced at the distance b in the direction making the angle β with the x-axis (Fig. 6.2), then the pattern is multiplied by $\exp[ikb\cos(\varphi - \beta)]$.

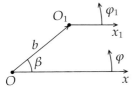

Fig. 6.2 Shift of the coordinate system

In the case of the elementary dipole we have $N = N_z = 1$, so that $N_\vartheta = -\sin\vartheta$ and (6.27a) gives the expression coincident with (6.11a). For the arbitrary antenna, N_ϑ, N_φ are the certain complex functions of the angles ϑ, φ. The total radiated power equals

$$S^{\text{rad}} = \frac{k^2}{8\pi c} \int \left| \vec{N} \right|^2 d\Omega, \quad d\Omega = \sin\vartheta \, d\vartheta d\varphi. \tag{6.28}$$

The integral is taken over the whole solid angle.

6.1.3
The elementary magnetic dipole and multipole

Another example of the simplest radiators is a plane ring with a radius much smaller than the wavelength, and azimuthal current flowing on it, which does not depend on the angle. If the ring radius a tends to zero whereas the linear current density J infinitely grows so that the quantity $\pi a^2 J$, that is, the product of the ring area by the current, remains finite and constant, then such a radiator creates the field similar to that created by the magnetic dipole with the magnetic current

$$j^{(m)} = ik\pi a^2 J. \tag{6.29}$$

The magnetic dipole has been already defined by equations (3.116) and (3.117) as a virtual quantity introduced into the Maxwell equations for solving certain boundary problem. The formulas describing the field created by the magnetic dipole can be obtained from those for the field created by the electric one with the same momentum, by substituting $\vec{E} \to \vec{H}$, $\vec{H} \to -\vec{E}$. This follows from the fact that the system of Maxwell equations (1.29) transforms into the artificial system (3.117) after such a substitution. In the far field zone, the field of the magnetic dipole differs from that for the electric one, given by (6.11), in substituting $E_\theta \to H_\varphi$, $H_\varphi \to -E_\theta$.

We show that this field is the field of a current on a small ring, that is, the magnetic dipole is realized as a small ring. Let a ring of the radius a lie in the plane (x, y). The current components are $J_x = -J\cos\varphi'$, $J_y = J\sin\varphi'$, the angle γ is defined by the relation $\cos\gamma = \sin\vartheta'\cos(\varphi - \varphi')$. Substituting these values into (6.26) and assuming $ka \ll 1$ we obtain

$$N_x = -j^{(m)}\sin\vartheta\cos\varphi, \quad N_y = j^{(m)}\sin\vartheta\sin\varphi. \tag{6.30}$$

Consequently, the vector $\vec{N}$ has only the spherical component $N_\varphi = j^{(m)}\sin\vartheta$. It follows from (6.25) that in the far field zone, the field of the small ring contains two components H_ϑ and E_φ having the form described above.

The field of the ring is an example of partial compensation of the fields created by different current segments. The fields of the current segments oppositely located at the ends of any diameter are inversely directed at any point. They cancel each other not completely, due to the different distances from this point to the current segments. For the points lying on the z-axis these distances are the same and the field equals zero at $\vartheta = 0, \pi$. The distance difference is the largest for the points lying in the ring plane; the field is maximal at $\vartheta = \pi/2$.

The field of the elementary magnetic dipole is proportional to the squared frequency. It can be easily shown that the radiation resistance is

$$\mathcal{R} = 10\pi^4 \, (ka)^4 \, \text{ohm}. \tag{6.31}$$

It is proportional to the fourth power of frequency. The magnetic dipole is similar to the electric quadrupole (not to the dipole). The field of quadrupole (6.17) is also proportional to the square of frequency and, analogously, its radiation resistance is proportional to the fourth power of the frequency. The fields of circular current segments cancel each other not completely, nearly in the same way as the fields of two inversely directed electric dipoles, that is, of the electric quadrupole does.

Fields of elementary radiators of the magnetic type, the so-called *magnetic multipoles*, can be created by the same scheme based on formula (6.16) as the fields of the electric multipoles.

It can be shown that in the far field zone, the field of any current system is expressible as a sum of fields of multipoles located at the coordinate origin. However, in this case the multipoles creating the fields, dependent not only on ϑ but also on the azimuthal angle φ, should be introduced.

6.1.4
The half-wave vibrator

The field created by a given current is found from (6.19), (1.43). This scheme of finding the field by current remains also valid in the case when the current is not given but when it is induced on the metallic surface of a certain body by a given ("extrinsic") field. In this case the most complicated part of the problem is finding the induced current. In the above problems on the excitation of waveguides and resonators (see Sections 3.4, 4.1), first the field was found. The current induced on the walls can be calculated as a limiting value of the tangential component of the magnetic field. In this subsection, another technique is applied. First, the induced current is found and then the field created by it (considered as the *diffracted field*) is calculated by the same formulas (6.19), (1.43).

The current induced on the metallic surface of a body is found from the condition that the field created by it (in absence of the body) and denoted as $\vec{E}^{\text{diff}}$, being summed with the incident field $\vec{E}^{\text{inc}}$ (which would be on the surface in absence of the body), makes up the field, tangential components of which are zero on this surface

$$E_{t_i}^{\text{inc}} + E_{t_i}^{\text{diff}} = 0, \quad i = 1, 2. \tag{6.32}$$

Here t_1 and t_2 are the two tangential directions (see (1.79)). This condition leads to an integral equation for the current flowing on the body surface. The right-hand side of this equation is a given field E_t^{inc}.

When the current is found, the diffracted field can be calculated in the whole space, in particular, the radiation pattern can be found. This is a general scheme of solving the diffraction problem on the metallic body. It will be used in the next chapter.

We apply the above scheme to the problem about radiator in the form of a thin metallic cylinder. We explain the quantities involved in equation (6.19) for this case. The integration is made over the cylinder surface. Denote the cylindrical coordinates of points on this surface by ξ, φ. Then the surface element is $a d\xi d\varphi$, where a is the cylinder radius. At $ka \ll 1$ the surface current density may be treated as a function independent of φ. The function $\vec{j}(\vec{r})$ in (6.19) and the linear current $J(\xi)$ are connected by the formula $j_z(\vec{r}) dV = J(\xi) d\xi d\varphi/(2\pi)$. The Hertz potential should be calculated at the points on the cylinder surface, at which condition (6.32) is imposed. At $a \ll L$, where $2L$ is a vibrator length, the z-component of $\vec{\Pi}^{(e)}$ is much larger than others. It equals

$$\Pi_z^{(e)}(z) = -\frac{i}{2\pi\omega} \int_{-L}^{L} J(\xi) \int_0^{2\pi} \frac{\exp(-ik\rho)}{\rho} d\varphi d\xi, \qquad (6.33)$$

where $\rho = |\vec{r} - \vec{r}'|$ is the distance between the integration point and the point at which $\Pi_z^{(e)}(z)$ is calculated. The coordinates of these points are $(z, a, 0)$ and (ξ, a, φ), respectively, so that

$$\rho = \sqrt{(z-\xi)^2 + a^2(1-\cos\varphi)}. \qquad (6.34)$$

At $a = 0$, the inner integral in (6.33) is divergent because in this case the multiplier $1/\rho$ has a nonintegrable singularity at $\xi = z$. At $a \neq 0$, this multiplier is large in the neighborhood of $\xi = z$. At $a/L \ll 1$ the entire integral being a function of z, is approximately proportional to the value of $J(\xi)$ at $\xi = z$, that is, $\Pi_z^{(e)}(z)$ is approximately proportional to $J(z)$ and it mainly depends not on the value of the current on the whole vibrator, but only on $J(z)$. Estimation of integral (6.33) at $a \to 0$ shows that the proportionality coefficient is $(-2i/\omega)\ln(L/a)$. Extracting the main term, we obtain

$$\Pi_z^{(e)}(z) = -\frac{2i}{\omega}\ln\frac{L}{a}J(z) + U[J(z)]. \qquad (6.35)$$

Here U denotes the operator determining the nonlocal field created by the current flowing in the area not adjacent to point $\xi = z$. This operator can be written in the explicit form by comparing formulas (6.33) and (6.35). It contains the integral of the function $[\exp(-ik\rho)-1]/\rho$ having no singularity at $\rho = 0$ and, therefore, we can put $a = 0$, that is, $\rho = |z-\xi|$ in the operator.

The above operator depends not only on the vibrator length but also on its shape. It is different for a prolate ellipsoid of revolution and for a circular cylinder of the same length and strongly depends on the vibrator length. The function $J(z)$ participates in this operator linearly.

In the higher order of the small parameter a/L, the field $\vec{E}$ created by the current $J(z)$ on the cylinder surface contains only the z-component

$$E_z = \left(\frac{d^2}{dz^2} + k^2\right) \Pi_z^{(e)}(z).$$ (6.36)

According to (6.32), the sought current $J(z)$ induced on a thin metallic vibrator located in the field $\vec{E}^{inc}$, satisfies the one-dimensional integro-differential equation

$$\frac{d^2 J}{dz^2} + k^2 J = -i\omega\chi \left\{ E_z^{inc} + \left(\frac{d^2}{dz^2} + k^2\right) U\left[J(z)\right] \right\},$$ (6.37)

where

$$\chi = \frac{1}{2\ln(L/a)}.$$ (6.38)

The current must vanish at the vibrator ends, that is, the function $J(z)$ should fulfill the ending condition

$$J(\pm L) = 0.$$ (6.39)

The approximate explicit solution to the problem (6.37), (6.39) can be obtained only for very thin vibrators, such that $\chi \ll 1$. In this case the sought function can be expressed by powers of the parameter χ. We represent this function in the form of the series

$$J(z) = J_0(z) + \chi J_1(z) + \chi^2 J_2(z) + \cdots$$ (6.40)

and equate the terms with the same powers of χ on both the sides of equation (6.37). The demand (6.39) relates to each term in (6.40).

We outline this solution keeping in mind that it is qualitatively also valid for the vibrators which fulfill not the condition $\chi \ll 1$, but a much weaker one $ka \ll 1$.

Substituting series (6.40) into (6.37) and (6.39), we obtain an infinite system of ordinary differential equations. Together with the ending conditions these equations make up an iterative set of the boundary value problems. The first two of them are

$$\frac{d^2 J_0}{dz^2} + k^2 J_0 = 0, \quad J_0(\pm L) = 0,$$ (6.41a)

$$\frac{d^2 J_1}{dz^2} + k^2 J_1 = -i\omega\left[E_z^{inc} + B(z)\right], \quad J_1(\pm L) = 0,$$ (6.41b)

where $B(z) = (d^2/dz^2 + k^2) \, U \, [J_0 \, (z)]$. The right-hand sides of all the equations are known functions found at the preceding step of the iterative process.

For the half-wave dipole ($2L = \lambda/2$), that is, at

$$kL = \pi/2, \tag{6.42}$$

problem (6.41a) has the solution

$$J_0 \, (z) = A \cos \left(\frac{\pi z}{2L} \right). \tag{6.43}$$

The current distribution along the vibrator is a cosine one; it does not depend on the excitation field $\vec{E}^{\mathrm{inc}}$. The amplitude A is found from the demand of problem (6.41b) solvability. Since the homogeneous system (6.41a) with the same left-hand side has a nonzero solution, then the corresponding nonhomogeneous system is solvable only if the right-hand side of the first equation in (6.41b) is orthogonal to solution (6.43) of the homogeneous problem. This demand leads to the equation

$$\int_{-L}^{L} \cos \frac{\pi z}{2L} B \, (z) \, dz = - \int_{-L}^{L} E_z^{\mathrm{inc}} \, (z) \cos \frac{\pi z}{2L} \, dz. \tag{6.44}$$

Since the operator $B(z)$ contains the current $J_0(z)$ in a linear form, the left-hand side of (6.44) is proportional to the factor A in (6.43), so that

$$A = \frac{1}{Z} \int_{-L}^{L} E_z^{\mathrm{inc}} \, (z) \cos \frac{\pi z}{2L} \, dz. \tag{6.45}$$

Here Z (similarly as the operator $U[J(z)]$) characterizes the vibrator and depends only on its shape and the precise length. The factor Z has the same dimension as the resistance. It is called the *entrance resistance* of the vibrator. It depends on the operator U and can be calculated from (6.44).

For the vibrator of not a half-wave length, the homogeneous problem (6.41a) has only the zero solution $J_0(z) = 0$. In this case the first equation in (6.41b) becomes

$$\frac{d^2 J_1}{dz^2} + k^2 J_1 = -i\omega E_z^{\mathrm{inc}}. \tag{6.46}$$

This equation supplied by the ending condition $J_1(\pm L) = 0$ is always solvable. Its solution can be easily found in the explicit form. In such a vibrator the current has the order of χ and its distribution depends on the function $E_z^{\mathrm{inc}}(z)$.

System (6.41) also allows us to investigate the intermediate case when the vibrator length is such that the difference $|kL - \pi/2|$ is small in comparison with χ. Then the current is almost the same as in the case of the vibrator for which condition (6.42) holds, that is, the current distribution is similar to (6.43). If the difference is much larger than χ, then the current is small and has the order χ.

In the case of the half-wave vibrator when the current is distributed by (6.43), the far field can be easily obtained as

$$E_\vartheta = -H_\varphi = \frac{2i}{c} A \frac{\cos\left((\pi/2)\cos\vartheta\right)}{\sin\vartheta} \frac{\exp\left(-ikR\right)}{R}. \tag{6.47}$$

The radiation pattern of the half-wave vibrator is a little narrower than that of the elementary dipole (6.11).

The notion of the radiation resistance defined by formula (6.14) can also be introduced for the half-wave vibrator. In this formula, J is equal to the value of the current at the vibrator center, that is, to the factor A. After substituting $1/c = 30$ ohms, the calculation gives $\mathcal{R} = 73$ ohms (see, e. g., (6.15) and the text before it).

Note that $\mathcal{R}$ calculated by (6.15) with $l_1 = \pi/k$, is 2.7 times larger than the above one. This is explained not only by the fact that (6.15) is valid only at $kl_1 \ll 1$ but also that the factor A in (6.43) is a maximal (not an average) value of the current. If we use the ratio of the radiated energy and the average square of the current, then at $kl_1 = \pi$, formula (6.15) gives the value of $\mathcal{R}$ erroneous only by one third. This formula can also be used for the rough estimation of the radiation resistance in the case of the vibrators with a finite length.

The radiation resistance $\mathcal{R}$ of the vibrator and its entrance resistance Z are related as $\mathcal{R} = \mathrm{Re}\, Z$, so that Z can be presented in the form

$$Z = \mathcal{R} + iY. \tag{6.48}$$

This formula follows from (1.56). Integrating both sides of (1.56) over a sphere of a large radius, encircling the vibrator, we found that the energy flux radiated by the vibrator is

$$\frac{1}{2}\mathcal{R}|A|^2 = \frac{1}{2}\mathrm{Re}\int_{-L}^{L} J(z)E_z^{\mathrm{inc}}\,dz. \tag{6.49}$$

For the current in the form of (6.43), the integral in (6.49) is the same as that participating in (6.44), which gives (6.48).

The real quantity Y depends both on the vibrator shape and its length. It can be calculated by (6.44). For instance, $Y = 0$ for a prolate ellipsoid of revolution with the larger axis equal to $\lambda/2$. For a cylinder, Y is zero if the cylinder length is a little smaller than $\lambda/2$. According to (6.16), in the vibrator for which $Y = 0$, the current is maximal (*tuned-up*) at any given E_z^{inc}. In the similar problem of the elasticity theory, the tuned-up string must have the length $\lambda/2$: the tuning does not require any shortening. The distinction of the three-dimensional problem on the vibrator from the one-dimensional one on string, manifests itself in the above fact.

6.1.5
The multivibrator antenna

The main result of the tuned-up vibrator theory is given by formulas (6.43) and (6.44). It is sufficient for constructing the theory of a system consisting of several tuned-up vibrators. Such a theory should account the mutual influence of the vibrators. An example of such antenna is the phased-array antenna in which the excitation is provided to all vibrators by feeders, and the current in each of them depends on the field on it. However, in this antenna, the mutual influence of elements should be accounted as well.

Let a system consist of M tuned-up vibrators. Denote the current amplitude in the mth vibrator by $A_m (m = 1, 2, \dots, M)$. The field acting on the mth vibrator and being the incident one for it is made up by the sum of fields of all other vibrators and the field $\vec{E}_m^{\text{inc}}$ which would act on this vibrator in absence of all others. For the receiving multivibrator antenna, $\vec{E}_z^{\text{inc}}$ is a field which would exist at the place of the mth vibrator if the antenna were absent. For the transmitting one, it is a field created by the feeder.

The field acting on the mth vibrator from the nth one is proportional to the current amplitude A_n of the nth vibrator. It can be written as $A_n \vec{E}_{nm}$, where elements of the matrix $\vec{E}_{nm}$ ($n, m = 1, 2, \dots, M$) depend only on the mutual location of the nth and mth vibrators. Below, $E_{mn,z}(z)$ denotes the component of the field generated by the nth vibrator in the case when the cosine current of the unit amplitude flows in it, namely, the component directed along the mth vibrator. Then, according to (6.45), for each of the M vibrators, we have

$$A_m = \frac{1}{\mathcal{R} + iY_m} \int \left[\sum_{n \neq m} A_n E_{mn,z}(z) + E_{m,z}^{\text{inc}}(z) \right] \cos(kz)\, dz. \tag{6.50}$$

Denoting

$$\int E_{mn,z}(z) \cos(kz)\, dz = T_{nm} (n \neq m),$$

$$\mathcal{R} + iY_m = T_{mm}, \quad \int E_{m,z}^{\text{inc}}(z) \cos(kz)\, dz = e_m, \tag{6.51}$$

we obtain

$$\sum_{n=1}^{M} A_n T_{nm} = e_m, \quad m = 1, 2, \dots, M. \tag{6.52}$$

The coefficients T_{nm} ($n \neq m$) depend only on the mutual orientation of the vibrators and distances between them. The vibrator field must be defined in the near field zone for calculating these coefficients. The coefficients T_{mm} depend on the shape and length of the vibrators. The quantities e_m are determined by the magnitude and type (incident or fed) of the excitation fields. They have a meaning of *electromotive forces* acting on the vibrators.

Calculating of the current amplitudes in the vibrators of the multivibrator antenna is reduced to solving the system of linear algebraic equations (6.52). When the currents are found, then the radiation pattern and radiated power can be calculated.

6.2
Aperture antennas

6.2.1
Radiation from aperture

The field created by the radiating system can be determined not only by currents flowing on metallic conductors but also by the electric field on a certain closed auxiliary surface inside which all conductors are located. The latter way is more preferable in the cases when this surface field (its tangential component) can be found more easily than the currents, or when for finding the exterior field with a given accuracy, it is sufficient to calculate the surface field less accurately than the currents. This way of the radiation source representation has been applied in Section 5.3, as well as in Sections 3.3, 3.4 where the fields inside the volume were determined by the field on the hole in the wall of this volume.

Below, the field created by the antenna is determined by the field on its aperture, that is, on a part of plane adjacent to the antenna. We accept that the electric field is zero on this plane outside the aperture. For the antenna with flange, this assumption is not an approximation.

The problem is similar to that solved in Section 5.3, in which, however, the field of the plane antenna was determined only in the domain where the spherical wave had not been formed yet. In this subsection the formal solution is constructed in a similar way as in Section 5.3. The field given at $Z = 0$, is continued into the domain $Z > 0$ using formula (5.76). Here, however, restriction (5.78) imposed in Section 5.3 is not used when investigating the far field. The field is calculated in another way.

Introduce the spherical coordinate system (R, ϑ, φ) with the origin lying on the aperture and the z-axis directed perpendicularly to the aperture plane. In the domain where ϑ is not small, integral (5.76) should be estimated using the stationary phase method.

We pass from the Cartesian coordinates to the spherical ones

$$X = R \sin \vartheta \cos \varphi, \quad Y = R \sin \vartheta \sin \varphi, \quad Z = R \cos \vartheta \qquad (6.53)$$

in (5.76). Then (5.76) becomes

$$U(R, \vartheta, \varphi) = \int\!\!\!\int_{-\infty}^{\infty} f(\gamma, \beta) \exp(-ikR\psi) \, d\gamma d\beta, \qquad (6.54)$$

where the phase ψ is

$$\psi(\gamma, \beta) = \frac{\gamma}{k} \sin \vartheta \cos \varphi + \frac{\beta}{k} \sin \vartheta \sin \varphi + \left[1 - \left(\frac{\gamma}{k} \right)^2 - \left(\frac{\beta}{k} \right)^2 \right]^{1/2} \cos \vartheta. \qquad (6.55)$$

In (6.54) the exponent contains the multiplier kR. In the domain

$$kR \gg 1, \qquad (6.56)$$

we apply the stationary phase method. This method has already been used in Subsection 5.1.3 when calculating the field of the spherical wave, excited by a dipole located inside the dielectric waveguide. This field did not depend on the azimuthal angle and the integration was done over the wave number h only. In (6.54), the integration is performed in the plane of two wave numbers γ, β. The idea of the method is that the higher term of the asymptotic expansion of integral (6.54) by powers of the small parameter $1/kR$, can be obtained by taking out the factor $f(\gamma, \beta)$ at $\gamma = \gamma_s$, $\beta = \beta_s$, where γ_s and β_s are the roots of the equation system

$$\frac{\partial \psi}{\partial \gamma} = 0, \quad \frac{\partial \psi}{\partial \beta} = 0. \qquad (6.57)$$

In this case, the phase $\psi(\gamma, \beta)$ should be replaced with the first two terms of its expansion into the Taylor series near these roots. The phase (6.55) is stationary at the points $(\gamma, \beta) = (\gamma_s, \beta_s)$; it is slow varying in its neighborhood. The mutual compensation of different elements of the integral is weakened near this point.

In order to make these derivations for the phase (6.55), it is convenient to introduce the new integration variables ν and μ in (6.54) instead of γ and β, by the formulas

$$\gamma = k \sin \vartheta \cos \mu, \quad \beta = k \sin \vartheta \sin \mu \qquad (6.58)$$

suggested by (6.53). In these variables, the phase equals

$$\psi(\nu, \mu) = \sin \nu \sin \vartheta \cos(\mu - \varphi) + \cos \nu \cos \vartheta. \qquad (6.59)$$

The Jacobian of transformation (6.58) is $k^2 \sin v \cos v$, so that

$$d\gamma d\beta = k^2 \sin v \cos v\, dv\, d\mu. \tag{6.60}$$

The stationary phase points, that is, the roots of the equations

$$\frac{\partial \psi}{\partial v} = 0, \quad \frac{\partial \psi}{\partial \mu} = 0, \tag{6.61}$$

similar to (6.57), are

$$v_s = \vartheta, \quad \mu_s = \varphi. \tag{6.62}$$

Values of the function $\psi(v, \mu)$ and its second derivatives near this point are

$$\psi\left(v_s, \mu_s\right) = 1, \quad \left.\frac{\partial^2 \psi}{\partial v^2}\right|_s = -1, \quad \left.\frac{\partial^2 \psi}{\partial \mu^2}\right|_s = -\sin^2 \vartheta. \tag{6.63}$$

Near the stationary phase point, the function $\psi(v, \mu)$ can be represented as

$$\psi(v, \mu) = 1 - \frac{1}{2}(v - \vartheta)^2 - \frac{1}{2}\sin^2 \vartheta\, (\mu - \varphi)^2. \tag{6.64}$$

Taking the function $f(\gamma_s, \beta_s)$ and the factor $\exp(-ikR)$ out of the integral in (6.54), we obtain

$$U\left(R, \vartheta, \varphi\right) = \exp\left(-ikR\right) k^2 \sin \vartheta \cos \vartheta \int_{-\infty}^{\infty} \exp\left[\frac{ikR\,(v - \vartheta)^2}{2}\right] dv \times$$

$$\int_{-\infty}^{\infty} \exp\left[\frac{ikR \sin^2 \vartheta\,(\mu - \varphi)^2}{2}\right] d\mu \cdot f\left(k \sin \vartheta \cos \varphi, k \sin \vartheta \sin \varphi\right). \tag{6.65}$$

The first integral equals $\sqrt{-2i\pi/(kR)}$, the second one is $\sqrt{-2i\pi/(kR)} \cdot 1/\sin \vartheta$, so that

$$U\left(R, \vartheta, \varphi\right) = \frac{\exp\left(-ikR\right)}{kR} \cos \vartheta \cdot k^2 f\left(k \sin \vartheta \cos \varphi, k \sin \vartheta \sin \varphi\right). \tag{6.66}$$

Substituting the expression of the function $f(\gamma, \beta, 0)$ by the field $U(\widehat{X}, \widehat{Y}, 0)$ (denoted further as $U(\widehat{X}, \widehat{Y})$), we obtain the sought expression for the pattern of wave created by the field $U(\widehat{X}, \widehat{Y})$ distributed on the aperture D:

$$F\left(\vartheta, \varphi\right) = k^2 \cos \vartheta \iint_D U(\widehat{X}, \widehat{Y}) \exp\left[ik \sin \vartheta \cos \varphi \widehat{X} + \right.$$

$$\left. ik \sin \vartheta \sin \varphi \widehat{Y}\right] d\widehat{X}\, d\widehat{Y}. \tag{6.67}$$

Recall that $U(\widehat{X}, \widehat{Y})$ is a Cartesian component of the field on the aperture, $U(R, \vartheta, \varphi)$ is the same component in the spherical wave.

The above expression can be obtained from formula (5.81) for the field in the beam, using substitution (6.53), replacing Z with R (not quite correctly) in the phase and factor $1/Z$, and dropping the phase $\psi_z(X, Y)$. The above derivations show that this procedure would give the expression for the first term of the asymptotic expansion of the pattern at large kR.

6.2.2
The Green function method

Formula (6.67) is the main one in the theory of aperture antennas. In this subsection, we outline another way of its derivation which allows us to understand its structure from a little different point of view. This subsection has only a methodical purpose; there are no new results here. It will be used in the next section.

The method is based on the connection between two solutions to the Maxwell equations with different extrinsic currents. The first solution is the sought field $\{\vec{E}, \vec{H}\}$, created by a field given on the aperture. The second one is the so-called Green function, namely, the field $\{\vec{E}^g, \vec{H}^g\}$ created by a certain electric dipole. These fields are connected by (3.76). The field $\{\vec{E}^g, \vec{H}^g\}$ is subjected to the boundary conditions dependent on the type of the surface on which the aperture is located.

The Green function method has already been used for finding the field in the waveguide and resonator excited by a field given on the wall hole. The fields of the eigenmodes in the waveguide and fields of the eigenoscillations in the resonator, both having no holes, were chosen as the Green functions. These fields satisfied the boundary conditions on the walls without holes. In the problem of this subsection, the field $\{\vec{E}^g, \vec{H}^g\}$ is also chosen such that they fulfill the conditions on the "metallized" aperture.

The surface S surrounding the domain in (3.76) is chosen in the form of a circle lying on the aperture plane and containing the aperture, and a half-sphere supported by this circle. The radius a of the circle and half-sphere is so large that the field $\{\vec{E}, \vec{H}\}$ satisfies the radiation condition (1.33) on the half-sphere. On the aperture, $E_{\tan} = e$, where e is a field given on D; it was denoted as $U(\widehat{X}, \widehat{Y}, 0)$ in Subsection 6.2.1. In the rest of the circle, that is, outside the aperture, $E_{\tan} = 0$.

The field $\{\vec{E}^g, \vec{H}^g\}$ is created by the elementary dipole, located inside the half-sphere at the point $\vec{r}^g$ with the coordinates $(R^g, \vartheta^g, \varphi^g)$. This field satisfies the condition $E_{\tan} = 0$ on the whole plane $Z = 0$ and the radiation condition on the half-sphere.

It can be easily seen that if both the fields in (3.76) have the form (1.33) on the half-sphere, then the difference under the integral on the right-hand side of (3.76), decreases faster than $1/a^2$ with a increasing. The surface element is proportional to a^2, and the integral over the half-sphere must decrease with a increasing. The right-hand side of the formula, however, does not depend on a. Quantity which decreases with a increasing and does not depend on a, is zero. Hence, the integral in (3.76) over the half-sphere equals zero.

The integral over the part of the plane $Z = 0$ lying outside D is also zero, because both $\vec{E}_{tan}$ and $\vec{E}^g_{tan}$ are zero there. Since $\vec{E}^g_{tan} = 0$ on the aperture, only the integral of the first term taken over the aperture remains on the left-hand side. Under the volume integral in (3.76) $\vec{j}^{ext} = 0$, and only the integral of the δ-function remains. In a similar way, formula (1.52) was obtained as well.

At such a choice of the Green function, it follows from (3.76) that

$$E\left(\vec{r}^g\right) = -\frac{c}{4\pi} \iint_D \left(\vec{e} \times \vec{H}^g\right)_N dS, \tag{6.68}$$

where N is the outward normal to the plane $Z = 0$, $N_z = -1$. Formula (6.68) expresses the component of the field $\vec{E}$, parallel to the direction of the dipole located at the point $\vec{r}^g$, creating the field $\vec{H}^g$.

On the metallic surface, the tangential component H^g_{tan} involved in (6.68) is equal to the doubled value of this component created by the same source in vacuum. With accuracy to a nonessential factor, this field equals $\cos\alpha \cdot \exp(-ikb)/b$, where b is the distance from the point $\vec{r}^g$ to an integration point in (6.68). The angle α depends on how the momentum of the elementary dipole creating the field $\{\vec{E}^g, \vec{H}^g\}$ is directed with respect to the Z-axis.

Let a and R^g tend to infinity so that the condition $R^g \ll a$ holds. In other words, we perform two consecutive passages to the limit: first $ka \to \infty$ and then $kR^g \to \infty$. In this case H^g_{tan} is the same on the aperture as in the field of the plane wave coming onto the aperture from the direction of the dipole, that is, from $\vartheta = \vartheta^g$, $\varphi = \varphi^g$.

Substitute the value of the field H^g_{tan} on the aperture into (6.68). The factor $1/b$ can be taken out of the integral by replacing it with $1/R$. In the exponent we put $b = R^g - \rho\cos\gamma$, where ρ is a distance between the origin and the integration point, γ is an angle made up by the Z-axis and the direction (ϑ^g, φ^g) of the plane wave incoming. From (6.21), $\cos\gamma = \sin\vartheta^g \cos(\varphi - \varphi^g)$ for the points lying on the aperture ($\vartheta = \pi/2$). Hence,

$$E\left(R^g, \vartheta^g, \varphi^g\right) = \frac{\exp\left(-ikR^g\right)}{kR^g} \sin\alpha \iint_D e\left(\rho, \varphi\right) \times$$

$$\exp\left(ik\rho\sin\vartheta^g \left(\cos\vartheta^g \cos\varphi + \sin\vartheta^g \sin\varphi\right)\right)\rho d\rho d\varphi. \tag{6.69}$$

Since $\rho\cos\varphi = \hat{X}$, $\rho\sin\varphi = \hat{Y}$ in notations (6.67), then (6.69) coincides with (6.67). The dependence of the field polarization in the spherical wave on that on the aperture is not interpreted in (6.67).

Formulas (6.67), (6.69) have the same meaning as (6.18). The field radiated by the aperture is made up by the waves radiated by the aperture segments; it is proportional to the value of the electric field on these segments. The summation is performed in the same way as for the fields created by the current segments, that is, with accounting the difference of distances covered by the waves radiated from different segments.

The analogy between the boundary problems with a given value of E_{tan} on a part of the boundary and those about the field of currents flowing on the metallic surfaces, prompts the usage of the notion magnetic currents. The vector like $\vec{e}$ in (6.69) or in similar boundary problems, is called the *magnetic current*. Formulas of the type (6.67), (6.69) can be treated as the generalization of the formulas (6.19) for the field created not by the electric (i. e., really existing) currents, but by the conventional magnetic ones. In the next subsection, some principal distinction of these formulas will be discussed.

6.2.3
The radiation pattern

With accuracy to a nonessential factor, the angular dependence of the field in the far field zone is described by the function

$$F(\vartheta, \varphi) = \cos \vartheta \iint_{D} U(\widehat{\rho}, \widehat{\varphi}) \exp \left[ik\widehat{\rho} \sin \vartheta \cos (\varphi - \widehat{\varphi}) \right] \widehat{\rho} \, d\widehat{\rho} \, d\widehat{\varphi} \qquad (6.70)$$

(which is the function (6.67) rewritten in cylindrical coordinates). Here $\widehat{\rho}$, $\widehat{\varphi}$ are the cylindrical coordinates in the aperture plane, and angles do not have the index g in contrast to (6.69). The function $U(\widehat{\rho}, \widehat{\varphi})$ is one of the components of the field $\vec{e}$ given on the aperture.

The point in the direction $\vartheta = 0$ in the far field zone (see (6.23)) can be considered as equally distanced from all the aperture elements (segments), so that

$$F(0, \varphi) = \iint_{D} U(\widehat{\rho}, \widehat{\varphi}) \widehat{\rho} \, d\widehat{\rho} \, d\widehat{\varphi}. \qquad (6.71)$$

Of course, the pattern does not depend on φ in this direction. At the given field modulus $|U(\widehat{\rho}, \widehat{\varphi})|$, the function (6.71) is maximal if the field is in-phase and does not change the sign. Then the antenna focuses on the infinite point located on the line perpendicular to the aperture plane. The main lobe of the pattern is directed along this line. The direction of the main lobe can be turned by introducing the linear phase factor $\exp(-i\alpha\widehat{X})$ into (6.71), without moving the antenna. Further in this subsection we put the function $U(\widehat{\rho}, \widehat{\varphi})$ to be a real one.

The pattern depends on the shape and size of the aperture and on the field distributed on it, that is, on the function $U(\widehat{\rho}, \widehat{\varphi})$. To illustrate this dependence, we consider a case when the antenna possesses the cylindrical symmetry, that is, the domain D is a circle and U does not depend on the angle $\widehat{\varphi}$. We do not account the symmetry violation caused by the fact that the vector $\vec{e}$ is a two-dimensional vector on the aperture, because this violation is not essential for our problem. The pattern also does not depend on the angle φ and

$$F(\vartheta) = \cos \vartheta \int_0^a U(\widehat{\rho}) \exp\left(-ik\widehat{\rho}\sin\vartheta\right) \widehat{\rho} \, d\rho, \qquad (6.72)$$

where a is a circle radius.

The larger the value ka, the faster the exponent varies with ϑ increasing and the faster the pattern decreases in comparison with its maximal value $F(0)$. This means that the larger ka, the narrower the main lobe of the pattern. The value of ϑ at which the pattern becomes zero for the first time, depends also on the function $U(\widehat{\rho})$. However, this value is such that the product $ka\sin\vartheta$ is about 2π. The width of the main lobe is about λ/a radian.

The next (side) lobes are smaller than the main one. The rule by which they decrease, depends on the function $U(\widehat{\rho})$ behavior near the aperture edge, namely, on $U(a)$, $dU/d\widehat{\rho}|_{\widehat{\rho}=a}$ and so on. This follows from the first terms of the asymptotic series by the inverse powers of $ka\sin\vartheta$ which can be obtained after calculating integral (6.72) by parts. The first term of this series is

$$\frac{\exp\left(-ika\sin\vartheta\right)}{ka\sin\vartheta} f(a). \qquad (6.73)$$

Just as any asymptotic series, the above one is divergent, but its several first terms give a good approximation of the pattern at nonsmall values of $ka\sin\vartheta$.

The series of terms like (6.73) describes the envelope of the side-lobe maxima. The pattern is fast decreasing when moving away from the main lobe if $U(a) = 0$, that is, if the field vanishes on the aperture edge. This effect is similar to that described in Subsection 5.3.5: in the wave beam, the field decreases faster with moving away from the rectenna edge in the case when the field vanishes on the antenna edge. It can be approximately accepted that in the beam theory, the rectenna plays a role of the main lobe in terms of the antenna theory.

The side lobes decrease even more faster with ϑ increasing if not only the field vanishes on the aperture boundary but also its normal derivative does, that is, if $U(a) = 0$ and $dU/d\widehat{\rho}|_{\widehat{\rho}=a} = 0$. Since the theory is constructed for the model in which the field is zero outside the aperture, the above implies that with ϑ increasing the side lobes decrease faster, the "more analytical" the field near the aperture boundary is. It should be kept in mind, however, that in this case the utilization factor of the aperture surface decreases: at a part of the aperture the field is significantly smaller than its maximal value.

At small ka, that is, for small antennas, the pattern is mainly determined not by integral (6.72), but by the factor $\cos\vartheta$. If the field $U(\widehat{\rho},\widehat{\varphi})$ has the same direction on the whole aperture, then the aperture radiates as an electric dipole in this case.

6.2.4
Inverse problem

The inverse problem consists in finding a field on the antenna aperture, which creates a given radiation pattern. The problem is similar to that considered in Subsection 5.3.4 about the field on the antenna creating a given field in the rectenna plane. This similarity can be shown when comparing formulas (5.95) and (6.67) for the field in the rectenna plane and radiation pattern, respectively. In both problems, the field on the radiating antenna differs from zero only on a finite part of the plane, and therefore the antenna can create certain bounded class of fields in the rectenna plane (or, respectively, of patterns). If the given field (or pattern) does not belong to this class, then the problem has no solution. Similarly as in the long-beam theory, the problem on finding a field on the antenna, which creates a pattern maximally close to the given one, has the practical sense in the antenna theory.

There is a difference between the above problems, consisting in the fact that in the antenna problem, the functional to be minimized could be naturally formulated as mean-square difference between the given and obtained patterns in the real ("visible") angle range. This would correspond to the limitation of the integration domain in integral (5.109), by the rectenna area. In the beam theory such an approach leads to absence of restriction on the radiated energy passing the rectenna. Similar situation also occurs in the antenna theory: minimization of the functional with integration over the real angles does not restrict the energy of nonpropagating waves localized in the near field. In mathematical terms, such a statement can lead to the situation when the functional has no minimum, or the function minimizing it is fast oscillating and thereby nonrealizable. Such solutions are called *super-directive*.

There are two ways to avoid this effect. The first one consists in supplying the integration domain with the "complex" angles and setting the desired pattern to be zero there. This approach leads (in appropriate notations) to the same formulation of the problem as in the beam theory (5.109), and, correspondingly, to the same solution (5.114). We explain it for the case of the so-called *linear antenna*: the thin wire of the length $2a$. The current with linear density $J(x)$, flowing on such an antenna, creates the radiation pattern

$$F(\vartheta) = \int\limits_{-a}^{a} J(x)\exp(-ikx\sin\vartheta)\,dx \tag{6.74}$$

(a constant factor is omitted). This formula can be easily obtained from (6.27), (6.26). Similar formula for the pattern created by the "magnetic current," that is, by the electric field on the aperture, can be obtained from (6.70). According to (6.74), the pattern treated as a function of the variable ("generalized coordinate")

$$\xi = \sin \vartheta \qquad (6.75)$$

is the Fourier transformation of the current, taken in the finite limits. In this coordinate, the real angle range $-\pi/2 \leq \vartheta \leq \pi/2$ corresponds to the interval $-1 \leq \xi \leq 1$. For simplicity, we use for the pattern in this coordinate the same notation $F(\xi)$.

Let the given pattern be $F_0(\xi)$, $-1 \leq \xi \leq 1$. Formally, substituting $F_0(\xi)$ into the left-hand side of (6.74), we obtain the integral equation of the first kind

$$\int_{-a}^{a} J(x) \exp(-ikx\xi)\, dx = F_0(\xi). \qquad (6.76)$$

It is known that this equation has a solution only if $F_0(\xi)$ belongs to a certain class of functions, and even in this case the solution is unstable.

Similarly as in Subsection 5.3.5.5, we find the function $J_\sigma(x)$, minimizing the functional

$$\sigma[u(x)] = \frac{\int_{-\infty}^{\infty} |F(\xi) - F_0(\xi)|^2\, d\xi}{\int_{-1}^{1} |F_0(\xi)|^2\, d\xi}. \qquad (6.77)$$

Here we assume that $F_0(\xi)$ is expanded into the complex angles as $F_0(\xi) = 0$, $|\xi| > 1$. The function $F(\xi)$ is expanded according to (6.74). The function $J_\sigma(x)$ has an explicit form as

$$J_\sigma(x) = \frac{1}{2\pi} \int_{-1}^{1} F_0(\xi) \exp(ikx\xi)\, d\xi. \qquad (6.78)$$

This expression is analogous to (5.113), $J_\sigma(x)$ does not solve equation (6.76).

Another way of avoiding super-directive solutions is to use functionals containing the energy of the sought field (current) together with the difference between the given and obtained directivity patterns. One of such functionals, written for the above example is

$$\sigma_\alpha[u(x)] = \frac{\int_{-1}^{1} |F(\xi) - F_0(\xi)|^2\, d\xi + \alpha \int_{-a}^{a} |J(x)|^2\, dx}{\int_{-1}^{1} |F_0(\xi)|^2\, d\xi}. \qquad (6.79)$$

The weight factor α can be determined from the condition $\int_{-a}^{a} |J(x)|^2\, dx = N$ with given N, or from a compromise between the summands in the numerator of (6.79). The Lagrange–Euler equation for this functional has the form

$$\alpha J(x) + \frac{2}{k} \int_{-a}^{a} \frac{\sin[k(x - x')]}{(x - x')} J(x')\, dx' = \int_{-1}^{1} F_0(\xi) \exp(ikx\xi)\, d\xi. \tag{6.80}$$

This is the integral equation of the second kind. It has a unique solution at any $F_0(\xi)$.

Similarly as in the analogous problem of the beam theory, in the case if not the complex pattern F_0, but only its modulus $|F_0|$ is given in both functionals σ and σ_α, then the difference between the given and obtained patterns can be decreased at a fixed antenna size, or, which is equivalent, the antenna size can be decreased at a fixed difference between the given and obtained patterns.

6.2.5
The near field zone, Fresnel zone, and far field zone

Near a large aperture, the field has a ray-like structure; in the far distance, it turns into a spherical wave. There exists an intermediate region in which the field changes its structure (see Subsection 6.1.2). We specify this statement.

In the near field zone, the field is described by (5.79). First, we assume that the field is real and does not change its sign on the aperture, that is, the antenna "focuses onto infinity." This means that the antenna should create a good pattern, but not a "good" field in a finite distance from the aperture. We denote the linear size of the aperture by a. The integral in (5.79) describing the field at the point (X, Y, Z) is proportional to the integral

$$\int_D U(x, y, 0) \exp \left\{ -\frac{ika^2}{2Z} \left[\left(\frac{X}{a} - x \right)^2 + \left(\frac{Y}{a} - y \right)^2 \right] \right\} dx dy, \tag{6.81}$$

where the dimensionless integration variables $x = \widehat{X}/a$, $y = \widehat{Y}/a$ are introduced. Recall that $\widehat{X}, \widehat{Y}$ are the Cartesian coordinates in the plane $Z = 0$. The integral depend on X, Y, Z as parameters.

The range of values of Z, at which the factor ka^2/Z is large, that is,

$$Z \ll ka^2 \tag{6.82}$$

is called the *near field zone*. In this zone, the integral in (6.81) can be estimated by the stationary phase method. The x- and y-derivatives of the phase (i. e., of the square bracket in (6.81)), equals $x - X/a$ and $y - Y/a$, respectively. The stationary phase points are

$$x_s = \frac{X}{a}, \quad y_s = \frac{Y}{a}. \tag{6.83}$$

In nonnormalized coordinates, $\widehat{X}_s = X$, $\widehat{Y}_s = Y$. Consequently, the first term of the asymptotic expansion of the field by the inverse powers of the ratio ka^2/Z in the plane $Z = const$, equals

$$U(X,Y,Z) = \exp(-ikZ)\, U(X,Y,0).$$ (6.84)

This formula implies that at any point of the plane $Z = const$, the field coincides (with accuracy to the phase factor $\exp(-ikZ)$) with its value at the point on the plane $Z = 0$, having the same coordinates X, Y. The field does not depend on its values at other points of the plane $Z = 0$. The field has a ray structure. At any point, it depends only on its values at the points of a ray coming to this point. According to (6.84), the rays are perpendicular lines to the plane $Z = 0$. We remind that in our case, this plane is an equiphase surface.

We repeat this derivation assuming that the aperture focuses onto the point located in the finite distance d from it. In this case, the field $U(\widehat{X}, \widehat{Y}, 0)$ in (5.79) has the phase factor $\exp\left[ik/(2d)(\widehat{X}^2 + \widehat{Y}^2)\right]$ (see text below formula (5.82a)). The phase in (5.79) equals

$$-\frac{ika^2}{2Z}\left[\left(\frac{X}{a} - x\right)^2 + \left(\frac{Y}{a} - y\right)^2 - \frac{Za^2}{d}\left(x^2 + y^2\right)\right].$$ (6.85)

The stationary phase points are

$$x_s = \frac{X}{a}\frac{1}{1 - Z/d}, \quad y_s = \frac{Y}{a}\frac{1}{1 - Z/d}.$$ (6.86)

For the first term of the asymptotic expansion, formula (5.79) gives the expansion

$$U(X,Y,Z) = \exp(-ikZ)\, U\left(\frac{X}{1 - Z/d}, \frac{Y}{1 - Z/d}, 0\right).$$ (6.87)

In this case, the near field has the ray structure, either. The field at the point (X, Y, Z) depends only on the field at one point of the plane $Z = 0$. The rays are parallel lines connecting the points (X, Y, Z), $(X/(1 - Z/d), Y/(1 - Z/d), 0)$ and being orthogonal to the equiphase surface given by (5.82a). In the next subsection, we will give a connection between the equiphase surface and rays in a general form.

In the domain, where $kR \gg 1$, the field has the structure of a spherical wave. The intermediate range in which Z has the order ka^2

$$Z \approx ka^2,$$ (6.88)

is called the *Fresnel zone*. The ratio

$$c_F = \frac{ka^2}{Z}$$ (6.89)

is called the *Fresnel coefficient*. In a rough approximation, $c_F \gg 1$ in the near field zone, $c_F \approx 1$ in the Fresnel zone. The zone where $c_F \ll 1$, that is, the inequality opposite to (6.82), is called the *far field zone*. Of course, the field structure and, therefore, the location of these three zones depends not only on the coefficient c_F but also on the field created on the aperture.

6.2.6
Geometric optics

Condition (6.82) means that the wavelength $\lambda = 2\pi/k$ is small in comparison with all linear sizes involved in the problem. This demand can be written as a symbolic equality

$$\lim k = \infty. \tag{6.90}$$

In this case the solution to the Maxwell equations has a form of the product of a slow varying (in the λ-scale) function and a fast varying one. The fast-varying function is conveniently expressed as the exponent of a function of coordinates, multiplied by k. The form of this function depends on the shape of the equiphase surface and on the type of the medium where the waves propagate. For instance, solution (5.83) has this form if the equiphase surface is the plane $Z = const$.

We consider the qualitative structure of the field in the case when condition (6.90) holds. Similarly as in Section 5.3, we confine ourselves to investigation of the scalar wave equation (5.73). Substituting the solution in the form (5.83) into this equation yields equation (5.84) for the function $W(\vec{r})$. In the example of Section 5.3, approximation (5.86) stronger than (6.90), has been accepted, leading to the parabolic approximation (5.88). The geometric approximation is more rough in this example. It implies that only the higher term with respect to the large parameter k is kept in equation (5.84) for $W(\vec{r})$ so that this function satisfies the equation

$$\frac{dW}{dZ} = 0. \tag{6.91}$$

In this example, the rays are lines $\{X = const, Y = const\}$. They are perpendicular to the equiphase surfaces $Z = const$. Equation (6.91) means that at any point of a ray, the field depends only on the fields at other points of this ray.

In the neighborhood of each ray, the so-called *ray tube*, that is, a cylinder having this ray as an axis and consisting of the nearby rays can be constructed. In the geometric approximation, the energy flux inside the ray tubes is kept. The tubes do not exchange the energy. The diffusion process is described by a more precise parabolic equation. In general, the tube cross-section is not constant along the ray. In this meaning, the set of parallel rays is an exception.

For instance, the ray structure of the field considered at the end of the preceding subsection, is a set of lines perpendicular to the concentric spherical surfaces being the equiphase surfaces. When approaching the center of these spherical surfaces (i. e., the focus of the ray set), the cross-section of the ray tubes infinitely decreases; density of the energy transferred in each tube tends to infinity. The geometric theory does not describe the field near the focus. In this domain, the density of the energy in the neighboring tubes is strongly different; significant transverse diffusion appears. If the ray set has an envelope (caustics), that is, a surface tangent to a ray at any its point, outside which there are no rays, then the geometric approximation is not applicable in its neighborhood, as well.

We consider a more general case of the usage of geometric approximation. Let the medium be nonhomogeneous, that is, the product $\varepsilon\mu$ depends on coordinates, $n = n(\vec{r})$, $n = \sqrt{\varepsilon\mu}$. We confine ourselves to the scalar wave equation

$$\frac{\partial^2 U}{\partial X^2} + \frac{\partial^2 U}{\partial Y^2} + \frac{\partial U^2}{\partial Z^2} + k^2 n^2 U = 0. \tag{6.92}$$

We seek the solution in the form

$$U = W \exp(-ikS). \tag{6.93}$$

The geometro-optical approximation implies that the functions $W(\vec{r})$ and $S(\vec{r})$ are found by equating to zero the first two terms of the expansion of the left-hand side of (6.92) by the powers of k, that is, the terms proportional to k^2 and k. This gives the equations for S and W

$$(\operatorname{grad} S)^2 = n^2, \tag{6.94}$$

$$\frac{1}{W} \operatorname{grad} W \operatorname{grad} S = -\frac{1}{2} \operatorname{div} \operatorname{grad} S. \tag{6.95}$$

The function S (*eikonal*) is defined by equation (6.94) (*eikonal equation*) and by its given value on one of the equiphase surfaces $S = const$. The ray is a line at each point of which the tangential direction coincides with the vector $\operatorname{grad} S$. By definition, the rays are orthogonal to the equiphase surfaces. For instance, if $n = 1$ and these surfaces are planes $Z = const$, then $S = Z$ solves (6.94). In this case W does not depend on Z.

The function $W(\vec{r})$ is slow varying in the direction at which $S(\vec{r})$ varies fast. This fact follows from (6.95), since the right-hand side of this equality is usually not large. If $W = 0$ at a certain point of the equiphase surface, then the energy of a ray passing through this point is zero; it is small at all points of this ray.

Equation (6.94) appears not only in the geometric optics. Efficient methods are developed for its solving in the case of an arbitrary function $n(\vec{r})$. If the function $S(\vec{r})$ is known, then the ray direction, that is, the vector grad S tangential to the ray, can be found at any point of it. In the medium with $n = 1$, the rays are straight lines.

If $n \neq const$, then it can be shown that the vector grad S changes its direction, it turns toward grad n, so that the angle between these two vectors diminishes. In particular, when refracting on the interface of two media, the ray approaches the medium with larger n, as it follows from the refraction law (Section 2.2). In the optical fiber theory (Section 5.1), where the ray path was investigated in the nonhomogeneous medium without using the general theory (i. e., formula (6.94)), it was also shown that the ray refraction consists in its deflection toward the larger n.

The rays possess the following property: integral of the refraction index $n(P)$ taken over a line connecting any two points, reaches its maximal value if this line is a ray passing through these points (the Fermat principle). This principle states that a ray passing through the points P_1 and P_2, provides the extremum to the integrals $\int_{P_1}^{P_2} n ds$; taken over all lines connecting these points. The integral over the ray, as a function of the upper limit, is the eikonal $S(P_2)$. It is easy to show that the integral satisfies equation (6.94). It is called the *optical length* of the ray from the point P_1 to P_2 and is proportional to the time of wave propagation along the ray. This follows from the fact that the wave velocity is c/n, that is, the time needed for passing the ray element ds is nds/c.

In this context, the extremum always means minimum, but there exist cases when it is maximum. The assertion that the time needed for the wave propagation between two points is minimal if the wave propagates along the ray, is almost always true.

We show how one of the simplest laws of geometric optics, namely, the law of refraction at the interface of two media, follows from the Fermat principle. Let the points P_1 and P_2 with the coordinates x_1, y_1 ($y_1 < 0$) and x_2, y_2 ($y_2 > 0$), respectively, lie in the corresponding media with the refraction indices n_1 and n_2 (Fig. 6.3). The x-axis is the media interface; at $n = n_1, y < 0$ and at $n = n_2$, $y > 0$. The optical length of the broken line connecting the two points and intersecting the interface at the point $x = a, y = 0$ ($x_1 < a < x_2$), equals

$$S = n_1 l_1 + n_2 l_2, \quad l_1 = \sqrt{(a - x_1)^2 + y_1^2}, \quad l_2 = \sqrt{(x_2 - a)^2 + y_2^2}. \quad (6.96)$$

If the coordinate a is such that $dS/da = 0$, then this length is minimal among the optical lengths of all broken lines. This gives

$$\frac{a - x_1}{l_1} n_1 = \frac{x_2 - a}{l_2} n_2. \quad (6.97)$$

It is a usual refraction law which implies that the cosines of the glancing angles of the incident and refracted rays relate as the refraction indices of both the media.

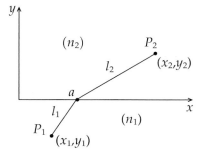

Fig. 6.3 Illustration of the Fermat principle

In the next chapter it will be noted that the geometro-optical approach in the formulation of the Fermat principle is also valid in the shadow domain.

6.3
Volume antennas

6.3.1
Radiation from holes

Antenna can radiate not only from the large hole (e. g., from the mirror antenna aperture, see Section 6.2) but also from several small ones, for instance, from slots in walls of the waveguide or resonator. Of course, the radiated field of such antennas can also be expressed by currents on the walls. However, these currents should be known with a very high accuracy, accounting into their small alternation caused by the slot cutting (before cutting, the fields of the currents were mutually canceled in the exterior space and the total field was zero). In this case, it is more convenient first to find the tangential components of the electric field on the holes, and then use them for calculation of the field radiated by the antenna.

In this section, formulations of other boundary value problems are considered. Since only the tangential components of the fields participate in these problems, below we omit the words "tangential components" and write "the field on the surface" meaning just these components.

If the fields on the surface are known, then the exterior field created by these fields can be calculated. For the slot antennas, it is convenient to consider the slots (not the metallic surfaces) as radiating elements.

The electric field between the slots on the antenna surface (i. e., on metallic surface) equals zero. Therefore, the field is known on the whole closed antenna surface if it is known on the slots. The problem of finding the exterior field by the known field on the slots consists in determining the field subject to the radiation condition, by its value on a certain closed surface. This boundary value problem is always uniquely solvable. The uniqueness follows from the fact that if the electric field equals zero on a closed surface (ideal metallic surface) and there are no sources outside it (in particular, no wave incident from infinity), then the exterior field equals zero (see Subsection 1.4.1). The existence of the solution is physically obvious.

The solution to the above boundary value problem can be easily obtained in the way outlined in Subsection 6.2.2, that is, by using the Green function. The direct continuation of the field from the surface into space is possible only for the surfaces coincident with a coordinate one of a simple coordinate system. For instance, in Section 5.3 and Subsection 6.2.1 such a surface is a plane. In this subsection, we derive formulas for the far field only, that is, for the radiation pattern. In this case we can choose the simple Green function and avoid two passages to the limit (as in Section 6.2). It is assumed that the field on the surface is created by certain inner sources.

The Green function technique is based on formula (3.76). It connects the sought field $\{\vec{E}, \vec{H}\}$ with the Green function $\{\vec{E}^g, \vec{H}^g\}$. We apply (3.76) to the domain bounded by the surface S and a spherical surface A encircling it. The radius a of the sphere is so large that the field $\{\vec{E}, \vec{H}\}$ on it has the asymptotic form (1.33). As the Green function, we choose the field arisen when the plane wave falls onto the "metallized" surface S. By definition, the electric field of the Green function $\vec{E}^g$ equals zero on S. The field $\{\vec{E}^g, \vec{H}^g\}$ consists of the field of the plane (incident) wave and of that of the scattered one. The below formula for the radiation pattern contains only the value of the field $\vec{H}^g$ on S, that is, the current arisen on the metallic surface coinciding with S, when the plane wave falls onto it.

The approach, used in Subsections 3.4.3, and 3.4.4 when solving the problem about a diaphragm with hole in the rectangular waveguide, is a particular case of the above technique. The mentioned problem was simpler than that considered here, since the Green function was the field of the standing wave in the waveguide with the metallic diaphragm and it was easily found in the explicit form.

There are no sources in the domain bounded by the surfaces S and A, and formula (3.76) is an equality of two surface integrals

$$\int_A \left[\left(\vec{E}^g \times \vec{H} \right)_N - \left(\vec{E} \times \vec{H}^g \right)_N \right] dS$$

$$= -\int_S \left[\left(\vec{E}^g \times \vec{H} \right)_N - \left(\vec{E} \times \vec{H}^g \right)_N \right] dS. \quad (6.98)$$

The normal $\vec{N}$ has different sense in these two integrals.

First, calculate the integral over A. The surface element is $dS = a^2 \sin\theta d\theta d\varphi$ in it. Denote by θ^g, φ^g the angles defining the direction from where the plane wave forming the Green function comes. The integral over A depends on these angles as on parameters. All the components of the plane wave have the phase $ik(x \sin\theta^g \cos\varphi^g + y \sin\theta^g \sin\varphi^g + z \cos\theta^g)$. Expressing the Cartesian coordinates x, y, z of a point on the sphere by spherical ones a, θ, φ, we obtain the expression $ika\psi(\theta, \varphi)$ for the phase, where

$$\psi(\theta, \varphi) = ika\left[\sin\theta^g \sin\theta \cos(\varphi - \varphi^g) + \cos\theta^g \cos\theta\right]. \tag{6.99}$$

The field $\{\vec{E}^g, \vec{H}^g\}$ consists of the fields of both the plane wave and the scattered one. First we keep only the field of the plane wave in the integral over the sphere A. Since the field $\{\vec{E}, \vec{H}\}$ has the phase $(-ika)$, all the terms in the integrand have the phase $ika(\psi - 1)$. Since $ka \gg 1$, the integral can be calculated by the stationary phase method. Similarly to (6.62), the coordinates θ_s, φ_s of the stationary phase point, that is, the angles satisfying the equation system

$$\frac{\partial\psi}{\partial\theta} = 0, \qquad \frac{\partial\psi}{\partial\varphi} = 0, \tag{6.100}$$

equal

$$\theta_s = \theta^g, \qquad \varphi_s = \varphi^g. \tag{6.101}$$

System (6.100) has also another solution

$$\theta_s = \pi + \theta^g, \qquad \varphi_s = \pi + \varphi^g, \tag{6.102}$$

that is, there exists the second stationary phase point. It lies on the opposite end of the diameter, on which the first point (6.101) lies. As it will be shown below, the point (6.102) does not contribute to the value of the calculated integral.

At the point (6.101), the second derivatives $\partial^2\psi/\partial\theta^2, \partial^2\psi/\partial\varphi^2$ have the same values as in (6.63), and the value of the phase itself equals zero. Replace the phase of integrand in the integral over A with the first terms of its Taylor series, that is, with

$$\psi = ika\left[1 - \frac{1}{2}(\theta - \theta^g)^2 - \frac{1}{2}(\varphi - \varphi^g)^2 \sin^2\theta\right]. \tag{6.103}$$

The factor $\sin\theta$ in dS cancels with the same function in the expression obtained by the stationary phase method. Consequently, a linear combination of the functions $F_1(\theta^g, \varphi^g)$, $F_2(\theta^g, \varphi^g)$ determining the sought pattern of the antenna, appears on the left-hand side of (6.98). The coefficients of this combination depend on the polarization of the incident plane wave defining the

Green function. Substituting successively the incoming Green waves of two polarizations into the right-hand side of (6.103), we obtain there a system of two linear combinations of the functions F_1 and F_2.

As it has been pointed out, the Green function contains not only the incident plane wave but also the one arising when this wave falls onto the "metallized" surface S. On the sphere A, this term in the Green function has the same structure of the divergent spherical wave as the sought field $\{\vec{E}, \vec{H}\}$. The integrand under the integral over A vanishes (in the higher order with respect to $1/a$) if both the fields $\{\vec{E}, \vec{H}\}$ and $\{\vec{E}^g, \vec{H}^g\}$ are the divergent spherical waves. Therefore, only the incident plane wave should be accounted in the Green function when calculating this integral.

By the same reason, the second stationary phase point (6.102) should not be accounted. Near this point on the sphere $R = a$, the Green function (both the plane wave and the divergent one) has the structure of the divergent spherical wave. More precisely, the outgoing plane wave is asymptotically close to the convergent spherical one.

If the two waves under the integral over A in (6.98) differ in the direction of propagation, then the bracket in this integral is not zero. If their structure is the same (both are divergent waves), then the integrand is asymptotically zero. If the radial dependence of the fields is $\exp(-ikR)$, then the components of $\vec{E}, \vec{H}$ are related similarly as in (1.43). If this dependence is $\exp(ikR)$, then $E_\theta = H_\varphi$, $E_\varphi = -H_\theta$, that is, the relation has different signs. Due to this fact, only the opposite-directed waves participate in (6.98).

Consider the integral over S in (6.98). The integration is made only over the slots: the integral over the metallic surface equals zero since both the fields $\vec{E}$ and $\vec{E}^g$ are zero on it. We have chosen the complicated Green function in order that the value of the magnetic field $\vec{H}$ on the metal is not involved in the expression for the sought field. If the field of the plane wave in vacuum were chosen as the Green function, then the integral over S would contain the value of $\vec{H}$ on the metal, that is, the current flowing on the antenna walls between the slots. This current cannot be given arbitrarily, it should be found after the problem about the field outside S is solved. This field is completely defined by the electric field on S. Earlier, when solving the problems about the fields in the waveguide or resonator, created by the electric field on the slots in the wall, we also chosen the Green function as a field, for which $E = 0$ on the walls.

Consequently, the integral over S in (6.98) is

$$\int_\Sigma \left(\vec{e} \times \vec{H}^g \right)_N dS, \tag{6.104}$$

where Σ is the hole, and $\vec{e}$ is a two-dimensional vector describing the electric field given on Σ.

The integrand of (6.104) contains the tangential component of the field H^g, equal to the current flowing on the "metallized" surface S when the plane wave of the unit amplitude, incoming from the direction (θ^g, φ^g), falls on it. The existence of the metallic surface between the slots, on which $E_{\text{tan}} = 0$, is accounted in the field H^g_{tan}. The necessity to account this surface causes the Green function $\{\vec{E}^g, \vec{H}^g\}$ to be a solution of a certain diffraction problem. In formula (6.19) analogous to (6.104), for the pattern created by the real (electric) currents, the function playing a role of $\vec{H}^g$ is significantly simpler. It does not depend on the shape of the surface, on which the current flows. In fact, the additional condition for the electric field to be zero on the metallic surface (outside the slots) is connected with the fact that the "magnetic current," equal to e_{tan}, does not really exist. It cannot "flow" in vacuum, as a real current, a metallic surface should always be near it.

Integral (6.104) should be calculated for both polarizations of the incoming Green waves. Then equation (6.98) leads to the system of two linear equations for the two functions $F_1(\theta^g, \varphi^g)$, $F_2(\theta^g, \varphi^g)$. The system solves the problem of this subsection, consisting in finding the pattern scattered from the holes cut in a closed metallic surface, by the given electric field on them.

6.3.2
Field in narrow long slot

In this subsection, a method for finding the field $\vec{e}$ on the hole is qualitatively described. The field is created by the source located inside the antenna (waveguide or resonator) and directed across the slot. Its second component, parallel to the long sides of the slot, is small, since it becomes zero on these sides, that is, on the close parallel lines.

In many aspects, the problem about the field on the narrow long slot is similar to that about the current on a long thin vibrator (see Subsection 6.1.4). A scalar function $e(\xi)$ (the electric field on the slot) depending on the longitudinal coordinate ξ directed along the slot is to be found. The function is analogous to the current J on the vibrator. At the slot ends, $e(\xi)$ vanishes, similarly as the current $J(\xi)$ does at the vibrator ends. The magnetic field on both the sides of the slots, created by the field $e(\xi)$, is calculated by this function. The magnetic field is mainly directed along the slot, its second component is small. On both the sides of the slot, at $z = +0$ and $z = -0$ (Fig. 6.4), the sums of this field and the magnetic fields created by the same sources on both the sides before cutting the slot, should be the same. This condition leads to a first-kind integral equation for the function $e(\xi)$. The free term of this equation is the difference between the magnetic fields at $z = +0$ and $z = -0$, created by the sources when metallizing the slot. If the sources exist only on the one side, for instance at $z < 0$, as in the antenna, then the free term is a current flowing on the metal in the place where the slot is cut, that is, the current is cross cut by the slot. This scheme has already been realized when solving the problem about the diaphragm with a slot in the rectangular waveguide (Section 3.3).

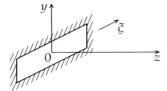

Fig. 6.4 The slot in the screen

   The electric field distributed on a certain surface Σ, creates the same field outside it, as the surface magnetic current, equal to this field and perpendicular-directed to it. This statement is based on the following consideration. Along with the Maxwell equation system (1.29), the system (3.117) is introduced which differs from (1.29) in having certain vector $\vec{j}^{(m)}$ as a source instead of the current with the density $\vec{j}$. The vector $\vec{j}^{(m)}$ is tangent to the surface Σ and differs from zero only on it. The field source is the two-dimensional vector $\vec{I}^{(m)}$ being the surface density of the magnetic current. On the surface Σ, the electric field has a jump, equal to $I^{(m)}$ and perpendicular to the vector $\vec{I}^{(m)}$. This fact can be proven using the Stokes theorem, similarly as the fact that if a current with the surface density $\vec{I}$ flows on a certain surface, then the magnetic field has the jump (1.80) on this surface. When calculating the field created by the field $\vec{E}$ distributed on the one side of the surface Σ, we can accept that on another side of the surface the field is zero, that is, $\vec{E}$ equals the jump of the electric field on the surface. This fact justifies the statement made in the first sentence of this paragraph.

   If the magnetic current is known, then the magnetic Hertz potential $\vec{\Pi}^{(m)}$ can be found. This vector satisfies the nonhomogeneous wave equation

$$\Delta\vec{\Pi}^{(m)} + k^2\vec{\Pi}^{(m)} = \frac{4\pi i}{\omega}\vec{j}^{(m)}. \tag{6.105}$$

It is easy to check that the field $\{\vec{E}, \vec{H}\}$ is expressed in terms of $\vec{\Pi}^{(m)}$ by formulas (3.24). This fact follows also from the analogy between the systems (1.29) and (3.117).

   The conditions for $\vec{\Pi}^{(e)}$ and $\vec{\Pi}^{(m)}$ differs only in the following. The potential $\vec{\Pi}^{(e)}$ as well as the field created by the electric current, does not satisfy any conditions except for the radiation one. The field created by a magnetic current, should additionally satisfy the usual conditions on the metallic surfaces, in which the slots are cut. Therefore, the kernel K of the integral expression

$$\vec{\Pi}^{(m)}(\vec{r}) = \int_{\Sigma} K(\vec{r}, \vec{r}') I^{(m)}(\vec{r}') \, dS_{r'} \tag{6.106}$$

is not $\exp(-ik\rho/(k\rho))$, as in (6.33). The kernel is

$$K(\vec{r},\vec{r}') = \frac{\exp(-ik\rho)}{k\rho} + Q(\vec{r},\vec{r}'), \tag{6.107}$$

where $\rho = |\vec{r} - \vec{r}'|$, and $Q(\vec{r},\vec{r}')$ depends on the shape of metallic surface. However, at $\rho \to 0$ the term in the kernel, denoted by Q, remains finite. Only the first term in (6.107) has the singularity at $\vec{r} = \vec{r}'$. This term is the same as in integral (6.33) in the theory of thin vibrator. Therefore, the main property of the thin vibrator that the value of $\vec{\Pi}^{(e)}$ on the surface is proportional to the value of the linear density of the current $I^{(m)}$ at the same point, remains valid for the relation between $I^{(m)}$ and $\vec{\Pi}^{(m)}$. Similarly to (6.35),

$$\vec{\Pi}_z^{(m)}(z) = -\frac{2i}{\omega} \ln \frac{L}{a} I^{(m)}(z) + \cdots, \tag{6.108}$$

where a is the slot width, and $2L$ is its length. The summands omitted in (6.108) differ from the integral summand in (6.35), but the term highest with respect to the small parameter $\chi = [2\ln(L/a)]^{-1}$ is the same.

The field H_z is obtained from $\vec{\Pi}_z^{(m)}$ by applying the operator $d^2/dz^2 + k^2$. At the slot ends, H_z should be zero.

The main results of the thin vibrator theory are transferred onto the theory of the narrow slot. In the slot, the length of which is close to the half-wavelength, the distribution of the transverse component of electric field is cosinusoidal and does not depend on the extrinsic force. The role of such a force is played by the current created by the same sources in the absence of the slot and cross cut by it. The amplitude of the electric field on the slot is proportional to the integral of this extrinsic force multiplied by the sine of the angle between this current and the slot direction, taken along the slot. If the slot length differs from the half-wavelength and is not multiple of it (*nonresonant* slot), then the distribution of the electric field depends on that of the extrinsic force, and the amplitude is small, of the order χ.

The theory of antennas consisting of several adjusted slots is constructed by the same scheme as the theory of several adjusted vibrators. The mutual influence of the slots is described by the coefficients depending both on their mutual location and on the shape of the antenna. These coefficients are involved in the linear algebraic equation system for the amplitudes of the magnetic fields in the slots. The system is similar to (6.52).

6.3.3
Resonant antennas

The volume resonator with slots in the walls, inside which a certain radiating element is located, is called the *resonant antenna*. The field arising in it is large in a narrow frequency range. The currents in the walls are also large, therefore, the electric field excited by these currents on the slots cross-cutting them, is large as well. The field on the slots is especially large if the slots are also tuned-up to this frequency, that is, if their length is equal or proportional to $\lambda/2$. The frequency, at which the field in the resonator is large, that is, the real part of the complex eigenfrequency of the open resonator (see Section 4.2), is close to the eigenfrequency of the closed (without slots) resonator. The shift of the resonant frequency, caused by the slot cutting, is larger if the slots are located more densely and their width is larger. We will return to this question at the end of Subsection 6.3.5. The radiation pattern depends on both the resonator shape and the slot locations.

The resonant antenna can radiate not only through the narrow slots but also through the walls segments made as the dense arrays or grids. The electric field exists on these segments of the antenna surface. It is proportional to the difference of the magnetic fields on both the sides of the surface. The proportionality coefficient (ratio of the electric field to above difference of the magnetic ones) is a characteristic of the array or grid, and it does not depend on the fields. It is a tensor of the second order, called the *transparency* of the surface segment.

The expression "the field on the short-periodical array" means the period-average value of the field or, which is the same, the field in a distance from the array much larger than its period and much smaller than the wavelength. In detail, this notion was defined in Subsection 5.2.3. This electric field $\vec{e}$ participates in the formulas for the external field of the antenna, in particular, for the radiation pattern.

The pattern created by the antenna of the form of a closed surface with holes is proportional to integral (6.104) taken over the whole surface. The vector $\vec{H}^g$ describes the magnetic field arisen on the metallic body of the same shape, illuminated by the plane wave of the unit amplitude, incoming from the direction, for which the pattern is calculated. If the slots are widely spaced, then the integral is taken only over the slots surface Σ, since $\vec{e} = 0$ on metal. On the array, the average field should be meant by $\vec{e}$, as it is explained in the preceding paragraph.

The transparency of the resonator walls must be small, otherwise the radiation losses would be large, the resonance would be weakly exposed and the field would be small even in the resonant frequency range. At the small transparency, the magnetic field inside the resonator is close to that in the closed one. The field $\vec{e}$ can be accepted to be proportional to the current flowing on the wall in the eigenoscillation of the closed resonator.

The contribution of any surface segment into the pattern depends not only on the field $\vec{e}$ on this segment but also on the shape of the whole surface S. For instance, the surface can shadow to some extent a certain direction from the radiation of this segment. This property of the antenna radiating through the semi-transparent surface or the slots, is accounted in formula (6.104) by the multiplier $\bar{H}^g$. This multiplier (the Green function) depends on the shape of the whole surface, it is not a local function of the given segment. If the plane wave coming from a certain direction almost does not illuminate the segment, then the field on it almost does not participates in creating the pattern in this direction. In fact, this is one of the formulations of the reciprocity principle.

6.3.4
Transparency of short-periodical array

The simplest theory of the short-periodical array, explained in Subsection 5.2.3, was sufficient for investigation of the helix line. However, for determination of the array transparency, this theory must be specified. We confine ourselves to the theory of infinitely thin strips.

On both the sides of the array, the electric field is the same, and the magnetic one has a jump. As in Subsection 5.2.6, by the field on the array we imply the limit, to which the field tends in the distance from the array, much larger than the array period p and much smaller than the wavelength λ. It follows from (5.68) that

$$E_t^{(1)} - E_t^{(2)} = 0, \qquad E_s^{(1)} - E_s^{(2)} = 0. \tag{6.109}$$

Here s and t are the directions along and across the strips, respectively; both directions lie in the array plane. The indices (1) and (2) refer to different sides of the array. The normal directed from (1) to (2) and the directions s and t make up the right triple.

Formula (6.109) remains valid in the more strict theory. However, instead of (5.68), the boundary conditions on the array contain, besides for (6.109), the relations

$$E_s = -iP_s(H_t^{(1)} - H_t^{(2)}), \tag{6.110a}$$

$$E_t = iP_t(H_s^{(1)} - H_s^{(2)}). \tag{6.110b}$$

The coefficients P_s, P_t describe transparencies of the strip array. Formulas (6.110) refer to the case when the field does not vary along the strips, that is, $\partial/\partial s \equiv 0$. If this derivative differs from zero, additional terms are present on the right-hand side of (6.110).

For the strip array, the transparencies are expressed explicitly by p, λ and the filling factor q equal to the ratio of the strip width to the period p, as follows:

$$P_s = -\frac{p}{\lambda} \ln \sin \frac{\pi q}{2}, \tag{6.111a}$$

$$P_t = \frac{\lambda}{4p} \frac{1}{\ln \cos(\pi q/2)}. \tag{6.111b}$$

These formulas are obtained from the solution of two independent scalar problems for the function $\varphi(t, N)$ satisfying the equation $\partial^2 \varphi / \partial t^2 + \partial^2 \varphi / \partial N^2 = 0$ with the boundary conditions $\varphi = 0$ or $\partial \varphi / \partial N = 0$, respectively, on the strip surface. Here t and N are the local coordinates on the strip, the t-axis lies in the strip plane, perpendicular to the strips, N is perpendicular to the strip plane, $\partial / \partial N$ is the normal derivative. The function φ satisfying the Dirichlet condition on the strip, is the electric field e_s; the function satisfying the Neumann condition is the magnetic field h_s. The fields e_s, h_s are the real fields, not the average ones E_s, H_s. The fields e_s, h_s are the periodical functions of the variable t with the period p. At $N \gg p$, they become the average ones E_s, H_s participating in formulas (6.109), (6.110).

The above Laplace equation can be solved, in particular, by the *conformal mapping method*. For the array constructed from the infinitely thin strips, the plane (t, N) contains the pieces of the straightforward lines (Fig. 6.5). For this geometry the conformal mapping is expressed in an explicit form. In such a way we obtain the explicit formulas for the transparency coefficients in (6.110). We omit these cumbersome calculations here.

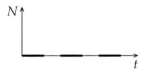

Fig. 6.5 The strip array

In the simpler theory, sufficient for describing the eigenmode in the helix line (Subsection 5.2.6), we put $P_s = 0$, $P_t = \infty$. These values can be obtained from (6.111) at $p / \lambda \to 0$. The array described by formulas (5.68) obtained for this case is completely nontransparent for the field in which $\vec{E}$ is directed along the strips, and completely transparent for the field with $\vec{E}$ directed across them.

Formulas (6.110), (6.111) specify this result. For instance, the transmission coefficients of the plane wave falling perpendicularly to the array, equal

$$D_s = \frac{2iP_s}{1 + 2iP_s}, \tag{6.112a}$$

$$D_t = \frac{2iP_t}{1 + 2iP_t} \tag{6.112b}$$

for the corresponding polarizations. According to these formulas, the array is partially transparent ($D_s \neq 0$) for the s-polarization and not completely transparent ($|D_t| < 1$) for the t-polarization.

Since $p \ll \lambda$ (the array is short-periodical), then if q is not too small and not too close to unity, then $|P_s| \ll 1$, $|P_t| \gg 1$. The array properties are close to those at which $D_s = 0$, $D_t = 1$. However, according to (6.111a), $P_s \to \infty$ at

$q \to 0$. If the strip width is much smaller than the period, then $D_s \to 1$ and the array ceases to reflect the s-polarized wave. In the second limiting case, when $q \to 1$, we have $P_t \to 0$, $D_t \to 0$. The array completely reflects the t-polarized wave if the slots between the strips are much smaller than the period. Both these results are physically obvious.

Since the transmission coefficients (6.112) are different for different polarizations, the wave polarized not along or across the strips becomes elliptically polarized after passing through the short-periodical array. This effect manifests itself the most of all when the wave falls, in which the vector $\vec{E}$ makes a small angle with the s-direction (along the strips). Denote this angle by α; $\alpha \ll 1$. In the incident wave, $E_t = E_0 \sin \alpha \approx E_0 \alpha$, $E_s = E_0 \cos \alpha \approx E_0$, where E_0 is the incident wave amplitude. If q and $1 - q$ are not small, then $|P_s| \ll 1$, $|P_t| \gg 1$. According to (6.112), in the transmitted wave, we have

$$E_s = E_0 \cdot 2iP_s, \qquad E_t = E_0\alpha. \tag{6.113}$$

The wave is elliptically polarized. In particular, if both the small values α and $2P_s$ coincide, that is, if $\alpha = -2\pi/\lambda \cdot \ln \sin(\pi q/2)$, then the transmitted wave is circularly polarized.

These elementary properties following from formulas (6.110), (6.111), are inherent to the short-periodical array not only at the normal incidence of the plane wave onto it but also when the transmitted wave has the amplitudes (5.87) (at $E_0 = 1$). These formulas describe the general relations between the electric field on the array and the current flowing on the wall when substituting the array with the solid metal. These relations are also valid when placing the array into a much complicated field. It is only needed to add the terms in (6.110), containing the derivative $\partial/\partial s$. These formulas give the fields arisen on the resonant antenna surface, which create the radiated field.

6.3.5
Dielectric antennas

Any body made from the material with $\varepsilon \gg 1$, is an open resonator, that is, an antenna. Its Q-factor has the order $\sqrt{\varepsilon}$. The advantage of the dielectric antennas is their small sizes (in comparison with λ). The resonant antennas considered in Subsection 6.3.3 can be small only if the interior volume has a complicated structure, for instance, its shape is a spiral or meander.

The field excited in the dielectric body by a certain source located inside it weakly penetrates into the exterior space. The interface of two media having $\varepsilon \gg 1$ and $\varepsilon = 1$, respectively, is a small-transparent screen. This fact follows, for instance, from formulas (2.51), (2.52) for the reflection coefficient in the simplest problem about the normal incidence of the plane wave from the medium with $\varepsilon > 1$ onto the interface with the medium with $\varepsilon = 1$. If the field in the first medium is represented as

$$E = \exp(-ik\sqrt{\varepsilon}z) + R\exp(ik\sqrt{\varepsilon}z), \tag{6.114}$$

then the reflection coefficient R equals

$$R = \frac{\sqrt{\varepsilon}-1}{\sqrt{\varepsilon}+1}. \tag{6.115}$$

It is assumed that $\mu = 1$ in both the media. At $\varepsilon \gg 1$,

$$R = 1 - \frac{2}{\sqrt{\varepsilon}} + \cdots . \tag{6.116}$$

The wave almost completely reflects from the interface. Inside the medium with $\varepsilon > 1$, the magnetic field amplitude is $\sqrt{\varepsilon}$ times larger than that of the electric one, but $|\vec{H}| = |\vec{E}|$ on the interface, $|\vec{H}|$ has the order $|\vec{E}|$, since such a relation between the amplitudes exists in the exterior medium, and the fields on the interface are continuous. Therefore, at $\varepsilon \gg 1$, the reflection from the interface with $\varepsilon = 1$ proceeds almost in the same way as from the surface with the condition

$$H_{\text{tan}} = 0. \tag{6.117}$$

The reflection coefficient from such a surface would be $R = 1$.

The virtual medium on the surface of which condition (6.117) holds, is the ideal magnetic. This condition holds if $\mu = \infty$ in the medium. Note that the reflection coefficient from ideal metal, that is, from the medium with $\varepsilon = \infty$ has the opposite sign, it equals -1.

We find the eigenoscillation of the simplest model of the dielectric resonator. Let the dielectric plate be bounded by the planes $z = -a$ and $z = a$. The dielectric permittivity of its material is $\varepsilon \gg 1$. The eigenoscillation is considered in which the field depends only on the z-coordinate. Only the two components, E_x and H_y differ from zero in the field. They are connected by the Maxwell equations

$$\frac{dH_y}{dz} = -ik\varepsilon E_x, \qquad \frac{dE_x}{dz} = -ikH_y. \tag{6.118}$$

At $|z| \to \infty$, the radiation conditions must be fulfilled: the field must be an outgoing wave. For the even oscillation, the field on the plate is

$$E_x = \cos(k\sqrt{\varepsilon}z), \quad H_y = -i\sqrt{\varepsilon}\sin(k\sqrt{\varepsilon}z), \qquad |x| < a. \tag{6.119}$$

At $z > a$, the field represents an outgoing wave, $H_y / E_x = -1$ in it. The equation

$$\tan(k_N\sqrt{\varepsilon}a) = \frac{i}{\sqrt{\varepsilon}}, \qquad N = 1, 2, \ldots \tag{6.120}$$

for the complex eigenfrequency follows from the field continuity at $z = a$. At $\varepsilon \gg 1$, the roots of this equations are approximately equal to

$$k_N = \frac{\pi N}{a\sqrt{\varepsilon}} + \frac{i}{a\varepsilon}, \qquad N = 1, 2, \ldots \tag{6.121}$$

They are close to the roots of the equation

$$\tan(k_N^0 \sqrt{\varepsilon}a) = 0 \tag{6.122}$$

for the eigenfrequencies k_N^0 of the closed resonator with condition (6.117). These roots are

$$k_N^0 = \frac{\pi N}{a\sqrt{\varepsilon}}, \qquad N = 1, 2, \ldots \tag{6.123}$$

According to (6.121), the plate of the material with $\varepsilon \gg 1$ is a high-quality resonator. Its Q-factor, that is, the ratio $Q = k_N'/(2k_N'')$ equals

$$Q = \frac{N\pi}{2}\sqrt{\varepsilon}. \tag{6.124}$$

On the boundary of this open resonator, the electric field equals $\cos(k_N \sqrt{\varepsilon}a)$. It differs from the electric field $\cos(k_N^0 \sqrt{\varepsilon}a)$ on the walls of the closed resonator with the same ε and the boundary condition (6.117), only by the value of the order $1/\varepsilon$ (under the condition that the fields are the same at large distances from the boundaries of the compared resonators). In fact, the electric field on the boundary of the open resonator with $\varepsilon \gg 1$ coincides with that of the closed resonator with the boundary condition (6.117), filled with the same dielectric.

This result is valid for the dielectric resonator of arbitrary shape. The relation between the amplitudes of the electric and magnetic fields on any closed surface S is defined only by the shape of this boundary and by the bodies located outside the resonator (if they exist). The field $\vec{E}$ on S defines the field in the whole space outside S, in particular, the field $\vec{H}$ on S. This field has the same order as $\vec{E}$; it does not depend on the medium located inside the surface. The field $|\vec{H}|$ is about $\sqrt{\varepsilon}$ times larger than $|\vec{E}|$ there. This fact follows from formula (4.17), which is valid, as it can be easily shown, also under condition (6.117) on the walls of the closed resonator. Hence, the demand for the amplitude of $\vec{H}$ on the surface to be of the same order as the amplitude of $\vec{E}$, is close to condition (6.117).

The general theory of closed resonators with the boundary condition (6.117) is constructed by the same scheme as the theory of closed resonators with the boundary condition (4.10b). There exists a countable sequence of the real eigenfrequencies k_N^0, $N = 1, 2, \ldots$ One or several eigenoscillations with the fields $\{\vec{E}^N, \vec{H}^N\}$ correspond to each eigenfrequency. The fields corresponding to different eigenfrequencies are orthogonal (see (4.16)).

With small corrections, the variational technique for calculating the eigenfrequencies also remains the same. The functional (4.24) for the problem with the boundary condition (11.8b) is stationary on the eigenoscillations, and the above condition should not be imposed on the admissible functions. In contrast to (4.24), the functional (4.33) is stationary on the eigenoscillations only if the admissible functions satisfy condition (11.8b). Under condition (6.117), the properties of these functionals are interchanged.

The closeness of the condition on the interface between the medium with $\varepsilon \gg 1$ and vacuum, and the boundary condition (6.117) leads to the closeness of the electric field distributions on the boundaries of the closed and open resonators with the same surface S. The radiation pattern of the antenna can be calculated if the electric field distribution on its surface is known. However, in order to find the amplitude of the field excited by a certain source and the frequency dependence of this amplitude, we must know the complex eigenfrequency of the open resonator.

For calculating the imaginary part k_N'' of the eigenfrequency, it is sufficient to know the radiation pattern. The energy carried away by the spherical wave radiated by the antenna is proportional to k_N''

$$k_N'' = \frac{\int_S \mathrm{Re}\left(\vec{E} \times \vec{H}^*\right)_N dS}{\int_V \left(|H|^2 + \varepsilon |E|^2\right) dV}. \tag{6.125}$$

The integrals in the numerator and denominator are taken over the antenna surface and its volume, respectively. This formula can be obtained by integration of the expression $\mathrm{div}(\vec{E} \times \vec{H}^*)$ over the volume, as it was made in Section 4.1. For this purpose, it is sufficient to replace rotors in (1.55) by those from the homogeneous Maxwell equations and take the real part of the obtained equality. This formula does not depend on the field normalization. The integral in the numerator equals (with accuracy to the factor $c/(8\pi)$) the Poynting vector flux through the surface S. This flux is equal to the same vector flux through the infinitely removed sphere and hence it is expressible by the radiation pattern.

The above consideration relates to both the dielectric antenna and the resonant antenna, considered in Subsection 6.3.3. At a small transparency of the walls, the electric field on the resonant antenna surface equals the boundary value of the magnetic field in the closed resonator, multiplied by the transparency. Similarly as for the dielectric antenna, the radiation pattern (and hence the imaginary part of the eigenfrequency) are calculated by the field on the exterior boundary.

More complicated calculations are required for determining the shift of the real part k_N' of complex eigenfrequency of the open resonator with respect to eigenfrequency k_N^0 of the closed one. Similarly as k_N'', this value depends on the objects located outside the surface S. However, in contrast to k''_N, the value k'_N cannot be expressed by the radiation pattern. For its determination, the field $\vec{H}$ on S must be preliminary calculated by the known field $\vec{E}$ on the same surface. The latter problem is much complicated than the calculation of the radiation pattern. For the case when $\vec{E}$ differs from zero only on the narrow long strip, this problem was solved in Subsection 6.3.2 in the slot theory. We do not give the appropriate cumbersome formulas for the general case.

6.3.6
The inverse problem for volume antennas

The inverse problem for the volume antennas consists in finding the electric field on the given antenna surface which creates the given radiation pattern. After finding this field, we can then solve the inverse problem for the reso-nant antenna in the following formulation: to find the surface S transparency distribution, at which the antenna has the given radiation pattern.

Formally, the problem about the field on the surface S can be reduced to the integral equation, the symbolic form of which is

$$\int_S e(S)K(S,\Omega)\,dS = F(\Omega). \tag{6.126}$$

Here S is the coordinate of the point on the antenna, dS is a surface element, and $\Omega = \{\theta, \varphi\}$ is the aggregate angular coordinate on the infinitely removed sphere. The function $e(S)$ is the sought electric field, $F(\Omega)$ is the given pat-tern. According to (6.104), the kernel $K(S,\Omega)$ is the magnetic field at the point S, arising on the surface of an ideal-conducting body with the same boundary, at the incidence of the plane wave of unit amplitude coming from the direction Ω. In the detailed form, (6.126) should be replaced by the system of two equa-tions for the two polarizations, the function e and kernel K should be replaced by the corresponding vectors.

We explain a method for solving this problem on the two-dimensional (and hence scalar) model. Let an electric field E_z creating the pattern $F(\varphi)$ exist on the cylinder, cross-section of which is bounded by the contour C, and the director is parallel to the z-axis. The contour C and function $F(\varphi)$ are given, the function E_z on C is to be found.

Denote the function $E_z(r, \varphi)$ by $u(r, \varphi)$. This function satisfies the equation

$$\frac{\partial^2 u}{\partial r^2} + \frac{1}{r}\frac{\partial u}{\partial r} + \left(k^2 - \frac{1}{r^2}\frac{\partial^2}{\partial \varphi^2}\right)u = 0 \tag{6.127}$$

and the radiation condition

$$u|_{r \to \infty} \simeq F(\varphi)\frac{\exp(-ikr)}{\sqrt{kr}}. \tag{6.128}$$

The magnetic field has only components H_r, H_φ, and can be calculated by $u(r, \varphi)$ according to (1.83).

The problem of finding the function $u|_C$ can be reduced to the first-kind integral equation

$$\int_C u(s)K(s,\varphi)\,ds = F(\varphi). \tag{6.129}$$

Here s is the coordinate on the contour C. The kernel $K(s,\varphi)$ is the magnetic field component, tangential to the contour C, which would arise on the ideal-conducting cylinder with the same contour of the cross-section, in which the electric field is parallel to the z-axis, at the incidence of the plane wave of unit amplitude coming from the direction φ.

The problem is similar to that formulated in Section 5.3, about the field on a part of the plane, creating the given field on another plane. In general case, the problem has no solution. Similarly, the problem about the field on C creating the given field at infinity has also no solution in general case, that is, for an arbitrary contour C and function $F(\varphi)$. In this subsection we investigate this question, using formulas (6.127), (6.128), without solving the integral equation (6.129).

There exist two main reasons by which the above problem may have no solution. Firstly, the function $F(\varphi)$ may not be sufficiently smooth to be a pattern. Secondly, the contour C can be too small for the antenna bounded by it to be able to create the given pattern. We specify these reasons and point out the way for finding the solution if it exists.

We find the function $u(r,\varphi)$ solving problems (6.127), (6.128). For simplicity, we assume that $F(\varphi)$ is an even function. Express it into the Fourier series

$$F(\varphi) = \sum_{n=0}^{\infty} A_n \cos(n\varphi). \tag{6.130}$$

Partial solutions to the wave equation (6.127), proportional to $\cos(n\varphi)$ are

$$\sqrt{\frac{\pi}{2i}} \exp\left(-\frac{i\pi n}{2}\right) H_n^{(2)}(kr)\cos(n\varphi), \qquad n = 0,1,2,\ldots \tag{6.131}$$

The asymptotics of these solutions are $\exp(-ikr)/\sqrt{kr} \cdot \cos(n\varphi)$. This follows from the asymptotic of the Hankel function

$$H_n^{(2)}(kr) \approx \sqrt{\frac{2i}{\pi}} \exp\left(\frac{i\pi n}{2}\right) \frac{\exp(-ikr)}{\sqrt{kr}}, \tag{6.132}$$

valid for $kr \gg 1$, $kr \gg n$. Consequently, a formal solution to equation (6.127), having the asymptotic (6.128), (6.130) is

$$u(r,\varphi) = \sqrt{\frac{\pi}{2i}} \sum_{n=0}^{\infty} A_n \exp\left(-\frac{i\pi n}{2}\right) H_n^{(2)}(kr)\cos(n\varphi). \tag{6.133}$$

This function exists if the series converges.

The series convergence is defined by the rate of its term decreasing at $n \to \infty$, that is, by the asymptotic of the far terms of the series. At $n \gg kr$, the Debye asymptotic

$$H_n^{(2)}(kr) \approx i\sqrt{\frac{2i}{\pi n}}\left(\frac{2n}{ekr}\right)^n, \qquad e = 2.71\dots \tag{6.134}$$

of the Hankel function holds. The far terms of series (6.133) have the form

$$\frac{A_n}{\sqrt{n}}\left(\frac{2n}{ekr}\right)^n \cos(n\varphi), \tag{6.135}$$

where a common factor is omitted. The coefficient at $\cos(n\varphi)$ can be written as $(a_0/r)^n$, where

$$a_0 = \frac{2}{ek}\lim_{n\to\infty}\left(n|A_n|^{1/n}\right). \tag{6.136}$$

If this limit exists, then the general term of series (6.133) at $n \gg kr$, $n \gg 1$ is subject to the estimation $(a_0/r)^n \cos(n\varphi)$ (in fact, it is sufficient that the upper limit exits in (6.136)). The series with such an asymptotic of the general term converges at $r > a_0$ and diverges at $r < a_0$. At $r = a_0$, the series converges to the generalized function $\delta(\varphi - \widehat{\varphi})$ with certain $\widehat{\varphi}$ or to a sum of such functions.

If the coefficients of series (6.130) decrease so slowly that the limit (6.136) does not exist ("equals infinity"), then there is no function $u(r,\varphi)$ satisfying equation (6.126) and condition (6.128). This means that the function $F(\varphi)$ cannot be a radiation pattern. For instance, if $A_n = O(n^{-p})$ at some $p > 1$, then series (6.130) converges, but a_0 does not exist, and therefore there is no antenna, the pattern of which is $F(\varphi)$. At such asymptotic of A_n, one of the derivatives $\partial^t F/\partial \varphi^t$ (and all the next) does not exist. The Fourier coefficients of any pattern should decrease faster than n^{-p}.

If the limit in (6.136) exists but a_0 is so large that the contour C lies inside the circle of radius a_0, centered in the origin (dashed lines in Fig. 6.6), then the inverse problem has no solution at the given C and $F(\varphi)$. Really, in this case a ring (shaded area in Fig. 6.6(a)) exists inside which the contour C lies, and the ring itself lies inside the circle of the radius a_0. The series for the field in this ring cannot contain the terms with $H_n^{(1)}(kr)$, because this fact would contradict the demand (6.128), and the series containing only $H_n^{(2)}(kr)$ does not converge. Consequently, there is no field on the contour C, satisfying (6.127) and (6.128).

The Gauss function

$$F(\varphi) = \exp\left(-M\sin^2\frac{\varphi}{2}\right) \tag{6.137}$$

with a given M is an example of such a function. The coefficients A_n for this function are proportional to the modified Bessel function $I_n(M/2)$. At the

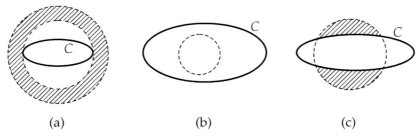

Fig. 6.6 Illustration of the inverse problem solvability

large n, this function asymptotically equals $[eM/(2n)]^n$, so that $a_0 = M/k$. The larger the number M, that is, the sharper the pattern to be created, the larger must be the contour C. The circle of the radius M/k must not contain this contour inside itself.

In the case when the contour C is so large that the circle of the radius a_0 completely lies inside it (Fig. 6.6(b)), then the inverse problem has a solution which can be found in a very simple way. The contour C lies in the domain where the field $u(r, \varphi)$ is known. It is expressed by the convergent series (6.133). The value $u|_C$ is the sought field on C, at which the antenna creates the pattern $F(\varphi)$. This statement follows from the fact that if there are no objects outside C, then the value $u|_C$ determines the field everywhere outside C. If two different values of the field existed at a certain point, then the difference field, equal to zero on C, would not be identical to zero, which would violate the uniqueness of the solution to the Maxwell equations.

The inverse problem is also solvable in the case if the circle of radius a_0 intersects the contour C so that a part of C lies outside the circle and the rest inside it (Fig. 6.6(c)). Outside the circle the field $u(r, \varphi)$ is expressed by the convergent series (6.133). The field exists also in the domain inside the circle but outside C (shaded area in Fig. 6.6(c)), but it is not expressed by (6.133). The series for $u(r, \varphi)$ in this domain should contain also the terms with $H_n^{(1)}(kr) \cos(n\varphi)$. In this domain the field also contains the waves propagating to the antenna surface ("back waves"). The presence of these waves does not contradict the radiation condition, since in this form the series should represent a continuous function not for all values of φ; inside C this series has no physical sense. After constructing the field $u(r, \varphi)$ everywhere outside C, we obtain the value $u|_C$ by "embedding" the contour in this field.

The radius a_0 of the circle is given by (6.136) as a limit. The value of the limit is defined by the behavior of the coefficients A_n in (6.130) at large n. It is obvious that a small alteration of $F(\varphi)$ can change these far terms of the Fourier series in such a way that the limit in (6.136) will have any value from 0 to ∞. For instance, by truncating the series, we obtain $a_0 = 0$. On the other hand, if we add the function $\sum_{n=N}^{\infty} n^{-2} \cos(n\varphi)$ to the series, then at $N \gg 1$ the function $F(\varphi)$ varies a little, but $a_0 = \infty$. Small alterations of the function $F(\varphi)$ may essentially change the solvability conditions for the inverse problem, and the sought function $u|_C$ itself. This instability of the problem is also connected with the fact that $u|_C$ satisfies the first-kind integral equation (6.126) with the smooth kernel and integration over the bounded medium. It is known that this problem is ill-posed and its solution is unstable. By this reason, for instance, the results obtained above that the pattern (6.137) cannot be created by the antenna completely lying inside the circle of the radius a_0, do not contradict the fact that a pattern arbitrary close to (6.137) can be created by an arbitrary small antenna. However, this antenna can turn out to be superdirective, that is, the function $u|_C$ can have large amplitude jumps and it cannot be realized practically. But, formally, the solution to the stated inverse problem exists. As a rule, all inverse problems are ill-posed.

Considerations of this subsection transfer on the three-dimensional vector problems almost automatically. The pattern given by the two functions $F_1(\theta, \varphi)$, $F_2(\theta, \varphi)$ can be created by the electric field distributed over a surface if (and only if) this surface is so large that it cannot be located inside a sphere of the radius depending on these functions. The sharper the pattern, the larger the radius and the larger antenna must be, to create this pattern. However, there always exist patterns close to the functions F_1, F_2 for which this radius is small. These patterns approximating the given one can be, in principle, created by a small antenna.

When transferring the method described above for the two-dimensional scalar problem into the three-dimensional vector problem, the singlefold Fourier series by the trigonometric functions is substituted with the double one by the much complicated functions of the angles. The radial functions are complicated as well. The fields are expressed by the two (not one) scalar functions. There are also other technical difficulties. This technique will be partially described in the next chapter.

6.3.7
Analytical continuation of the field

In this subsection, another statement of the problem about the volume antenna is considered. In this statement, only the desired pattern is given, whereas the antenna surface is unknown. The problem is to find a surface, as small as possible, which could be the surface of the antenna creating this pattern. Above, we have shown that this surface cannot wholly lie inside a certain circle (sphere in three-dimensional case). Now we establish several more exact conditions, which must be satisfied by the surface. The constructions and derivations are illustrated on the two-dimensional model.

We construct the field in the whole space having the given pattern. It is assumed that there are no bodies in space. Outside the circle of the radius a_0 the field is described by series (6.133). Continue this field inside the circle analytically, that is, in such a way that the field components and their derivatives are continuous on the circle. This continuation obligatory has singularity points, at which the field is infinite. If such points were absent, then the field should satisfy the Maxwell equation in the whole space and the radiation condition, but such a field is zero if the current is absent. The key task in the problem is to localize these points.

The antenna surface must enclose all the singularity points of the continued field. This follows from the fact that the field outside the antenna has no such points. The field on the constructed surface must be the same as the field in which the antenna is "embedded." The antenna with such a surface and such field distribution on it creates the given radiation pattern.

After the antenna is really constructed, the field inside it is, of course, not connected with the field found by the analytical continuation of the external field under assumption that the antenna is absent. This domain is called *nonphysical*.

Further, in Subsection 7.4.8, we describe an application of the analytical continuation method to the diffraction problems. The structure of the field continued into the nonphysical domain is discussed there. Here we only mention two methods of localization of the singularity points. The more precise this localization is, the more exact the condition on the surface of antenna, creating the given pattern. All singularity points must lie inside the antenna or on its surface. The first method uses the fact that at least one singularity point lies on the circle of radius a_0 (6.136), centered in the coordinate origin. Of course, this point lies on a part of the circle, located inside the contour C (Fig. 6.6(c)), that is, on the nonphysical domain boundary. The method consists in introducing a new coordinate systems shifted with respect to the old one. This shift changes the pattern by the factor $\exp[ikb\cos(\varphi - \tilde{\varphi})]$, where b is the length of the segment by which the origin is shifted, the angle $\tilde{\varphi}$ describes the shift direction. The Fourier coefficients are changed as well, and hence the value a_0 is changed, either.

If there is only one singularity point on the circle, then the origin should be moved directly to this point. Then the radius a_0 decreases, that is, the localization of the domain where the singularity points lie, is improved. If there are two singularity points on the circle, then the origin should be shifted to center of the chord connecting these points. If the points lie on the one diameter, then the singularity point localization cannot be improved by a circle, since the minimal circle is already constructed, outside which the singularity points are absent. In this case, the area of the domain containing all singularity points can be decreased by substituting the minimal circle by two arcs of larger circles passing through the two given points.

The second method for localization of singularity points of the field given by its asymptotic at $r \to \infty$ is based on introducing the plane of the complex angle $\bar{\varphi} = \alpha + i\beta$. The field in any point (r, φ) can be expressed as the integral

$$u(r, \varphi) = \int_{-i\infty}^{\pi+i\infty} F(\bar{\varphi}) \exp\left[-ikr\cos(\bar{\varphi} - \varphi)\right] d\bar{\varphi} \qquad (6.138)$$

taken over the contour on the complex plane $\bar{\varphi}$, shown in Fig. 6.7 (*Sommerfeld contour*). This function satisfies equation (6.127) in the whole plane and has the asymptotic (6.128) at infinity.

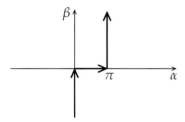

Fig. 6.7 The Sommerfeld contour

It can be shown that this integral has no singularities in the half-plane

$$r\sin\varphi > C. \qquad (6.139)$$

The number C is found by the given function $F(\varphi)$ in a certain way based on the theory of entire functions. Condition (6.139) bounds the domain where the singularity points are located by a straight line.

Turning the coordinate system changes the function $F(\varphi)$. The position and direction of the straight line on one side of which singularity points lie is changed as well. Varying the turning angle from 0 to 2π, we can construct the minimal domain bounded by these lines, outside which the singularity points are absent. In this way we can localize the singularity points in the domain, smaller than the minimal circle (or domain bounded by two circle arcs) constructed by the first method.

7
Diffraction on Metallic and Dielectric Objects

7.1
Diffraction of the plane wave on circular waveguide

7.1.1
Scalar potentials

The simplest diffraction problem is the problem of the plane wave scattering on an infinite circular cylinder (as a model). Expressions for the fields in not very large distances, obtained in this problem, are approximately also valid for a cylinder with length much larger than both the wavelength and its cross-section size.

Below, we consider the case when the plane wave propagates perpendicular to the cylinder axis. In this case the incident field does not depend on the z-coordinate of the cylindrical coordinate system (z, r, φ), axis of which coincides with the cylinder one (Fig. 7.1). Since the cylinder surface does not depend on z, the diffracted field does not depend on it as well.

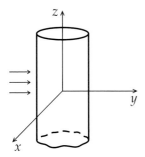

Fig. 7.1 The circular cylinder

At $\partial/\partial z \equiv 0$, the Maxwell equations are divided into two systems, for the components E_z, H_r, H_φ and H_z, E_r, E_φ, respectively. The components H_r, H_φ of the first system can be expressed by E_z, the components E_r, E_φ of the second one by H_z. These two functions play a role of potentials; the field components are expressed by their derivatives.

High-frequency Electrodynamics. Boris Z. Katsenelenbaum
Copyright © 2006 WILEY-VCH Verlag GmbH & Co. KGaA, Weinheim
ISBN: 3-527-40529-1

It follows from the Maxwell equations that

$$H_r = \frac{i}{k}\frac{\partial U}{r\partial\varphi}, \quad H_\varphi = -\frac{i}{k}\frac{\partial U}{\partial r}, \quad E_z = U. \tag{7.1}$$

The function U is a Cartesian component satisfying the wave equation (1.41)

$$\frac{\partial^2 U}{\partial r^2} + \frac{1}{r}\frac{\partial U}{\partial r} + \left(k^2\varepsilon + \frac{\partial^2 U}{r^2\partial\varphi^2}\right) = 0. \tag{7.2}$$

The dielectric permittivity is kept in this equation in order to use it in the problem of diffraction on the dielectric cylinder (see Subsection 7.1.4). Equations (7.1), (7.2), as well as equations (7.3) are valid for any dependence of ε on the transverse coordinates r, φ. In this chapter, it is assumed that $\mu = 1$.

In the second equation system connecting the components H_z, E_r, E_φ, the role of the potential function is played by $H_z = V(r, \varphi)$. The electric field components are expressed by V as

$$E_r = -\frac{i}{k\varepsilon}\frac{\partial V}{r\partial\varphi}, \quad E_\varphi = \frac{i}{k\varepsilon}\frac{\partial V}{\partial r}. \tag{7.3}$$

The wave equation for V is more complicated than (7.2); it involves the derivatives $\partial\varepsilon/\partial r$ and $\partial\varepsilon/\partial\varphi$. We do not write it here. In domains where $\varepsilon = const$, the wave equation for V coincides with (7.2).

The equations for U and V are not connected. The functions U and V satisfy the independent equations. If the polarizations are not connected by the excitation conditions, then the functions are not connected at all. If $V \equiv 0$, that is, $E_r \equiv 0$, $E_\varphi \equiv 0$ in the incident wave, so that the wave is linearly polarized along the z-axis, then the diffracted field is also linearly polarized along this axis. This assertion is also valid for the second polarization, as well as for cylinders with arbitrary cross-sections.

7.1.2
The variable separation method: the Rayleigh series

Consider a case when $V \equiv 0$, that is, the field $\vec{E}$ has the only component $E_z = U(r, \varphi)$. The total field U is a sum of the incident field U^0 and the diffracted field U^{diff}, $U = U^0 + U^{\mathrm{diff}}$. The field U^{diff} satisfies (7.2), the radiation condition and the boundary one

$$U^{\mathrm{diff}} = -U^0, \quad r = a \tag{7.4}$$

on the cylinder surface, where a is the cylinder radius.

At any $r > a$, $U^{\mathrm{diff}}(r, \varphi)$ is expressible into the Fourier series by the functions $\cos(n\varphi)$ in the form

$$U^{\mathrm{diff}}(r, \varphi) = \sum_{n=0}^{\infty} A_n Z_n(r) \cos(n\varphi). \tag{7.5}$$

Here we omit the terms with $\sin(n\varphi)$, assuming temporary that the waveguide cross-section as well as the incident field is symmetrical with respect to the axis $\varphi = 0$. Substituting (7.5) into (7.2) and accounting the orthogonality of the functions $\cos(n\varphi)$, we obtain that the functions Z_n should satisfy the equation

$$\frac{d^2 Z_n}{dr^2} + \frac{1}{r}\frac{dZ_n}{dr} + \left(k^2\varepsilon - \frac{n^2}{r^2}\right) Z_n = 0. \tag{7.6}$$

This is equation for the cylindrical functions. It follows from the radiation condition for the function U^{diff} that $\dot{Z}_n(r) = H_n^{(2)}(kr)$. Therefore, (7.5) becomes

$$U^{\text{diff}} = \sum_{n=0}^{\infty} A_n H_n^{(2)}(kr) \cos(n\varphi). \tag{7.7}$$

The boundary condition (7.4) leads to the functional equation

$$\sum_{n=0}^{\infty} A_n H_n^{(2)}(ka) \cos(n\varphi) = -U^0(a,\varphi), \tag{7.8}$$

which should hold in the whole interval $0 \le \varphi \le 2\pi$. System of functions $\cos(n\varphi)$ is complete (for even functions) and orthogonal in this interval. Therefore, for the coefficients A_n of series (7.7), we have

$$A_n = -\frac{1}{(1+\delta_{0n})\pi}\frac{1}{H_n^{(2)}(ka)}\int_0^{2\pi} U^0(a,\varphi)\cos(n\varphi)\,d\varphi. \tag{7.9}$$

This formula holds for any incident field U^0. If the incident wave (7.10) is the plane one propagating along the x-axis, that is, its field is

$$U^0 = \exp(-ikr\cos\varphi), \tag{7.10}$$

then

$$A_n = -\frac{2(-i)^n}{(1+\delta_{0n})}\frac{J_n(ka)}{H_n^{(2)}(ka)}. \tag{7.11}$$

Series (7.7) with coefficients (7.11) converges at any $r \ne 0$. It also converges at $r < a$, that is, in the domain where the field is absent. In Subsection 6.3.7, the physical meaning of such series at $r < a$ was explained.

Series (7.7) is called the *Rayleigh series*. Each term of this series is a product of one-variable functions. The method in which the solution is represented as the series with such terms is called the *variable separation method*. It is applicable to the wave equation if surface on which the boundary conditions are imposed is a coordinate one. In the same form the solution of scattering problem on the elliptic metallic cylinder can be obtained. The surface of such a cylinder is a coordinate one of the elliptical coordinate system. Each term of the series is a product of the *radial* and *azimuthal Mathieu functions* (analogous to the Hankel and trigonometric functions, respectively) of elliptic coordinates similar to the cylindrical ones r and φ, respectively. However, for instance, this method is not applicable for the cylinder with the rectangular cross-section.

When the incident wave is polarized in the plane (r, φ), then $U \equiv 0$, and the potential V of the total field must satisfy, according to (7.3), the condition $\partial V / \partial r = 0$ on the cylinder surface. The potential V is the sum $V^0 + V^{\text{diff}}$, so that the function V^{diff} satisfies the boundary condition

$$\frac{\partial V^{\text{diff}}}{\partial r} = -\frac{\partial V^0}{\partial r}, \quad r = a. \tag{7.12}$$

The function V^{diff} can be obtained in the form of series similar to (7.8):

$$V^{\text{diff}}(r, \varphi) = \sum_{n=0}^{\infty} B_n H_n^{(2)}(kr) \cos(n\varphi). \tag{7.13}$$

This series satisfies the wave equation and radiation condition termwise. The coefficients B_n are obtained from the boundary condition (7.12), leading to the functional equation

$$\sum_{n=0}^{\infty} B_n H_n^{(2)\prime}(ka) \cos(n\varphi) = -\frac{1}{k}\frac{\partial V_0}{\partial r}, \quad 0 < \varphi < 2\pi. \tag{7.14}$$

When V^0 has the form (7.10), that is, the plane wave falls, then

$$B_n = \frac{(-i)^n 2}{(1+\delta_{0n})} \frac{J_n'(ka)}{H_n^{(2)\prime}(ka)}. \tag{7.15}$$

7.1.3
The Watson series

The terms of series (7.7) start decreasing only at n of the order $2ka$, so that many terms should be considered when summing the series for large cylinders ($ka \gg 1$). In this subsection, we obtain the solution of the problem in the form of a series fast convergent at $ka \gg 1$; the terms of this series have more complicated structure than those of (7.7). This series can be obtained from the Rayleigh one by applying the so-called *asymptotic summation method*. We give the qualitative description of this method, but then the *Watson series* will be obtained in a more simple way, immediately from the wave equation and the boundary condition at $r = a$.

The asymptotic summation method is based on the Cauchy theorem (3.89) from the theory of integration in the complex variable plane. Let $c(\nu)$ be a function of the complex variable v having no singularities on the half-line $\operatorname{Im} v = 0, 0 \leq \operatorname{Re} v < \infty$. Consider the integral

$$\int_{\mathcal{L}} c\left(v\right) \frac{1}{\sin(\pi v)} \, dv. \tag{7.16}$$

The contour $\mathcal{L}$ is a loop encircling the mentioned half-line (Fig. 7.2). The integrand has poles at the points $v = n \, (n = 0, 1, 2, \ldots)$ where the denominator $\sin(\pi v)$ vanishes. When contracting the contour into the infinitely removed point, we sweep all the poles. Integral (7.16) is equal to a sum of residues at these poles (see (3.89)), that is, it equals

$$2i \sum_{n=0}^{\infty} \left(-1\right)^n c\left(n\right). \tag{7.17}$$

We perform another deformation of the contour $\mathcal{L}$ by turning the lower side of the loop by 2π clockwise. The integral (7.16) is equal to the integral over the infinitely remote circle plus the sum of residues at the poles of function $c(v)$. The integral over the circle is zero if the function $c(v)$ does not increase at $|v| \to \infty$ faster than the denominator $\sin(\pi v)$ does, that is, not faster than $\exp(i\pi v)$. In this case (7.17) equals the sum of residues at the poles of the function $c(v)$. This sum is called the Watson series.

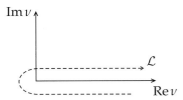

Fig. 7.2 Integration contour in (7.16)

This method of transforming the slowly convergent series to a fast convergent one is applied not to series (7.7) for the diffracted field but to an analogous series for the total field created by sources located on the plane $\varphi = 0$. The total field is zero at $r = a$. The φ-dependence of each of the series terms is described not by the symmetrical functions $\cos(n\varphi)$ but by more complicated functions $\Phi_n(\varphi)$ determined further. The terms of this series differ from those (7.9) of series (7.7); they involve the function $H_v^{(2)}(ka)$ in the denominator. The poles of the function $c(v)$ are roots of the equation

$$H_v^{(2)}(ka) = 0 \tag{7.18}$$

for the unknown v.

It follows from the cylindrical function theory that these roots are complex and lie in the first quadrant of the plane v, that is, $\operatorname{Re} v > 0$, $\operatorname{Im} v > 0$. We denote the nth root of equation (7.18) by ν_n ($n = 0, 1, 2, \ldots$). The Watson series for the total field has the form

$$U^0 + U^{\text{diff}} = \sum_{n=0}^{\infty} D_n \Phi_n(\varphi) H_{\nu_n}^{(2)}(ka). \tag{7.19}$$

It follows from the wave equation for the total field that the functions $\Phi_n(\varphi)$ satisfy the equation $\Phi_n'' + \nu_n^2 \Phi_n = 0$.

Now we construct the Watson series (7.19) in another way by finding the functions $\Phi_n(\varphi)$ and the coefficients D_n directly. Since the terms of this series are zero at $r = a$, the total field $U(r, \varphi)$ (being zero at $r = a$), should be expressed by it.

We subject each function $Z_n(z)$ of series (7.5) not only to the radiation condition but also to the condition $Z_n(a) = 0$. Then $Z_n(r) = H_{\nu_n}^{(2)}(kr)$. These functions satisfy equation (7.6) after replacing n by ν_n. The numbers ν_n are found from equation (7.18).

Series (7.19) satisfies the wave equation if $\Phi_n(\varphi)$ is a linear combination of the functions $\cos(\nu_n \varphi)$ and $\sin(\nu_n \varphi)$. Since ν_n is not a real integer number, then either the function $\Phi_n(\varphi)$ or its derivative takes different values at φ and $\varphi + 2\pi$. We choose the interval where $\Phi_n(\varphi)$ is a single-valued function of the point as $0 < \varphi < 2\pi$. Then, either $\Phi_n(\varphi)$ or $d\Phi_n(\varphi)/d\varphi$ has a jump when passing the ray $\varphi = 0$, that is, their values at $\varphi = 0$ and $\varphi = 2\pi$ are different. It is a price for the fact that $Z_n(r)$ satisfies the conditions at $r = a$ and at $r \to \infty$.

It is convenient to choose the function $\Phi_n(\varphi)$ to be continuous on the ray $\varphi = 0$ and its derivative has a jump. The linear combination of the functions $\cos(\nu_n \varphi)$ and $\sin(\nu_n \varphi)$ satisfying the conditions

$$\Phi_n(2\pi) - \Phi_n(0) = 0, \quad \left.\frac{d\Phi_n}{d\varphi}\right|_{\varphi=2\pi} - \left.\frac{d\Phi_n}{d\varphi}\right|_{\varphi=0} = 1 \tag{7.20}$$

is

$$\Phi_n(\varphi) = \frac{\cos[\nu_n(\pi - \varphi)]}{2\nu_n \sin(\nu_n \pi)}. \tag{7.21}$$

Series (7.19) with $\Phi_n(\varphi)$ having the form (7.21) and ν_n satisfying (7.18), satisfies the wave equation (7.2) together with the radiation condition and the condition $U(0, \varphi) = 0$ at any coefficients D_n for which (7.19) converges. Any function represented by this series is continuous in the whole interval $0 \leq \varphi \leq 2\pi$ and its φ-derivative has a jump when passing the ray $\varphi = 0$. According to (7.20), the jump equals

$$\left.\frac{dU}{d\varphi}\right|_{\varphi=2\pi} - \left.\frac{dU}{d\varphi}\right|_{\varphi=0} = \sum_{n=0}^{\infty} D_n H_{\nu_n}^{(2)}(kr). \tag{7.22}$$

The left-hand side of this equality is proportional to the surface density of the extrinsic current $I_z(r)$, distributed on the ray $\varphi = 0$. This follows from the fact that $U = E_z$, $\partial U / \partial \varphi = -ikrH_r$ and the jump of H_r is proportional to the surface current density: $H_r(2\pi) - H_r(0) = 4\pi/c \cdot I_z(r)$. Series (7.19) represents the total field created in presence of the cylinder, by the current having the surface density $I_z = ic/(4\pi k) \cdot F(r)$, where

$$F(r) = \frac{1}{r} \sum_{n=0}^{\infty} D_n H_{\nu_n}^{(2)}(kr), \quad a < r < \infty. \tag{7.23}$$

The coefficients D_n should be found from this functional equation when considering the function $F(r)$ to be known.

The functions $1/\sqrt{r} \cdot H_{\nu_n}^{(2)}(kr)$ make up complete orthogonal system in the interval $a < r < \infty$. They are eigenfunctions of the problem consisting of equation (7.6) (with the number n replaced by the eigenvalue ν_n), the boundary condition (7.18) and radiation one.

Let us derive the orthogonality conditions. Expanding $[rH_{\nu_n}^{(2)\prime} H_{\nu_m}^{(2)}]'$ by substituting the function $H_{\nu_n}^{(2)\prime\prime}$ from (7.6), gives

$$\left(rH_{\nu_n}^{(2)\prime} H_{\nu_m}^{(2)} \right)' = rH_{\nu_n}^{(2)\prime} H_{\nu_m}^{(2)\prime} - k^2 rH_{\nu_n}^{(2)} H_{\nu_m}^{(2)} + \nu_n^2 \frac{H_{\nu_m}^{(2)} H_{\nu_m}^{(2)}}{r}. \tag{7.24}$$

Subtracting the similar identity with the interchanged indices n and m from (7.24), and integrating the result from $r = a$ to $r = \infty$, we obtain the sought condition

$$\int_a^\infty H_{\nu_n}^{(2)} H_{\nu_m}^{(2)} \frac{dr}{r} = 0, \quad \nu_n \neq \nu_m. \tag{7.25}$$

According to this equality, the coefficients D_n are found from (7.23) as

$$D_n = \frac{1}{N_n} \int_a^\infty H_{\nu_n}^{(2)}(kr) F(r) \, dr, \tag{7.26}$$

where

$$N_n = \int_a^\infty \left[H_{\nu_n}^{(2)}(kr) \right]^2 \frac{dr}{r} \tag{7.27}$$

is the norm of the function $H_{\nu_n}^{(2)}(kr)$.

We prove that the norm N_n is proportional to $d[H_\nu^{(2)}(ka)]/d\nu \big|_{\nu=\nu_n}$. Denote this derivative (after replacing a by r) as L. Differentiating equation (7.6) with respect to ν and putting $\nu = \nu_n$, we obtain

$$L'' + \frac{1}{r}L' + \left(k^2 - \frac{\nu_n^2}{r^2}\right)L = \frac{2\nu_n}{r^2}H_{\nu_n}^{(2)}(kr).$$ (7.28)

Evaluate the expression $(rH_{\nu_n}^{(2)}L' - rH_{\nu_n}^{(2)\prime}L)'$ using equations (7.6) and (7.28):

$$(rH_{\nu_n}^{(2)}L' - rH_{\nu_n}^{(2)\prime}L)' = 2\nu_n\frac{\left[H_{\nu_n}^{(2)}\right]^2}{r}.$$ (7.29)

Integrating this expression from $r = a$ to $r = \infty$ and accounting condition (7.18) for $H_{\nu_n}^{(2)}(ka)$, we obtain

$$N_n = \frac{ka}{2\nu_n}H_{\nu_n}^{(2)\prime}(ka)\frac{d}{d\nu}H_{\nu_n}^{(2)}(ka).$$ (7.30)

Consequently, the denominator of expression (7.26) for D_n is proportional to the ν-derivative of $H_{\nu_n}^{(2)}$. This conforms with the fact that, according to (3.89), such denominator has the residue of the function $c(\nu)$ at $\nu = \nu_n$, that is, at the root of equation (7.18).

Formulas (7.19), (7.26), and (7.27) solve the problem about diffraction of the field created by any current flowing in the half-plane $\varphi = 0$, on the metallic cylinder $r = a$. If the current is not concentrated in one plane, then the field is obtained by integration over all currents.

If the field source is the current thread located at $\varphi = 0$, $r = r_0$, then $F(r) = I_0\delta(\vec{r} - \vec{r}_0)$ and, with accuracy to a nonessential factor, $D_n = 1/N_n \cdot H_{\nu_n}^{(2)}(kr_0)$. Then

$$U(r, \varphi) = I_0\sum_{n=0}^{\infty}\frac{1}{N_n}\Phi_n(\varphi)H_{\nu_n}^{(2)}(kr_0)H_{\nu_n}^{(2)}(kr).$$ (7.31)

Directing r_0 to infinity and increasing the amplitude of the current I_0, simultaneously, in such a way that its field be unity near the coordinate origin, from this formula we may obtain the expression for the total field arisen at the incidence of the plane wave onto the metallic cylinder.

By the same scheme, the Watson series is constructed for the function $V(r, \varphi)$ being the component H_z of the field polarized in the plane perpendicular to the cylinder axis. In this case the Watson series for the total field can also be obtained either by summation of the Rayleigh series (7.13) or immediately from the Maxwell equations. The coefficients of the Watson series are residues at the poles of the coefficients in (7.13). The poles are values of the parameter ν at which

$$H_{\nu}^{(2)\prime}(ka) = 0.$$ (7.32)

The eigenfunctions corresponding to these ν differ from those correspond-ing to the roots of equation (7.18). However, as it could be easily obtained from (7.24), they satisfy the same orthogonality condition (7.25). The same functions (7.21) are the angular functions in the Watson series. All the sum-mands of the Watson series for the total field $V(r, \varphi)$ fulfils the condition $\partial V / \partial r|_{r=a} = 0$ following from (7.12).

7.1.4
Dielectric cylinder

In this subsection, the Rayleigh series is constructed for the problem about the plane wave incidence onto the circular dielectric cylinder of radius a, with the dielectric permittivity $\varepsilon \neq 1$ at $r < a$. First, we consider the E-polarization case. Find the field $E_z = U(r, \varphi)$. In the cylinder exterior ($r \geq a$), it has the form

$$U^- = U^0 + \sum_{n=0}^{\infty} A_n H_n^{(2)}(kr) \cos(n\varphi), \tag{7.33}$$

where the incident field $U^0(r, \varphi)$ is given by (7.10). The field should have no singularities inside the cylinder (at $r \leq a$). It satisfies equation (7.2) and is expressible by the functions $J_n(k\sqrt{\varepsilon}r)$ only. Other cylindrical functions have a singularity at $r = 0$. We express the inner field as

$$U^+ = \sum_{n=0}^{\infty} B_n J_n(k\sqrt{\varepsilon}r) \cos(n\varphi). \tag{7.34}$$

Since this formula should not describe the field in large distances, it is possible not to separate the diffracted field satisfying the radiation condition.

The two tangential components E_z and H_φ should be continuous on the boundary $r = a$. According to (7.1),

$$U^+ = U^-, \quad \frac{\partial U^+}{\partial r} = \frac{\partial U^-}{\partial r}. \tag{7.35}$$

Consequently, the two functional equations

$$U^0(a, \varphi) + \sum_{n=0}^{\infty} A_n H_n^{(2)}(ka) \cos(n\varphi) = \sum_{n=0}^{\infty} B_n J_n(k\sqrt{\varepsilon}a) \cos(n\varphi), \tag{7.36a}$$

$$\frac{\partial U^0(r, \varphi)}{\partial r}\bigg|_{r=a} + k \sum_{n=0}^{\infty} A_n H_n^{(2)\prime}(ka) \cos(n\varphi)$$

$$= k\sqrt{\varepsilon} \sum_{n=0}^{\infty} B_n J_n'(k\sqrt{\varepsilon}a) \cos(n\varphi) \tag{7.36b}$$

should be satisfied in the interval $0 < \varphi < 2\pi$. Series by the same complete orthogonal system of functions $\cos(n\varphi)$ are involved on both the sides of the equations. For each pair of the coefficients A_n, B_n ($n = 0, 1, 2, \ldots$) we obtain the independent system of two algebraic equations

$$A_n H_n^{(2)}(ka) - B_n J_n\left(k\sqrt{\varepsilon}a\right) = -\frac{1}{\pi(1 + \delta_{0n})} \int_0^{2\pi} U^0(a, \varphi) \cos(n\varphi)\, d\varphi, \quad (7.37a)$$

$$A_n H_n^{(2)\prime}(ka) - \sqrt{\varepsilon} B_n J_n'\left(k\sqrt{\varepsilon}a\right)$$

$$= -\frac{1}{k\pi(1 + \delta_{0n})} \int_0^{2\pi} \left.\frac{\partial U^0(r, \varphi)}{\partial r}\right|_{r=a} \cos(n\varphi)\, d\varphi. \quad (7.37b)$$

The coefficients of series (7.33) and (7.34) are easily found from these systems.

The same variable separation method is used for finding the solution in the case of H-polarization. The function $H_z = V(r, \varphi)$ is sought in the form of series similar to (7.33), (7.34). The components H_z and E_φ, tangential to the cylinder surface, should be continuous at $r = a$. According to (7.3), this leads to the following boundary conditions at $r = a$:

$$V^+ = V^-, \quad \frac{1}{\varepsilon}\frac{\partial V^+}{\partial r} = \frac{\partial V^-}{\partial r}. \quad (7.38)$$

They differ from (7.35) only by the factor $1/\varepsilon$ at $\partial V^+/\partial r$. The functional equations for the coefficients in the series for V^+, V^- differ from (7.36) by the same factor only. The equation system for the series coefficients are independent for different numbers as well.

The simplicity of transferring the variable separation method from the case of metallic cylinder onto that of dielectric one is explained by the fact that in the cylindrical coordinate system, the angular functions $\sin(n\varphi)$, $\cos(n\varphi)$ do not depend on ε. Therefore, the independent systems of two algebraic equations independent for the coefficients of different numbers, follow from the functional equations (7.36). In the case of the dielectric cylinder with elliptical cross-section, although its surface coincides with a coordinate one, the variable separation method results in the infinite algebraic system which does not divide into the separate systems of two equations. This is caused by the fact that in elliptical coordinates, the functions $\sin(n\varphi)$, $\cos(n\varphi)$ are replaced by the angular Mathieu functions dependent on ε. Different angular functions are involved on the left- and right-hand sides of the functional equations of the type (7.36) and, therefore, the infinite algebraic system appears when passing to the algebraic equations. No such complication arises for the metallic cylinder of the elliptic cross-section.

7.2
Diffraction on metallic half-plane

7.2.1
Usage of the variable separation method

The equation of a half-plane in the Cartesian coordinates is $y = 0$, $x > 0$, $-\infty < z < \infty$ (Fig. 7.3). Assume that the incident field does not depend on z; then the total field is independent of z as well. Similarly as in the case of diffraction on the cylinder, the vector problem is reduced to two independent scalar ones. In these problems, the field has only the components E_z, H_r, H_φ or H_z, E_r, E_φ, and the functions $U(r, \varphi) = E_z$, $V(r, \varphi) = H_z$ describe the so-called *electric* (*E*-) or *magnetic* (*H*-) polarization, respectively. These functions satisfy the wave equation (7.2). The components H_r, H_φ in the first problem and E_r, E_φ in the second problem are calculated by formulas (7.1), (7.2).

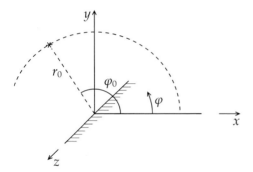

Fig. 7.3 Illustration to the diffraction problem on the half-plane

In cylindrical coordinates the half-plane is described by the equation $\varphi = 0$, $\varphi = 2\pi$. At these values of φ (i.e., on the upper and lower sides of the half-plane), the tangential components E_z, E_r of the electric field must be zero. This gives the boundary conditions

$$U = 0, \tag{7.39a}$$

$$\frac{\partial V}{\partial \varphi} = 0 \tag{7.39b}$$

at $\varphi = 0$ and $\varphi = 2\pi$. The tangential components H_z, H_r of the magnetic field discontinue on the metal; their values at $\varphi = 0$ and $\varphi = 2\pi$ are different. An induced current flows in the metal. Its surface densities are I_z or I_r for *E*- or *H*-polarization, respectively. With accuracy to nonessential factors these densities equal

$$I_z = \frac{\partial U}{r\partial \varphi}\bigg|_{\pi=2\pi} - \frac{\partial U}{r\partial \varphi}\bigg|_{\pi=0}, \tag{7.40a}$$

$$I_r = V(r, 2\pi) - V(r, 0). \tag{7.40b}$$

We find the solution to equation (7.2) using the variable separation method, that is, as the product $Z_{\nu_n}(kr)\{{\sin \atop \cos}\}(\nu_n\varphi)$, $n = 1, 2, \ldots$ of cylindrical and trigonometric functions. The indices ν_n are found from the boundary conditions (7.39). However, no cylindrical function exists, being finite at $r = 0$ and satisfying the radiation condition at $r \to \infty$. To apply the variable separation method, we first assume that the incident wave is not plane, but the cylindrical one with the source located at the finite distance r_0 from the origin (later we will put $r_0 \to \infty$). In this case, the field or its normal derivative has a jump on the circle $r = r_0$. The field inside the circle can be described by the Bessel functions and outside by the Hankel ones. The jump of the field or its normal derivative on the circle is caused by existence of an extrinsic current on it.

Let the function $U(r, \varphi)$ be continuous, and $\partial U/\partial r$ have a jump at $r = r_0$. This condition can be satisfied if the field is expressed by the series

$$U(r, \varphi) = \begin{cases} \sum_{n=1}^{\infty} A_n H_{\nu_n}^{(2)}(kr_0) J_{\nu_n}(kr) \sin(\nu_n \varphi), & r < r_0, \\ \sum_{n=1}^{\infty} A_n H_{\nu_n}^{(2)}(kr) J_{\nu_n}(kr_0) \sin(\nu_n \varphi), & r > r_0. \end{cases} \tag{7.41}$$

Since the coefficients A_n are the same in both the series, the function (7.41) fulfills the condition $U(r_0 + 0, \varphi) = U(r_0 - 0, \varphi)$. The jump of its derivative is

$$\frac{\partial U}{\partial r}\bigg|_{r=r_0+0} - \frac{\partial U}{\partial r}\bigg|_{r=r_0-0} = \frac{2i}{\pi r_0} \sum_{n=1}^{\infty} A_n \sin(\nu_n \pi). \tag{7.42}$$

Here a known expression for the Wronskian of the cylindrical functions is used.

According to (7.39a), the indices ν_n are found from equation $\sin(2\nu_n\pi) = 0$, which gives

$$\nu_n = \frac{n}{2}, \qquad n = 1, 2, \ldots \tag{7.43}$$

The left-hand side of (7.42) is proportional to the jump of the component H_φ on the circle $r = r_0$, that is, to the density of the extrinsic current creating the incident field. If this current is a thread located at the point (r_0, φ_0), then $j_z = I_0 \delta(\varphi - \varphi_0)$, and the coefficients A_n are proportional to $\sin(\nu_n \varphi_0)$. Therefore, the cylindrical wave radiated by the current thread creates the field

$$U(r, \varphi) = k r_0 \sum_{n=1}^{\infty} H_{n/2}^{(2)}(k r_0) J_{n/2}(kr) \sin(n\varphi_0/2) \sin(n\varphi/2), \quad r < r_0, \quad (7.44a)$$

$$U(r, \varphi) = k r_0 \sum_{n=1}^{\infty} H_{n/2}^{(2)}(kr) J_{n/2}(k r_0) \sin(n\varphi_0/2) \sin(n\varphi/2), \quad r > r_0. \quad (7.44b)$$

diffracted on the half-plane. According to the Debye asymptotic (6.134) of the cylindrical functions, series (7.44a) diverges as $(r/r_0)^{n/2}$, and series (7.44b) does as $(r_0/r)^{n/2}$.

7.2.2
Currents on metallic surface

In the case of the E-polarization, the induced current is parallel to the edge and the surface density I_z of the current differs from zero. The values of component H_r on the upper and lower sides, that is, at $\varphi = 0$ and $\varphi = 2\pi$ are different, because $\cos(n\pi) = -1$ for odd n and $\cos(0) = 1$. At $r < r_0$, these values are proportional to

$$H_r|_{\varphi=0} = I_0 \frac{r_0}{r} \sum_{n=1.}^{\infty} C_n J_{n/2}(kr), \qquad (7.45a)$$

$$H_r|_{\varphi=2\pi} = I_0 \frac{r_0}{r} \sum_{n=1}^{\infty} C_n J_{n/2}(kr)(-1)^n, \qquad (7.45b)$$

where $C_n = n H_{n/2}^{(2)}(k r_0) \cos(n\varphi_0/2)$. Their difference, that is, the surface density

$$I_z = I_0 \frac{r_0}{r} \sum_{n=1,3,5,\dots} C_n J_{n/2}(kr) \qquad (7.46)$$

is proportional to the sum involving the terms of odd numbers only. At $kr \ll 1$, that is, near the edge, with the accuracy to a nonessential factor, we have

$$I_z = \frac{1}{\sqrt{kr}} + A\sqrt{kr}, \qquad (7.47)$$

where the factor A depends on the point (φ_0, r_0). In particular, if $k r_0 \gg 1$, $\varphi_0 = \pi/2$, that is, when the source is located over the edge at the large distance from it, then $A = -9i/4$. Formula (7.47) specifies the result obtained in Subsection 1.3.5, according to which the current density grows as $(kr)^{-1/2}$ when approaching the edge. The second term in the approximate formula (7.47) has no universal form, it depends on the incident field.

The same variable separation method may by used for finding the formal solution (in the form of infinite series) in the case of the H-polarization. The function $V(p, \varphi)$ can be found as a series different from (7.41) only by replacing the factor $\sin(\nu_n \varphi)$ by $\cos(\nu_n \varphi)$. According to (20.1), the indices ν_n are found from the same equation $\sin(2\nu_n \pi) = 0$, that is, they are given by the same formula (7.43). The component E_φ has a jump on the circle $r = r_0$. This is in accordance with the fact that the magnetic currents $j_z^{(m)}$ are located on this circle. They may also have the form of a magnetic current thread.

On the different sides of the half-plane, the component H_z takes different values. A current flows on the metallic surface, directed perpendicular to the edge. The surface density I_r of this induced current is proportional to the jump of the component H_z. The formula

$$I_r = \sum_{n=1,3,5,...} D_n J_{n/2}(kr) \tag{7.48}$$

for I_r is similar to (7.46) for I_z in the E-polarization case, but it does not involve the factor $1/r$. The coefficients D_n differ from C_n in the absence of factor n.

At $kr \ll 1$, the density of induced current flowing perpendicular to the edge, approximately equals $I_z = \sqrt{kr} + B(kr)^{3/2}$, where the coefficient B depends on the location of the magnetic field creating the incident wave. At $kr_0 \gg 1$, $\varphi_0 = \pi/2$, we have $B = -3i/4$. As it follows from Subsection 1.3.5, the component I_r of the induced field has no singularity.

7.2.3
The far field

The above formulas for $U(r, \varphi)$, $V(r, \varphi)$ do not permit the direct evaluation of the field at $kr \gg 1$, that is, far from the edge. There exist other formulas which allow us to make the passage $kr \to \infty$, as well as $kr_0 \to \infty$ (i.e., when the incident field is the plane wave). These formulas represent the fields in the form of integrals taken in the complex variable plane. They are obtained from (7.44) by expressing the cylindrical functions as the contour integrals, and interchanging the summation and integration order. The formulas can be obtained immediately without separating the variables. For instance, the problem can be reduced to an integral equation with the half-infinite limits (similarly as in Subsection 3.4.6) and then solved by the factorization method. There are other ways for obtaining these formulas.

The formulas are very cumbersome. We do not present them here, only describe some results of their analysis, namely, the structure of the field arisen far from the edge, at the plane wave diffraction on the half-plane.

When moving away from the edge, the tangential components of the magnetic field on the metallic surface tend to the values which they obtain in the problem about the plane wave incident onto the infinite plane. On the illuminated surface side, the magnetic field tends to its twofold value in the incident wave. On the shadowed side, it tends to zero. Therefore, the current densities I_z, I_r tend to constant values.

Since the half-plane has the infinite "cross-section" ($0 < x < \infty$), the diffracted field at $kr \gg 1$ has complicated structure; it is not the cylindrical wave, as in the case of diffraction on arbitrary cylinder with a bounded cross-section. Qualitatively, it is the same for both polarizations.

When the plane wave falls onto the half-plane at the glancing angle φ_0, the field at $kr \gg 1$ can be divided into three summands: "optical" field, transient one (localized near the two lines $\varphi = \pi \pm \varphi_0$) and the cylindrical wave, outgoing from the edge.

The optical field is the part of the entire one not disturbed by the diffraction. It has different structure in the domains partitioned by the half-plane and the rays $\varphi = \pi \pm \varphi_0$ (Fig. 7.4). At $0 < \varphi < \pi - \varphi_0$ (domain *I*), this field consists of the incident and reflected plane waves. At $\pi - \varphi_0 < \varphi < \pi + \varphi_0$ (domain *II*), only the incident wave is present, the reflected one is absent. Domain *III* is shadowed, there is no optical field at $\pi + \varphi_0 < \varphi < 2\pi$. The optical field has a jump on the lines $\varphi = \pi \pm \varphi_0$.

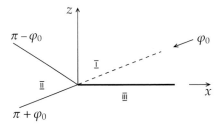

Fig. 7.4 Domains of different field structure

The transient field is concentrated near the above lines, dividing different domains of the optical one. The transient field is also discontinuous on these lines, so that the sum of two considered fields is continuous.

Consider the transient field concentrated near the limiting ray $\varphi = \pi - \varphi_0$. It does not depend on the coordinates r, φ apart, it depends on their combination

$$w = -\sqrt{2kr}\sin(\alpha/2) \tag{7.49}$$

only, where $\alpha = \varphi - (\pi - \varphi_0)$ is the angle between the ray with the coordinate φ and the limiting one. On the limiting ray, $w = 0$; it is positive in the domain *I* and negative in *II*. The transient field is proportional to the function (*Fresnel integral*)

$$\exp(-ikr\cos\alpha) \int_{\pm\infty}^{w} \exp(it^2)\,dt, \tag{7.50}$$

where the sign of the lower limit coincides with the sign of w. On the limiting ray, that is, at $w = 0$, the function (7.50) has a jump. The coefficient omitted in (7.50) is such that this jump is equal in modulus and opposite in sign to that of the optical field.

The lines $|w| = const$ are parabolas having the limiting ray $\varphi = \pi - \varphi_0$ as their axis. With r growing, the angle occupied by each parabola decreases. Any ray $\varphi = const$, except the boundary ray (i.e., $\alpha = 0$), transfers into the domain of large $|w|$ when r grows. At $|w| \gg 1$, the function (7.50) turns to be a field of the cylindrical wave and becomes proportional to

$$\frac{1}{\sin(\alpha/2)} \frac{\exp(-ikr)}{\sqrt{kr}}. \tag{7.51}$$

In the domain, bounded by the parabola $|w| = 1$, that is, at $|w| \le 1$, the transient field has a complicated structure described by formula (7.50). This field provides continuous passage between the domains I and II. In particular, on the limiting ray, the total field is the half-sum of the optical field values in both the domains. Outside the domain $|w| = 1$, at $|w| \gg 1$, the transient field transforms into the cylindrical wave outgoing from the edge. According to (7.51), the amplitude of this wave increases as $1/\sin(\alpha/2)$ when approaching the limiting ray. The smaller the α, the larger the r at which the field becomes the cylindrical wave.

Similar transient field is formed around the second limiting ray $\varphi = \pi + \varphi_0$. Near it, the field depends on r and φ by the same formula (7.50), with

$$w = -\sqrt{2kr}\sin(\beta/2), \tag{7.52}$$

where $\beta = \varphi - (\pi + \varphi_0)$. At $|w| \approx 1$, the field is described by the same formula (7.50). At $|w| \gg 1$, the field becomes the cylindrical wave having the amplitude proportional to $1/\sin(\beta/2)$.

Consequently, at $kr \gg 1$, the total field in the direction φ consists of the optical field, transient field and sum of fields of two cylindrical waves, having the amplitudes proportional to $1/\sin(\alpha/2)$ and $1/\sin(\beta/2)$, respectively, where α and β are the angles between the given direction and direction of the limiting lines $\varphi = \pi - \varphi_0$ or $\varphi = \pi + \varphi_0$.

The half-plane can be treated as a wedge with the zero angle between its faces. The field arisen at diffraction on the wedge with the finite angle lying in the interval $(0, \pi)$ has qualitatively the same structure as in the above case.

7.3

The Debye potentials: diffraction on a metallic sphere

7.3.1

The Debye potentials

In the spherical coordinates R, ϑ, φ, the fields $\vec{E}$, $\vec{H}$ satisfying the system of homogeneous Maxwell equations are expressible in terms of the two scalar functions $U(R, \vartheta, \varphi)$ and $V(R, \vartheta, \varphi)$. They are called the *electric* and *magnetic Debye potentials*, respectively. The spherical components of the fields are expressed as

$$E_R = \frac{\partial^2 (RU)}{\partial R^2} + k^2 RU, \quad H_R = \frac{\partial^2 (RV)}{\partial R^2} + k^2 RV, \tag{7.53a}$$

$$E_\vartheta = \frac{1}{R} \frac{\partial^2 (RU)}{\partial R \partial \vartheta} - \frac{ik}{R \sin \vartheta} \frac{\partial (RV)}{\partial \varphi}, \tag{7.53b}$$

$$E_\varphi = \frac{1}{R \sin \vartheta} \frac{\partial^2 (RU)}{\partial R \partial \varphi} + \frac{ik}{R} \frac{\partial (RV)}{d\vartheta},$$

$$H_\vartheta = \frac{1}{R} \frac{\partial^2 (RV)}{\partial R \partial \vartheta} + \frac{ik}{R \sin \vartheta} \frac{\partial (RU)}{\partial \varphi}, \tag{7.53c}$$

$$H_\varphi = \frac{1}{R \sin \vartheta} \frac{\partial^2 (RV)}{\partial R \partial \varphi} - \frac{ik}{R} \frac{\partial (RU)}{d\vartheta}.$$

Substituting (7.53) into the Maxwell equations, we obtain the equations for U and V as

$$\Delta U + k^2 U = 0, \tag{7.54a}$$

$$\Delta V + k^2 V = 0. \tag{7.54b}$$

If the functions U and V satisfy (7.54), then the fields $\vec{E}$, $\vec{H}$, calculated by them using (7.53), satisfy the Maxwell equations.

Equations (7.54a) and (7.54b) are independent, that is, there exist two solutions, with $V \equiv 0$ and $U \equiv 0$, respectively. In the first solution, $H_R \equiv 0$; in the second solution, $E_R \equiv 0$. The solutions are not connected by the equations, but they can be connected by the boundary or excitation conditions. Equations (7.54) have solutions represented as a product of three functions of the respective variable R, ϑ, and φ. Each function satisfies the ordinary differential equation of the second order with respect to the corresponding variable R, ϑ, or φ.

The system of functions dependent on φ is the simplest one. It consists of the trigonometric functions $\cos(m\varphi)$ and $\sin(m\varphi)$ ($m = 0, 1, 2, \ldots$). The functions dependent on R are $1/\sqrt{kR} \cdot Z_{n+1/2}(kR)$ ($n = 0, 1, 2..$), where Z

is a cylindrical function. Cylindrical functions of the half-integer index are expressed by the trigonometric or exponential ones. For instance, if $n = 0$ and the field is subjected to the radiation condition, then Z is the Hankel function $H_{1/2}^{(2)}(kR)$, so that the corresponding potential is $\exp(-ikR)/(kR)$.

The functions of angle ϑ depend on two indices n and m, coincident with the indices of functions of R and φ, respectively. They are denoted as $P_n^m(\vartheta)$ and called the *associated Legendre functions*. The functions with $m = 0$ are denoted as $P_n(\vartheta)$ and called the *Legendre polynomials*. With accuracy to normalizing factors, they are

$$
\begin{aligned}
&P_0(\vartheta) = 1, \quad P_1(\vartheta) = \cos\vartheta, \quad P_2(\vartheta) = 3\cos(2\vartheta) + 1, \quad m = 0, \\
&P_1^1(\vartheta) = \sin(\vartheta), \quad P_2^1(\vartheta) = \sin(2\vartheta), \quad m = 1.
\end{aligned}
\tag{7.55}
$$

Solution to equations (7.54) satisfying the radiation condition, is, in general, a double sum of the functions

$$
A_{nm}\frac{1}{\sqrt{kR}}H_{n+1/2}^{(2)}(kR)P_n^m(\vartheta)\left\{{\cos \atop \sin}\right\}(m\varphi), \quad n, m = 0, 1, \ldots, \quad m \leq n.
\tag{7.56}
$$

The functions $P_n^m(\vartheta)$ are orthogonal in the following sense:

$$
\int_0^\pi P_n^m(\vartheta)P_q^m(\vartheta)\sin\vartheta\,d\vartheta = 0
\tag{7.57}
$$

at $n \neq q$. The system of them is complete in the interval $0 < \vartheta < \pi$.

7.3.2
The Debye potentials for fields of elementary sources

In Section 6.1, elementary dipoles and multipoles were considered. Fields of such electric and magnetic sources are expressed in terms of the potentials U and V, respectively. For the elementary sources, each of these potentials coincides with the appropriate function in (7.56). In Section 6.1, only the sources independent of the azimuthal angle φ, were considered. Potentials describing fields of such sources correspond to $m = 0$. Potentials with $n = 1$ correspond to the dipoles, those with $n = 2$ correspond to the quadrupoles, and so on.

The field of the elementary electric dipole (6.5) can be expressed by the potential U (7.56) at $n = 1$, $m = 0$, using formulas (7.53). Substituting the expressions for $H_{3/2}^{(2)}(kR)$ and $P_1(\vartheta)$ into (7.56), we find the potential

$$
U = -k^2\mathcal{P}\left(1 - \frac{i}{kR}\right)\cos\vartheta\frac{\exp(-ikR)}{kR}.
\tag{7.58}
$$

According to (7.53), the field corresponding to this potential coincides with (6.5).

The field of the elementary electric quadrupole is obtained from the Debye potential $U(R, \vartheta, \varphi)$ (7.56) at $n = 2$. $m = 0$, that is, from the potential proportional to the function

$$\frac{1}{\sqrt{kR}} H_{5/2}^{(2)}(kR)\,(3\cos(2\varphi) + 1). \tag{7.59}$$

The fields of the elementary magnetic dipole and quadrupole are obtained by (7.53) from the Debye potential $V(R, \vartheta, \varphi)$, which also have the form (7.56). For the dipole, $n = 1$, $m = 0$, and the function V has the form (7.58); for the quadrupole, $n = 2$, $m = 0$, and the potential V is given by (7.59).

At the end of Subsection 6.1.3, we noted that at $kR \gg 1$ the field of an arbitrary current can be presented as a sum of the elementary source fields. This assertion is based on the fact that the field of each such source is described by the Debye potential in the form of one of the functions (7.56), and that an arbitrary field having no singularity at $R \neq 0$ and fulfilling the radiation condition can be expanded by these functions.

7.3.3
Fields of elementary sources located at point with complex coordinates

The functions (7.56) are the Debye potentials of the fields created by the elementary sources located at the origin. In this subsection, we introduce the functions formally obtained as the Debye potentials of the fields of sources located at a point with complex coordinates. These fields can be created by certain currents lying on a plane or on any other surface. Not these currents themselves, but the asymptotics of potentials, that is, the radiation patterns of the corresponding fields, are of the main interest.

The Laplace operator Δ is invariant with respect to the parallel coordinate shift along the z-axis by the constant value α. If a certain function $f(x, y, z)$ satisfies the wave equation, then the function $f(x, y, z + \alpha)$ satisfies it as well.

Let the values $\tilde{x} = x$, $\tilde{y} = y$, $\tilde{z} = z + \alpha$ be the Cartesian coordinates in the "shifted" coordinate system. Introduce corresponding coordinates $\widetilde{R}$, $\widetilde{\vartheta}$, $\widetilde{\varphi}$ in the spherical coordinate system with shifted origin and the same axis. New spherical coordinates are connected with the new Cartesian coordinates as

$$\tilde{x} = \widetilde{R}\sin\widetilde{\vartheta}\cos\widetilde{\varphi}, \quad \tilde{y} = \widetilde{R}\cos\widetilde{\vartheta}\sin\widetilde{\varphi}, \quad \tilde{z} = \widetilde{R}\cos\widetilde{\vartheta}. \tag{7.60}$$

Similar connection exists between the corresponding old coordinates

$$x = R\sin\vartheta\cos\varphi, \quad y = R\cos\vartheta\sin\varphi, \quad z = R\cos\vartheta. \tag{7.61}$$

Express the coordinates $\widetilde{R}$, $\widetilde{\vartheta}$, $\widetilde{\varphi}$ by R, ϑ, φ. Accounting that $\widetilde{x} = x$, $\widetilde{y} = y$, $\widetilde{z} = z + \alpha$, we eliminate the Cartesian coordinates from (7.60), (7.61) and obtain

$$\widetilde{R}\cos\widetilde{\vartheta} = R\cos\vartheta + \alpha, \quad \widetilde{R}\sin\widetilde{\vartheta} = R\sin\vartheta, \quad \widetilde{\varphi} = \varphi. \tag{7.62}$$

Consequently,

$$\widetilde{R} = \sqrt{R^2 + \alpha^2 + 2aR\cos\vartheta}, \quad \sin\widetilde{\vartheta} = \frac{R}{\widetilde{R}}\sin\vartheta. \tag{7.63}$$

In particular, at $R \gg |\alpha|$, $\widetilde{R} = R + \alpha\cos\vartheta$, $\widetilde{\vartheta} = \vartheta$.

If a certain function $f(R, \vartheta, \varphi)$ satisfies the wave equation written in the co-ordinates R, ϑ, φ, then the function $f(\widetilde{R}, \widetilde{\vartheta}, \widetilde{\varphi})$ satisfies the same equation written in the coordinates $\widetilde{R}$, $\widetilde{\vartheta}$, $\widetilde{\varphi}$ (for simplicity, we keep the same notation for the function f in new coordinates). The functions

$$\frac{1}{\sqrt{k\widetilde{R}}} H^{(2)}_{n+1/2}\left(k\widetilde{R}\right) P^m_n\left(\widetilde{\vartheta}\right) \begin{Bmatrix} \cos \\ \sin \end{Bmatrix} (m\varphi), \quad n, m = 0, 1, 2, \ldots, \quad m \le n, \tag{7.64}$$

where $\widetilde{R}$, $\widetilde{\vartheta}$ are given in (7.63) are the Debye potentials of the elementary sources located at the point with coordinates $x = 0$, $y = 0$, $z = -\alpha$. At $kR \gg 1$, $R \gg \alpha$, the asymptotic of these functions is

$$\exp\left(-ik\alpha\cos\vartheta\right) P^m_n(\vartheta) \begin{Bmatrix} \cos \\ \sin \end{Bmatrix} (m\varphi) \frac{\exp\left(-ikR\right)}{kR}. \tag{7.65}$$

Let $\alpha = ia$, where a is a real positive number. Then the arguments $\widetilde{R}$ and $\widetilde{\vartheta}$ in (7.64) are complex:

$$\widetilde{R} = \sqrt{R^2 - a^2 + 2iaR\cos\vartheta}, \quad \sin\widetilde{\vartheta} = \frac{R}{\widetilde{R}}\sin\vartheta. \tag{7.66}$$

At $kR \gg 1$, $R \gg a$, the asymptotic of the Debye potentials is

$$\exp\left(ka\cos\vartheta\right) P^m_n(\vartheta) \begin{matrix} \cos \\ \sin \end{matrix} (m\varphi) \frac{\exp\left(-ikR\right)}{kR}. \tag{7.67}$$

These formal derivations result in the assertion that the Debye potentials (7.64) with $\widetilde{R}$ and $\widetilde{\vartheta}$ being the complex functions (7.66) of R, ϑ, satisfy the wave equations. The fields calculated by formulas (7.53) satisfy the Maxwell equations and the radiation conditions. Asymptotic of these fields differs from that of the fields of elementary sources by the multiplier $\exp(ka\cos\vartheta)$. This multiplier is maximal at $\vartheta = 0$. The larger the parameter a is, the faster this multiplier decreases. The presence of this multiplier in the asymptotic of functions in (7.64), may simplify solving the antenna problems in which the connection between the field in the near field zone and the radiation pattern should be established.

If the pattern is calculated by the field measured in the near field zone of antenna, then this field should be expanded into the series by the variable ϑ with substituting each term of the series by its asymptotic. This series converges faster if the field (more exactly, the Debye potentials) is expanded not by the functions (7.56), but by (7.64) with appropriately chosen parameter a.

In the inverse problem, the field in the near field zone (from which the current is easily found) should be calculated by a given pattern. In this case, the pattern should be expanded into the Fourier series by ϑ. If the pattern is narrow directed, then the series converges faster if expanding it by the functions (7.67) and choosing the parameter a such that the multiplier $\exp(ka \cos \vartheta)$ is close to the envelope of this pattern. In this case the near field is also presented as the fast-convergent series by the functions (7.64).

7.3.4
Diffraction of the plane wave on metallic sphere: the Rayleigh series

The total field should satisfy the conditions

$$E_\vartheta = 0, \quad E_\varphi = 0 \tag{7.68}$$

on the sphere surface $R = a$. According to (7.53), the corresponding condition has the form

$$\frac{\partial (RU)}{\partial R} = 0, \quad V = 0 \quad \text{at } R = a \tag{7.69}$$

for the Debye potentials. We find the potentials U^{diff}, V^{diff} for the diffracted field. They should satisfy the conditions

$$\frac{\partial \left(RU^{\text{diff}} \right)}{\partial R} = -\frac{\partial \left(RU^0 \right)}{\partial R}, \quad V^{\text{diff}} = -V^0 \quad \text{at } R = a, \tag{7.70}$$

where U^0, V^0 are the potentials of the incident field calculated by its components.

Let a plane linearly polarized wave propagating along the z-axis fall. The wave has nonzero components $E_x^0 = \exp(-ikz)$, $H_y^0 = \exp(-ikz)$. Since (7.53a) does not involve the angular derivatives, the angular dependence of the potentials U^0 and V^0 is the same as that of the radial components E_R^0 and H_R^0. In the spherical coordinate system with the same z-axis, we have

$$\begin{aligned} E_R^0 &= \exp\left(-ikR \cos \vartheta\right) \sin \vartheta \cos \varphi, \\ H_R^0 &= \exp\left(-ikR \cos \vartheta\right) \sin \vartheta \sin \varphi. \end{aligned} \tag{7.71}$$

Following the variable separation method, we express the potentials in the form of series with terms similar to the products (7.56). The functions (7.71) have no singularities at $R = 0$ and do not satisfy the radiation conditions. Therefore, in these products, the Bessel functions should be involved instead of the Hankel functions. The components E_R^0 and H_R^0 depend on the angle φ by means of the multipliers $\cos \varphi$ and $\sin \varphi$ only. In similar way, the potential should depend on φ. Therefore, the upper index m of the function $P_n^m(\vartheta)$ should be unity; the potentials are expressed by a single, not double sum. They can be written as

$$U^0(R, \vartheta, \varphi) = \frac{1}{k^2 R} \sum_{n=1}^{\infty} A_n \bar{j}_n(kR) P_n^1(\vartheta) \cos \varphi,$$

$$V^0(R, \vartheta, \varphi) = \frac{1}{k^2 R} \sum_{n=1}^{\infty} A_n \bar{j}_n(kR) P_n^1(\vartheta) \sin \varphi.$$

(7.72)

Here the function

$$\bar{j}_n(t) = \sqrt{t} J_{n+1/2}(t)$$

(7.73a)

is introduced. Below we use the similar function

$$\bar{h}_n(t) = \sqrt{t} H_{n+1/2}^{(2)}(t)$$

(7.73b)

connected with the *spherical ones* $j_n(t)$ and $h_n(t)$ as $\bar{j}_n(t) = \sqrt{2/\pi t} \cdot j_n(t)$, $\bar{h}_n(t) = \sqrt{2/\pi t} \cdot h_n(t)$. Since both the components (7.71) equally depend on R and ϑ, then this property is inherent to the potentials U^0 and V^0 as well; the coefficients A_n of both the series (7.72) are the same. They can be found from the demand that the equality

$$\sum_{n=0}^{\infty} A_n \left(\frac{d^2}{d(kR)^2} + 1 \right) [\bar{j}_n(kR)] P_n^1(\vartheta) = \exp(-ikR \cos \vartheta) \sin \vartheta,$$

(7.74)

obtained after substituting (7.72) into (7.53a), holds at any ϑ ($0 \leq \vartheta \leq \pi$) and R ($0 \leq R < \infty$).

It follows from the completeness and orthogonality of the functions $P_n^1(\vartheta)$, that

$$A_n \left(\frac{d^2}{d(kR)^2} + 1 \right) [\bar{j}_n(kR)] = \frac{\int_0^\pi \exp(-ikR \cos \vartheta) P_n^1(\vartheta) \sin^2 \vartheta \, d\vartheta}{\int_0^\pi [P_n^1(\vartheta)]^2 \sin \vartheta \, d\vartheta}.$$

(7.75)

At any n, this equality holds for all R identically. The coefficients A_n can be calculated from (7.75) by setting any value to R. The simplest is to set $R = 0$. However, at $n > 1$, equality (7.75) becomes $0 = 0$ at $R = 0$. In order to find A_n at $n > 1$, one should first differentiate this equality $n - 1$ times with respect to kR and then only set $R = 0$. In this case the second term of the operator $d^2/d(kR)^2 + 1$ can be omitted since the $(n - 1)$th derivative of the function

$\bar{j}_n(kR)$ is zero at $R = 0$; the function $\bar{j}_n(kR)$ can be replaced with the first term of its expansion by kR:

$$\bar{j}_n(kR) \approx \frac{1}{2^{n+1/2}} \frac{1}{\Gamma(n + 3/2)} (kR)^{n+1}. \tag{7.76}$$

For the coefficients A_n ($n = 1, 2, \ldots$) we obtain the expression

$$A_n = (-i)^{n-1} \frac{2^{n+1/2} \Gamma(n + 3/2)}{\Gamma(n + 2)} \frac{\int_0^\pi (\cos \vartheta)^{n-1} P_n^1(\vartheta) \sin^2 \vartheta\, d\vartheta}{\int_0^\pi [P_n^1(\vartheta)]^2 \sin \vartheta\, d\vartheta}. \tag{7.77}$$

In contrast to the coefficients A_n, the product $A_n P_n^1(\vartheta)$ in (7.72) does not depend on the normalization of the functions $P_n^1(\vartheta)$. Expressing $P_1^1(\vartheta)$ and $P_2^1(\vartheta)$ as in (7.55), we obtain $A_1 = (3/2)\sqrt{\pi/2}$, $A_2 = -5/4\sqrt{\pi/2}i$ from (7.77). The integrals in (7.77) can be calculated in the explicit form for any n, since, according to the known formula, $P_n^1(\vartheta)$ is the polynomial of the $(n-1)$th degree of $\cos \vartheta$, multiplied by $\sin \vartheta$:

$$P_n^1(\vartheta) = \sin \vartheta \frac{d^{n+1}}{d(\cos \vartheta)^{n+1}} \left[\left(\cos^2 \vartheta\right) - 1 \right]^n. \tag{7.78}$$

The normalization different from (7.55) is used here.

If the expressions for the potentials U^0 and V^0 of the incident wave are known, then it is easy to derive those for the potentials U^{diff} and V^{diff} of the diffracted field. These potentials are expanded into the series similar to (7.72), but by the Hankel functions instead of the Bessel functions. Using notations (7.73b), these series can be written as

$$U^{\text{diff}}(R, \vartheta, \varphi) = \frac{1}{k^2 R} \sum_{n=1}^\infty B_n \bar{h}_n(kR) P_n^1(\vartheta) \cos \varphi,$$

$$V^{\text{diff}}(R, \vartheta, \varphi) = \frac{1}{k^2 R} \sum_{n=1}^\infty C_n \bar{h}_n(kR) P_n^1(\vartheta) \sin \varphi. \tag{7.79}$$

The coefficients are different in these series, since the boundary conditions at $R = a$ (7.70) are different for these functions. Comparing series (7.72) and (7.79) and accounting the completeness and orthogonality of the functions $P_n^1(\vartheta)$, we obtain

$$B_n = \frac{\bar{j}'_n(kR)\big|_{R=a}}{\bar{h}'_n(kR)\big|_{R=a}} A_n, \tag{7.80a}$$

$$C_n = \frac{\bar{j}_n(ka)}{\bar{h}_n(ka)} A_n. \tag{7.80b}$$

Formulas (7.79), (7.77), (7.73), and (7.80) solve the problem of the plane wave diffraction on the metallic sphere. The solution is represented in the form of Rayleigh series.

The incident field is expressed by both the potentials U^0 and V^0; the diffracted filed by U^{diff} and V^{diff}. Although U^{diff} and V^{diff} are expressed only by U^0 and V^0, respectively, in the diffracted field these two potentials are related differently than in the incident one. Therefore, in contrast to the incident field, the total one is not linearly polarized. The depolarization occurs at the diffraction on the metallic sphere. This does not contradict the fact that if $H_R \equiv 0$ in the incident field, then the total field possesses the same property. The definition "polarized wave" needs to be specified, which components are zero in the field.

7.3.5
Small sphere: dipole momenta

At $ka \ll 1$, the functions $\bar{j}_n(ka)$ and $\bar{h}_n(ka)$ in (7.80) can be replaced with the first terms of their expansions by the small parameter ka. In this case $\bar{j}_n$ and $\bar{h}_n$ have the orders $(ka)^{n+1}$ and $(ka)^{-n}$, respectively, so that the coefficients B_n and C_n have the order $(ka)^{2n+1}$. In the higher order, only the first terms, that is, the terms with $n = 1$, must be kept in series (7.80). The coefficients B_1 and C_1 have the order $(ka)^3$; $C_1 = (-1/2)B_1$. Expressing $P_1^1(\vartheta)$ by the functions $\sin\vartheta$ yields

$$
\begin{aligned}
U^{\text{diff}} &= \frac{1}{k^2 R} B_1 \bar{h}_1(kR) \sin\vartheta \cos\varphi, \\
V^{\text{diff}} &= \frac{1}{k^2 R} C_1 \bar{h}_1(kR) \sin\vartheta \sin\varphi.
\end{aligned}
\tag{7.81}
$$

Although both the potentials are symmetrical with respect to the plane $\vartheta = \pi/2$, the forward $(0 < \vartheta < \pi/2)$ and backward $(\pi/2 < \vartheta < \pi)$ radiations do not coincide. The components $E_\vartheta^{\text{diff}}$ and E_φ^{diff} consist of two terms, one of which does not depend on ϑ and the other is proportional to $\cos\vartheta$. Their sum is not symmetrical with respect to the plane $\vartheta = \pi/2$. Calculations show that for the small metallic sphere, the forward radiation is much smaller than the backward radiation.

In (7.81), the z-axis is selected, that is, it is assumed that the incident wave propagates along this axis. The angles ϑ and φ are coordinates in the spherical coordinate system having this axis. However, for the small sphere, the phase alteration in the incident wave is not large over the body. The x- and y-axes along which the fields $\vec{E}^0$ and $\vec{H}^0$ of the incident wave are directed are selected in the physical sense. Selection of these particular axes also follows from formulas (7.81) describing the angular dependence of the Debye potentials of the diffracted field. Denote the angles made by the radius-vector of a point with the axes x and y, by ϑ_x and ϑ_y, respectively. Then $\sin \vartheta \cos \varphi = \cos \vartheta_x$, $\sin \vartheta \sin \varphi = \cos \vartheta_y$, and formulas (7.81) can be written as

$$
\begin{aligned}
U^{\text{diff}}(R, \vartheta_x) &= \frac{1}{k^2 R} B_1 \bar{h}_1 (kR) \cos \vartheta_x, \\
V^{\text{diff}}(R, \vartheta_y) &= \frac{1}{k^2 R} C_1 \bar{h}_1 (kR) \cos \vartheta_y.
\end{aligned}
\tag{7.82}
$$

If the x-axis is taken as an axis of the spherical coordinate systems then the potential U^{diff} and fields expressed by it do not depend on the azimuthal coordinate. Analogous assertion is valid for the y-axis and potential V^{diff}. The dependence of potentials on the radial coordinate is given by the multiplier

$$
\frac{1}{kR} \bar{h}_1 (kR) = -\sqrt{\frac{2}{\pi}} \left(1 - \frac{i}{kR} \right) \frac{\exp (-ikR)}{kR}.
\tag{7.83}
$$

According to (7.58), formula (7.83) describes the radial dependence of the elementary dipole.

In the case of diffraction of the plane linearly polarized wave on a small metallic sphere, in the higher order with respect to the small parameter ka, the field is a superposition of the two fields of elementary electric and magnetic dipoles oriented along the fields $\vec{E}^0$ and $\vec{H}^0$ of the incident wave, respectively. Comparing the formulas of this subsections with (7.58) gives the expressions for the electric and magnetic polarization coefficients, ρ_E and ρ_H, respectively,

$$
\rho_E = a^3, \quad \rho_H = -\frac{1}{2} a^3.
\tag{7.84}
$$

Recall that the electric and magnetic dipole momenta equal $\rho_E E^0$ and $\rho_H H^0$, respectively. As will be shown in the next section, this result is valid for the case of diffraction of arbitrary field on any small metallic body. The momenta of both the dipoles are found from the two independent problems, namely, the electrostatic and magnetostatic ones. For the sphere, formulas (7.84) are elementarily found by this method.

7.3.6
Large sphere: the Watson series

The terms of the Rayleigh series (7.72) decrease starting from the number larger than the argument. For the large spheres, that is, at $ka \gg 1$, the numerical convergence for $R = a$ is observed only at the large term numbers. There exist other series fast convergent for the fields arisen at diffraction on the large sphere. Below we outline the technique of constructing such series (the Watson series).

The derivations of this subsection briefly repeat those used for constructing the Watson series in the diffraction problem on the cylinder (see Subsection 7.1.3). Two methods for obtaining this series were described there. In this subsection, we do not apply the first method, that is, the method of the asymptotic series summation based on transformation of the Rayleigh series into the contour integral in the plane of complex indices. We describe only the method consisting in direct constructing of series for the Debye potentials of the total field. The spherical functions of the coordinate R are involved into these series; satisfying both the radiation conditions at $R \rightarrow \infty$ and the boundary conditions (7.69) at $R = a$. The order of these functions is not the real number $n + 1/2$ ($n = 1, 2, \ldots$) as for the functions (7.69), but a complex one. Denote this index as $\nu_n + 1/2$ ($n = 1, 2, \ldots$). Then, ν_n is one of the roots of equations (7.86a) or (7.86b), similar to (7.18). In order that the Debye potentials satisfy the wave equation and, therefore, the fields expressed by these potentials by (7.53), satisfy the Maxwell equations, the complex values $\nu_n + 1/2$ should be the lower index of the Legendre functions. At the plane wave incidence onto the sphere, the upper index which determines the potential dependence on the azimuthal angle φ, equals unity, $m = 1$. The system of functions, by which the Watson series for the Debye potentials is constructed, is similar to (7.56). It has the form

$$\frac{1}{\sqrt{kR}} H_{\nu_n+1/2}^{(2)}(kR) P_{\nu_n}^1 (\vartheta) \left\{ \begin{array}{c} \cos \varphi \\ \sin \varphi \end{array} \right\}, \quad n = 1, 2, \ldots \tag{7.85}$$

The potentials $U(R, \vartheta, \varphi)$ and $V(R, \vartheta, \varphi)$ are proportional to $\cos \varphi$ and $\sin \varphi$, respectively.

The complex numbers ν_n for the functions by which the potential U of the total field is expanded, differ from those for the potential V. According to (7.69), these numbers should solve the equations

$$\frac{d}{dR} \overline{h}_{\nu_n} (kR) \Big|_{R=0} = 0, \tag{7.86a}$$

$$\overline{h}_{\nu_n} (kR) = 0 \tag{7.86b}$$

for the potentials U and V, respectively. According to (7.73b), the function $\overline{h}_{\nu_n}$ is

$$\overline{h}_{\nu_n} (kR) = \sqrt{kR} H_{\nu_n+1/2}^{(2)} (kR) . \tag{7.87}$$

Equations (7.86) coincide with those obtained by equating to zero the denominators in (7.80) for the coefficients of Rayleigh series and considering the obtained equalities as the equations for the index of the function $\overline{h}_n(kR)$. The above conforms with the following situation: if we obtained the Watson series by the asymptotic summation of the Rayleigh series (7.79), then, according to (7.80), the series for the potential U would be a series by the residues of the function $1/\overline{h}_\nu'(ka)$ treated as a function of the complex variable ν. The series for V would be a series by the residues of the function $1/\overline{h}_\nu(ka)$. The residues are taken into the roots of the denominators of these functions, that is, at the values of ν satisfying equations (7.86a) or (7.86b).

In the diffraction problem for the cylinder, the function $\Phi_{\nu_n}(\varphi)$ (7.26) with the complex index ν_n, involved into the Watson series for the total field is not analytical on the ray $\varphi = 0$. When passing this ray, the derivative of function Φ_{ν_n} discontinues and, therefore, so does the entire series (7.19) for the field. This situation can be treated as the existence of extrinsic currents on the ray $\varphi = 0$. The coefficients of series (7.19) are obtained from the condition that the jump of the series being the function of the r-coordinate equals these currents.

In the problem of diffraction on the sphere, the function $P_{\nu_n}^1(\vartheta)$ with the complex index ν_n, involved into the Watson series for the total field is singular on the ray $\vartheta = \pi$. It becomes infinite at $\vartheta = \pi$. This situation can also be treated as the existence of extrinsic currents on the ray $\vartheta = \pi$. The coefficients of the series for $U(R, \vartheta, \varphi)$ and $V(R, \vartheta, \varphi)$ by the functions (7.85) are found from the demand that at $\vartheta \to \pi$ the singularities of these series treated as the functions of R ($a < R < \infty$) be proportional to the given extrinsic currents. Replacing these currents by a dipole and moving it into infinity allows us to construct the Watson series for the problem of the plane wave diffraction on the metallic sphere.

The current distribution on the surface of large metallic sphere illuminated by a plane wave is described by several terms of the Watson series. According to (7.53c), the components H_φ and H_ϑ at $R = a$, that is, the meridian and azimuthal currents, are proportional to $\cos \varphi$ and $\sin \varphi$, respectively. The ϑ-dependence of these currents is described by the sum of several functions $P_{\nu_n}^1(\vartheta)$.

At $ka \gg 1$, the geometric optics terms are qualitatively applicable. There exist the illuminated side of sphere, the light-shadow border and the shadowed side. According to the geometrical optics theory, on the illuminated side, the current is proportional to the tangential component of the magnetic field of incident wave. There is no current on the shadowed side. At large but finite ka, all these assertions should be specified. The overall picture of the current distribution on the surface of metallic body with small curvature will be considered in the next section.

7.4
Small bodies; large bodies

7.4.1
Diffraction on the body with noncoordinate surface

In this subsection, the diffraction on small bodies with constant ε and μ, not equal (in general) to unity, is considered. If the largest linear size a of the body is small in comparison with the wavelength, that is, $ka \ll 1$, then the diffraction field is a superposition of the fields of electric and magnetic dipoles with the momenta $i\omega\rho_E\vec{E}^0$ and $i\omega\rho_H\vec{H}^0$, respectively. Here $\vec{E}^0$, $\vec{H}^0$ are the fields into which the body is placed, ρ_E, ρ_H are the electric and magnetic polarization coefficients, respectively. This assertion is also valid for the field in the distances larger than a, if assuming (which is made here and further) that there are no other bodies nearby and that the fields $\vec{E}^0$, $\vec{H}^0$ are nonzero and slow varying in distance of the order a.

We expand the total field $\{\vec{E}, \vec{H}\}$ into the series by powers of the small parameter k: $\vec{E} = \vec{E}_0 + k\vec{E}_1 + \cdots$, $\vec{H} = \vec{H}_0 + k\vec{H}_1 + \cdots$, and substitute these expansions into the Maxwell equations (1.29) with $\vec{j}^{\text{ext}} = 0$. Equating the terms with the same powers of k yields

$$\text{rot}\,\vec{E}_0 = 0, \quad \text{rot}\,\vec{H}_0 = 0, \tag{7.88a}$$

$$i\varepsilon\vec{E}_0 = \text{rot}\,\vec{H}_1, \quad -i\mu\vec{H}_0 = \text{rot}\,\vec{E}_1. \tag{7.88b}$$

The first-order equations (7.88b) subject the terms of the zero order $\vec{E}_0$, $\vec{H}_0$ to the existence conditions for the field $\{\vec{E}_1, \vec{H}_1\}$. Taking the divergence of equations (7.88b), we obtain, together with (7.88a), the system of four equations for $\vec{E}_0$, $\vec{H}_0$. Dropping the lower index 0, we write them as the two independent systems

$$\text{rot}\,\vec{E} = 0, \quad \text{div}\left(\varepsilon\vec{E}\right) = 0, \tag{7.89a}$$

$$\text{rot}\,\vec{H} = 0, \quad \text{div}\left(\mu\vec{H}\right) = 0, \tag{7.89b}$$

which are satisfied by the field $\{\vec{E}, \vec{H}\}$ in the higher order with respect to k. The systems (7.89a) and (7.89b) describe the electrostatic field $\vec{E}$ and magnetostatic field $\vec{H}$, respectively.

The permittivities ε and μ have jumps on the boundary S of the body. Equations (7.89) lead to the connection between the limiting values of the fields $\{\vec{E}^+, \vec{H}^+\}$ inside the body and $\{\vec{E}^-, \vec{H}^-\}$ outside it. Repeating considerations given for the Maxwell equations (see Subsection 1.3.1), from (7.89a) we obtain

$$\left. \left(E_t^+ - E_t^- \right) \right|_S = 0, \tag{7.90a}$$
$$\left. \left(\varepsilon E_N^+ - E_N^- \right) \right|_S = 0. \tag{7.90b}$$

Condition (7.90b) is not automatically fulfilled in contrast to the case when the fields satisfy the Maxwell equations on both the sides of the boundary. Although there is no magnetic field in the electrostatics, when passing to electrostatic equations (7.89a), the electric field must be subjected to the first equation of (7.88b), which is necessary for the magnetic field existence in the next order of k. This condition results in the boundary one (7.90b) (see remark at the end of Subsection 1.3.1).

In the magnetostatic, the boundary conditions have a similar form

$$\left. H_t^+ - H_t^- \right|_S = 0, \tag{7.91a}$$
$$\left. \mu H_N^+ - H_N^- \right|_S = 0. \tag{7.91b}$$

Equations (7.89) are valid in distances R much smaller than the wavelength, that is, at $kR \ll 1$. They are valid, in particular, in the distances larger than a, that is, at $R \gg a$. Since $ka \ll 1$, both these conditions are consistent. These distances are "infinitely large" for the electrostatic problem and, simultaneously, "infinitely small" for the electrodynamic problem. The limiting value of the field at $R \to \infty$ in the electrostatic problem is, simultaneously, the limiting value of that at $R \to 0$ in the electrodynamic problem. The similar situation occurs in the theory of short-periodical arrays (see Subsection 5.2.3).

The electrostatic problem must contain a condition at the "static infinity." This condition implies that the diffracted field should decrease as the dipole one, that is, according to (6.6), as $1/R^3$. After extracting the trigonometric multiplier, the coefficient in this asymptotic is equal to the dipole momentum for the component E_ϑ. For E_R, it is twice larger. Thus, the electrostatic problem consists of equations (7.89a), the boundary conditions (7.90), and the asymptotic condition

$$E_R = E_R^0 + \rho_E \frac{2\cos\vartheta}{R^3} E_R^0 + O\left(1/R^4\right) \tag{7.92}$$

at $R \to \infty$. Here we use the spherical coordinate system with origin inside the body and axis directed along the arisen dipole momentum. The coefficient ρ_E and the momentum direction can be found from the above conditions. If ρ_E is a tensor, then formula (7.92) has a more complicated form. The coefficient ρ_E has a dimension of volume. When the body is not very prolate or oblate and ε is not very close to unity, then ρ_E has order of the body volume.

For solving the electrostatic problem, the electrostatic potential is introduced as a scalar function Φ. The field is calculated by Φ as

$$\vec{E} = -\nabla\Phi. \tag{7.93}$$

According to (7.89a), (7.90), and (7.92), the potential Φ solves the problem

$$\Delta\Phi = 0, \tag{7.94a}$$

$$\Phi^+ - \Phi^-|_S = 0, \quad \varepsilon\frac{\partial\Phi^+}{\partial N} - \frac{\partial\Phi^-}{\partial N}\bigg|_S = 0, \tag{7.94b}$$

$$\Phi = -E^0(0)\left(R\cos\vartheta - \rho_E\frac{\cos\vartheta}{R^2}\right) + O\left(1/R^3\right). \tag{7.94c}$$

Similar to (7.92), formula (7.94c) is obtained under the assumption that either ρ_E is scalar or $\vec{E}^0$ is directed along one of the polarization tensor axes.

The magnetostatic problem from which ρ_H and direction of the magnetic momentum can be obtained, consists in finding the magnetostatic potential Ψ from the problem

$$\Delta\Psi = 0, \tag{7.95a}$$

$$\Psi^+ - \Psi^-|_S = 0, \quad \mu\frac{\partial\Psi^+}{\partial N} - \frac{\partial\Psi^-}{\partial N}\bigg|_S = 0, \tag{7.95b}$$

$$\Psi = -H^0(0)\left(R\cos\vartheta - \rho_H\frac{\cos\vartheta}{R^2}\right) + O\left(1/R^3\right). \tag{7.95c}$$

The field is calculated $\vec{H} = -\nabla\Psi$.

In the case of sphere, problem (7.94) is solved elementarily. The field inside the dielectric sphere placed into the homogeneous external field is also homogeneous and has the same direction. Outside the sphere, it is equal to the first two terms of its asymptotic expansion. The field potential equals

$$\Phi^+ = -A \cdot R\cos\varphi \cdot E^0(0), \quad R < a$$
$$\Phi^- = -\left(R\cos\vartheta - \rho_E\frac{\cos\vartheta}{R^2}\right) \cdot E^0(0), \quad R > a. \tag{7.96}$$

The numbers A and ρ_E are obtained from the two conditions (7.94b) at $R = a$

$$\rho_E = a^3\frac{\varepsilon - 1}{\varepsilon + 2}, \tag{7.97a}$$

$$\vec{E}^+ = \frac{3}{\varepsilon + 2}\vec{E}^0. \tag{7.97b}$$

In almost the same simple way, the problem for dielectric ellipsoid is solved. In this case the polarization coefficient is the tensor with axes coincident with those of the ellipsoid. If $\vec{E}^0$ is directed along one of these axes, then, similarly as for the sphere, the inner field is homogeneous and has the same direction. However, the formulas similar to (7.97), are more cumbersome for the ellipsoid. In the next subsection, we will give them only for the metallic ellipsoid of revolution.

It follows from the analogy between formulas (7.94) and (7.95) that

$$\rho_H = a^3 \frac{\mu - 1}{\mu + 2}, \tag{7.98a}$$

$$\vec{H}^+ = \vec{H}^0 \frac{3}{\mu + 2} \tag{7.98b}$$

for the sphere made from the material with $\mu \neq 1$. Formula (7.98a) implies, in particular, that if $\mu = 1$, then $\rho_H = 0$, that is, the small dielectric sphere made from the nonmagnetic material, does not create the magnetic momentum in the electromagnetic field. This assertion ($\Psi^+ = \Psi^-$ and $P_H = 0$ at $\mu = 1$) follows from (7.95) for an arbitrary body. However, it is not valid for the metallic sphere which has the magnetic momentum (see (7.84)). Passage from dielectric to metal will be considered in the next subsection.

7.4.2
Small metallic bodies

The field $\{\vec{E}^+, \vec{H}^+\}$ satisfies the static equations (7.89) inside the body only if the body size is small in comparison with the wavelength inside the body material, that is, if $k\sqrt{|\varepsilon\mu|}a \ll 1$. If this condition violates, but the weaker one $ka \ll 1$ holds, then the fields are statical only outside the body; they satisfy the Maxwell equations inside it. The electric and magnetic fields are connected and exist simultaneously.

The passage from the small body, material of which has finite ε and μ, to the metallic one is illustrated on the two-dimensional model. We consider a plane wave which falls perpendicular onto a circular cylinder having the axis coincident with the z-axis and propagates in the x-direction. The wave has the components $E_z^0 = -\exp(-ikx)$, $H_y^0 = \exp(-ikx)$. The cylinder radius is small, $ka \ll 1$, but no restrictions are imposed on the parameter $k\sqrt{|\varepsilon\mu|}a$. According to (7.32), the total field outside the cylinder (at $r > a$) can be expressed in the form as

$$E_z^- = -1 + ikr\cos\varphi + A_0 H_0^{(2)}(kr) + A_1 H_1^{(2)}(kr)\cos\varphi + \cdots, \tag{7.99a}$$

$$H_\varphi^- = \cos\varphi - iA_0 H_0^{(2)\prime}(kr) - iA_1 H_1^{(2)\prime}(kr)\cos\varphi + \cdots. \tag{7.99b}$$

Only the components involved into the boundary conditions at $r = a$ are given. According to (7.34), inside the cylinder (at $r < a$) these components are

$$E_r^+ = B_0 J_0(k\sqrt{\varepsilon\mu}r) + B_1 J_1(k\sqrt{\varepsilon\mu}r)\cos\varphi + \cdots, \tag{7.100a}$$

$$H_\varphi^+ = -i\frac{\sqrt{\varepsilon\mu}}{\mu}\left[B_0 J_0'(k\sqrt{\varepsilon\mu}r) + B_1 J_1'(k\sqrt{\varepsilon\mu}r)\cos\varphi + \cdots\right]. \tag{7.100b}$$

Here we account that, according to (1.32), $H_\varphi = -i/(k\mu) \cdot \partial E_z/\partial r$; in (7.1) $\mu = 1$ was put. It is easy to show that the term in (7.99b), proportional to $\cos \varphi$, is the field of magnetic dipole directed along the y-axis, that is, in the direction of the field $\vec{H}^0$. The angle φ measured from the x-axis in the plane (x, y), equals $\pi/2 - \vartheta$, where ϑ is measured from the y-axis. The component H_ϑ differs from H_φ only in sign. Consequently, the component H_ϑ^- contains the term decreasing proportional to $\sin \vartheta/r^2$, that is, as the field of magnetic dipole in the two-dimensional problems. Substituting $H_1^{(2)} = -2i/\pi \cdot 1/(kr)$ yields the expression $2/(k\pi) \cdot A_1 \sin(\vartheta/2)$ for the third term in (7.99b). It is the field of the magnetic dipole with the momentum $\rho_H = 2A_1/(k\pi)$ directed along the external field $\vec{H}^0$ (at $|\vec{H}^0| = 1$). Equating the boundary values of the tangential components E_z, H_φ at $r = a$ which are proportional to $\cos \varphi$, leads to the system of two nonhomogeneous linear equations for ρ_H and B_1. The sought expression for ρ_H is found from this system in the convenient form

$$\rho_H = \frac{\mu T - 1}{\mu T + 1} a^2, \tag{7.101a}$$

where

$$T = \frac{J_1\left(ka\sqrt{\varepsilon\mu}\right)}{ka\sqrt{\varepsilon\mu} J_1'\left(ka\sqrt{\varepsilon\mu}\right)}. \tag{7.101b}$$

At $ka\sqrt{|\varepsilon\mu|} \ll 1$, we have $T = 1$ and (7.101a) gives $\rho_H = a^2 \cdot (\mu - 1)/(\mu + 1)$. At $ka\sqrt{|\varepsilon\mu|} \gg 1$, $T \ll 1$ and (7.101) gives $\rho_H = -a^2$. These two limiting values of ρ_H obtained for the circle, are similar to those for a sphere made from the magnetic material (see (7.98a)) and the metallic one ($\rho_H = -a^3/2$). In the static approximation ($ka\sqrt{|\varepsilon\mu|} \ll 1$), the magnetic dipole momentum does not arise in a nonmagnetic sphere ($\mu = 1$). According to (7.97a), in the sphere with $\varepsilon = 1$, the electric dipole momentum does not arise in this approximation. However, the above results are valid only at $T = 1$.

Although formula (7.97a) is not valid at large $|\varepsilon|$, at $|\varepsilon| \to \infty$ it gives the correct value a^3 to the electric dipole momentum for the metallic sphere. If we put $\varepsilon = 0$ in this formula, then it gives the correct value $-a^3/2$ to the magnetic dipole momentum for the metallic sphere.

We show that for an arbitrary body, this property follows from the static conditions (7.94), (7.95) imposed on the potentials Φ and Ψ. If $\varepsilon = 1$, then Φ^+ continuously passes into Φ^-, which means that $\rho_E = 0$. At $\varepsilon \to \infty$, it follows from (7.94b) that $\partial\Phi^+/\partial N|_S = 0$. For Φ^+ satisfying the Laplace equation, the above is possible only if $\Phi^+ \equiv 0$, which provides that $\Phi^-|_C = 0$ follows from the first formula (7.94a). According to (7.93), this implies that $E_t^-|_S = 0$ which is the boundary condition on metallic surface.

Putting $\varepsilon = 0$ in the second condition of (7.94b) yields $\partial\Phi^-/\partial N|_S = 0$. Assuming that Φ is the magnetic field potential, we obtain $H_N|_S = 0$; this is the usual boundary condition imposed on the magnetic field on metallic surface (see text before (1.81)). Note that in this case the first condition in (7.95b) is not fulfilled: the tangential components of magnetic field are discontinuous on the metallic body surface.

If the coefficient ρ_E is found from system (7.94) as a some function of ε, $\rho_E = f(\varepsilon)$ for a certain body, then $f(1) = 0$. It follows from the symmetry of formulas (7.95) and (7.94), that $\rho_H = f(\mu)$ for the body made from the magnetic material. The polarization coefficients of metallic body are equal to $\rho_E = f(\infty), \rho_H = f(0)$, respectively.

At the end of this subsection, we give formulas for the polarization coefficients of ellipsoids of revolution. These formulas are obtained by solving the Laplace equation in elliptic coordinates. They generalize formulas (7.84) for the sphere.

If the field $\vec{E}^0$ or $\vec{H}^0$ is directed along the axis of revolution, then the dipole momenta are directed in the same direction. The polarization coefficients are expressed by the auxiliary parameter n ($0 < n < 1$) depending only on the axes ratio

$$\rho_E = \frac{ab^2}{3} \cdot \frac{1}{n}, \tag{7.102a}$$

$$\rho_H = -\frac{ab^2}{3} \cdot \frac{1}{1-n}. \tag{7.102b}$$

Here a is the half-axis directed along the axis of revolution, b is the second half-axis. For the prolate ellipsoid ($a > b$),

$$n = \frac{1-e^2}{e^3}\left(\frac{1}{2}\ln\frac{1+e}{1-e} - e\right), \quad e^2 = \frac{a^2 - b^2}{a^2}, \tag{7.103}$$

whereas for the oblate one ($a < b$),

$$n = \frac{1+t^2}{t^3}(t - \arctan t), \quad t^2 = \frac{b^2 - a^2}{a^2}. \tag{7.104}$$

For the sphere, both formulas give $n = 1/3$.

In the second case if $\vec{E}^0$ or $\vec{H}^0$ is directed perpendicular to the axis of revolution (the dipole momenta are directed in the same direction), then n should be replaced by $(1-n)/2$ in (7.102).

In this case we have

$$\rho_E = \frac{2ab^2}{3} \cdot \frac{1}{1-n}, \tag{7.105a}$$

$$\rho_H = -\frac{2ab^2}{3} \cdot \frac{1}{1+n}. \tag{7.105b}$$

We give one more formula obtained from (7.105a) for very oblate ellipsoid (disk) having the thickness $2a$ much smaller than its diameter $2b$. The electric field $\vec{E}^0$ lies in the disk plane. According to (7.104), if $a \ll b$, then $t \gg 1$. In such a field, according to (7.105a), the electric polarization coefficient of the disk is

$$\rho_E = \frac{4}{3\pi} b^3. \tag{7.106}$$

It does not depend on the disk thickness and is only $3\pi/4$ times, (i.e., approximately twice) smaller than for the sphere of the same radius b. Formula (7.106) is used in the next subsection.

7.4.3
The duality principle for the hole in plain screen

The duality principle establishes connection between the fields arisen in two problems of diffraction: on the hole in the plane infinite metallic screen (Fig. 7.5(a)) and on the finite screen having shape and location just as the hole in the first problem (Fig. 7.5(b)). Denote the field falling onto the hole in the first problem by $\{\vec{E}^0, \vec{H}^0\}$ and the arisen one by $\{\vec{E}, \vec{H}\}$. In the second problem another incident field $\{\tilde{\vec{E}}^0, \tilde{\vec{H}}^0\}$ different from that in the first one, falls onto the screen. The arisen field is $\{\tilde{\vec{E}}, \tilde{\vec{H}}\}$ in the second problem. Let the incident fields be related as

$$\tilde{\vec{E}}^0 = -\vec{H}^0, \tag{7.107a}$$
$$\tilde{\vec{H}}^0 = \vec{E}^0. \tag{7.107b}$$

Then the field $\{\tilde{\vec{E}}^0, \tilde{\vec{H}}^0\}$ obtained by this substitution from the field $\{\vec{E}^0, \vec{H}^0\}$ satisfying the Maxwell equations, satisfies these equations as well, that is, these equations are invariant to substitution (7.107).

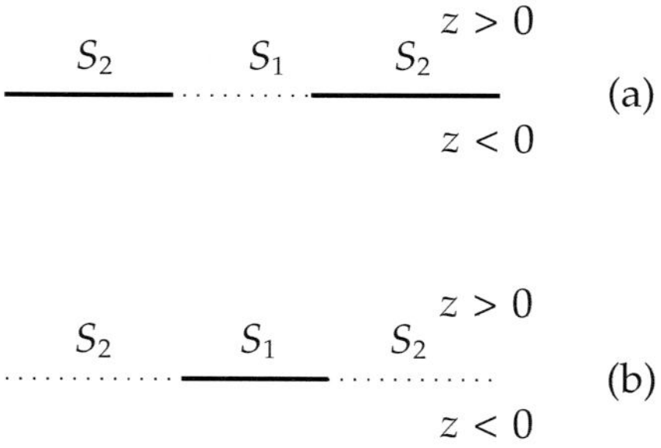

Fig. 7.5 Illustration of the duality principle

Here we confine ourselves to a simple case when the incident field $\{\vec{E}^0, \vec{H}^0\}$ is the plane wave falling perpendicular onto the hole. Then the field $\{\tilde{\vec{E}}^0, \tilde{\vec{H}}^0\}$ is a similar wave with another polarization direction. Let the screen be located in the coordinate plane $z = 0$, the wave fall from the half-space $z > 0$ and the field components normal to this plane be not present in the incident waves.

We accept that in the half-space $z < 0$, the field $\{\vec{E}, \vec{H}\}$ of the first problem coincides with the field $\{\vec{e}, \vec{h}\}$ obtained from the diffraction field $\tilde{\vec{E}}^{\text{diff}}$, $\tilde{\vec{H}}^{\text{diff}}$ of the second problem as

$$\vec{h} = \tilde{\vec{E}}^{\text{diff}}, \tag{7.108a}$$

$$\vec{e} = -\tilde{\vec{H}}^{\text{diff}}, \tag{7.108b}$$

where $\tilde{\vec{E}}^{\text{diff}} = \tilde{\vec{E}} - \tilde{\vec{E}}^0$, $\tilde{\vec{H}}^{\text{diff}} = \tilde{\vec{H}} - \tilde{\vec{H}}^0$. Since the field $\{\tilde{\vec{E}}^{\text{diff}}, \tilde{\vec{H}}^{\text{diff}}\}$ satisfies the Maxwell equations, the field $\{\vec{e}, \vec{h}\}$ defined by condition (7.108) satisfies them as well.

The proof of the fields $\{\vec{e}, \vec{h}\}$ and $\{\vec{E}, \vec{H}\}$ identity is based on comparison of the values of their tangential components on the plane $z = -0$. For the components E_t, H_t we have

$$H_t = H_t^0 \quad \text{on } S_1, \tag{7.109a}$$

$$E_t = 0 \quad \text{on } S_2. \tag{7.109b}$$

Formula (7.109a) follows from the fact that the field $\vec{H}$ differs from the incident field $\vec{H}^0$ by the field of currents induced on the screen S_2, that is, lying in the plane $z = 0$. Currents located on a plane create the magnetic field normal to this plane, that is, they have zero tangential components. Condition (7.109b) holds since the domain S_2 is occupied by the metallic screen.

In the second problem (Fig. 7.5(b)), the domain S_1 is occupied by the metallic screen, that is, $\tilde{E}_t = 0$ in this domain, hence, $E_t^{\text{diff}} = -\tilde{E}_t^0$. According to (7.108a), the left-hand side of this equality equals h_t, whereas the right-hand side equals H_t^0 in accordance with (7.107a). Consequently, $h_t = H_t^0$ on S_1. On S_2, the component $\tilde{H}_t$ differs from $\tilde{H}_t^0$ by the field created by currents flawing on S_1 (i.e., on the same plane), therefore, $\tilde{H}_t^{\text{diff}} = 0$. According to (7.108b), this implies that $e_t = 0$. Thus, the tangential components of $\vec{e}, \vec{h}$ are

$$h_t = H_t^0 \quad \text{on } S_1, \tag{7.110a}$$

$$e_t = 0 \quad \text{on } S_2. \tag{7.110b}$$

Therefore, the fields $\{\vec{E}, \vec{H}\}$ (7.109) and $\{\vec{e}, \vec{h}\}$ (7.110) coincide on the plane $z = 0$. Both these fields satisfy the radiation condition at $z \to -\infty$. We prove that these fields coincide in the whole half-space $z < 0$. Difference of these fields, that is, the field $\{\vec{E} - \vec{e}, \vec{H} - \vec{h}\}$, satisfies the homogeneous Maxwell equations and radiation conditions in the whole half-space. On the part S_1 of the plane $z = 0$, the tangential component of the difference magnetic field equals zero. On the other part S_2, this component of the difference electric field is zero, either. Therefore, the Poynting vector flux of this field through the plane $z = 0$ equals zero. Since there are no currents in the half-space $z < 0$, the Poynting vector flux through infinite half-sphere at $z < 0$ is zero as well. This implies that at infinity the difference field decreases faster than the distance from the origin. As it was already mentioned, such a field is identical to zero. Consequently, $\vec{E} = \vec{e}$, $\vec{H} = \vec{h}$ at $z < 0$.

It is proven that if we replace the incident field according to (7.107) in the diffraction problem on the hole and then replace the diffracted field in the lower half-space according to (7.108) in the problem of diffraction of this incident field on the finite screen, then the resulting field solves the diffraction problem on the hole.

The duality principle reduces the problem of diffraction of the plane wave falling perpendicular onto a small circular hole, to the above problem about electric dipole momentum of the plane disk of small radius. The diffracted field $\{\tilde{\vec{E}}^{\mathrm{diff}}, \tilde{\vec{H}}^{\mathrm{diff}}\}$ arisen when a small disk is placed into the electric field $\tilde{\vec{E}}^0$ parallel to its plane, is the field of electric dipole with the momentum $\mathcal{P} = i\omega\rho_E\tilde{E}^0_t$, where ρ_E is given in (7.106). By (7.107), $\mathcal{P} = -i\omega\rho_E H^0_t$. This formula seems to be somewhat paradoxical: the electric dipole momentum is proportional to the magnetic field of incident wave. It is connected with the fact that this expression gives the momentum for the screen replacing the hole.

If the field $\{\vec{E}, \vec{H}\}$ is known in the half-space $z < 0$, then the energy passing through the hole can be found. It is twice smaller than the energy radiated by elementary electric dipole in the whole space. According to (6.13), this energy is $k^2\omega^2 p^2/(6c)$. Substituting expression (7.106) for p and dividing by the energy $c/(8\pi) \cdot \pi b^2$ which falls onto the hole, we obtain the major characteristic of the process of diffraction on small circular hole at the normal incidence, namely, the relative part τ of the energy passing through the hole. It equals

$$\tau = C\,(kb)^4, \quad C = \frac{64}{27\pi^2} \approx 0.24. \tag{7.111}$$

Under the condition $kb \ll 1$, at which this formula is valid, very small part of the incident energy passes through the hole. Of course, under the inverse condition $kb \gg 1$, this ratio is close to unity.

Using (7.111), it is possible to find the electric field arising on a small circular hole. The energy flux through the hole is equal (with accuracy to the factor $c/(8\pi)$) to the integral over the hole area taken of the tangential component of $\vec{E}$ multiplied by the perpendicular to it tangential component of $\vec{H}^0$ of the incident field. Consequently, the average value of E_t on S_1 (Fig. 7.5(a)) equals the component E^0_t of incident field, multiplied by the coefficient (7.111). The field on the hole is much smaller than the incident one. On the hole border, that is, on metal, $E_t = 0$, and it cannot take large value at a small distance of the order b.

According to (7.108b), for the hole of arbitrary size and shape, the component E_t equals the magnetic field $\tilde{H}_t$ on the surface of metallic disk placed into the external electric field $\tilde{E}^0$ which equals, according to (7.107a), the magnetic field H_t^0 of the wave falling onto the hole.

The leakage of the field through the hole depends on how much the hole cutting disturbs the current induced by the incident wave on the solid screen. The larger this disturbance is, the larger the field E_t on the hole, and the more energy passes through it. The current easily flows around the small hole (Fig. 7.6(a)); a small electric field arises in the hole.

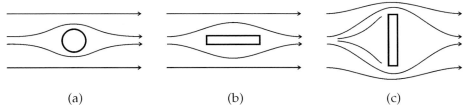

| (a) | (b) | (c) |

Fig. 7.6 The current lines deformation by different holes

If the hole has a shape of a long slot, then the magnitude of the electric field arising on it depends on the slot direction with respect to the current on metal. The direction of the induced current coincides with the direction of $\vec{E}^0$.

We consider two limiting cases. In the first case, the slot is parallel to the current, that is, to $\dot{E}_t^0$. In the complementary problem about the metallic strip, the external electric field $\tilde{E}_t^0$ is same directed as H_t^0, that is, across the strip. Then the current arising on the strip is small. In the initial problem about the slot, the electric field on it is small and directed parallel to the slot sides. The current easily flows around the slot parallel to it (Fig. 7.6(b)).

In the second limiting case, the slot is perpendicular to the field E_t^0. In the corresponding problem about strip, the external electric field $\tilde{E}_t^0$ is directed along the strip. The large current arises on the strip and the large magnetic field perpendicular to the strip arises on its surface. In the problem about slot this means that a large electric field perpendicular to the slot, arises on it. The slot cutting the current causes large perturbation of the current lines; they are significantly deformed (Fig. 7.6(c)).

We compare these results with the theory of the strip arrays, stated in Subsections 5.2.6 and 6.3.4. This theory is more complicated than that of isolated slot, since there is no infinite screen in the strip arrays. However, there exists an analogy between these two systems. The first limiting case (Fig. 7.6(b)) corresponds to the wave incidence onto the array with $\vec{E}^0$ parallel to the strips. In the approximation accepted in Subsection 5.2.6, according to which the field in slots is exactly zero, the array transparency equals zero (5.68a). More exact theory (Subsections 6.3.4) accounts the difference of this field from zero. Although the transparency P_S in (6.110a) is small, it differs from zero. In the array theory, the second limiting case (Fig. 7.6(c)) corresponds to the case when $\vec{E}^0$ is perpendicular to the strips. In the simplest theory, the field in slots is assumed to be equal to the incident one, which corresponds to formula (5.68b). In the more exact theory the field in slots is close to E_t^0 but smaller than it. The transparency P_S is large but smaller than unity.

7.4.4
Thin metallic cylinder of arbitrary cross-section: the H-polarization

Let the linearly polarized plane wave normally fall onto the thin metallic cylinder of an arbitrary cross-section. In the cylindrical coordinate system (z, r, φ), the cross-section does not depend on z. The field $\vec{H}^0$ has the only component H_z^0, the field $\vec{E}^0$ lies in the transverse plane. The entire field does not depend on the z-coordinate. Similarly as in Section 7.1, we denote $V = H_z(r, \varphi)$; the components E_r, E_φ are expressed by V according to formulas (7.3) with $\varepsilon = 1$; other components do not present from the field. The potential V satisfies the wave equation (7.2):

$$\Delta V + k^2 V = 0. \tag{7.112}$$

In this and the next subsections the symbol Δ denotes the two-dimensional Laplace operator.

The total field V consists of the incident field $V^0 = \exp(-ikr\cos\varphi)$ and diffracted field V^{diff}. On the cylinder surface, that is, on its cross-section contour C, the function V^{diff} satisfies the condition

$$\left.\frac{\partial V^{\text{diff}}}{\partial N}\right|_C = -\left.\frac{\partial V^0}{\partial N}\right|_C, \tag{7.113}$$

where N is the normal to C, lying in the cross-section plane, that is, N is the normal to the cylinder surface. Condition (7.113) means that $E_s|_C = 0$, where $\vec{E}$ is the total electric field, and s is tangent to the contour C. The current I_s induced on the surface is directed along C, its surface density differs from the value of H_z on metal (i. e., from $V|_C$) only by the factor $c/(4\pi)$, which we omit further. The induced current creates the diffracted field. Far from the cylinder, at $r \gg a$, where a is the characteristic size of the cross-section, V^{diff} tends to zero. The problem is to find the current I_s and field V^{diff} for the case $ka \ll 1$, that is, for thin cylinders.

In the case of the circular cylinder, the problem has an explicit solution in the form of the Rayleigh series (7.13). According to this solution, the current on the cylinder surface $r = a$ is

$$V|_C = V^0\Big|_C + B_0 H_0^{(2)}(ka) + B_1 H_1^{(2)}(ka)\cos\varphi + O\left((ka)^2\right). \tag{7.114}$$

At $ka \ll 1$, (7.15) gives

$$V|_C = 1 - 2ika \cos \varphi. \tag{7.115}$$

The current consists of two summands. The first (larger) one, independent of the angle φ, is called the *ring current*. For the circular cylinder, the ring current, that is, the first term in (7.115) is equal (in this approximation) to the field of the incident wave. The second summand, proportional to $\cos \varphi$, is called the *dipole current*. In fact, the dipole current is co-linearly directed on the whole line C, similarly as the current in the dipole. The elements of the ring current, located on the opposite sides of any diameter, are oppositely directed.

In the far field zone, at $r \gg a$, the field V^{diff} can be calculated by integration of the current over the contour C (which gives the additional factor a), with accounting both the current direction and the difference between distances to different current elements. The fields created by different elements of the dipole current are, in fact, in-phase. The diffracted field created by the dipole current, has the order $(ka)^2$. The phase difference between fields created by two elements of the ring current, located on opposite sides of a diameter (and different in the sign of $\cos \varphi$, only), is approximately equal to π. These fields do not cancel each other completely, since the distances from different elements of the current are different; they differ by values of the order a. The diffracted field created by the ring current has the same order $(ka)^2$, despite the ring current is larger than the dipole one by one order. The total diffracted field has the order $(ka)^2$. Of course, this fact also follows from (7.114), since, according to (7.15), the coefficients B_0 and B_1 have this order. The coefficients B_n, $n > 1$ have higher order of smallness.

The diffracted field arising when a thin cylinder is placed into the field in which $\vec{E}^0$ is perpendicular to the cylinder director, is small. Therefore, in particular, for this polarization of the incident field, the reflection coefficient from the array is small (see Subsection 6.3.4). A significant reflection occurs only from the array with very narrow slots. Physically, this is explained by the fact that the displacement current between the conductors ("capacity current") is essential in this case.

For the cylinder of arbitrary cross-section, in the higher order, the current (i. e., the value $V|_C$) consists also of the two summands: first is constant along C (ring current) and second changes its sign (dipole one). The field V satisfies the Laplace equation in the whole domain $kr \ll 1$, the boundary condition $\partial V / \partial N|_C = 0$ on the cross-section contour and the asymptotic condition

$$V(r, \varphi) = V^0(r, \varphi) + \rho_E \frac{\cos \varphi}{r} + O(r^{-2}) \tag{7.116}$$

at $r \gg a$. In the two-dimensional problem, the Laplace equation is satisfied not by the potential (as in the three-dimensional problem), but by the field V itself.

It decreases as r^{-1}, but not as R^{-2}, as the three-dimensional potential (compare (7.94) with (7.95)). The condition at the static infinity has such a simple form only in the case if the coefficient ρ_E is scalar (see the note below formula (7.94c)). Similarly as in the three-dimensional problem, the electric polarization coefficient ρ_E is found from the conditions formulated before (7.114).

Far from the cylinder the field is much smaller, than on the cylinder and in its neighborhood. In order to calculate the far field by the current, we must know the current very precisely. First of all, this relates to the ring current, since far from the cylinder the field created by it is $(ka)^2$ times smaller than the current itself.

The ring current is proportional to the area of the cylinder cross-section. For the circular cylinder this filed is proportional to a^2. If the cylinder is a strip, then this area is zero and the ring current does not create the external field. This result follows from the fact that the distance from any point outside the strip to both of its opposite sides is the same, and the ring currents on these sides are directed toward each other.

7.4.5
Thin metallic cylinder of arbitrary cross-section: the E-polarization

Here we consider the second case: the electric field of the wave normally falling onto the cylinder has the only component E_z^0. An induced current flowing along the long cylinder is large and disturbs the incident field significantly. Both these effects have some peculiar properties for the model of infinitely long cylinder. The main property is that the problem cannot be reduced to the electrostatic one for arbitrarily thin cylinder.

The field is expressed by the scalar function $U(r, \varphi) = E_z(r, \varphi)$. The other two nonzero components H_r, H_φ are calculated by U according to (7.1). The function U satisfies the wave equation

$$\Delta U + k^2 U = 0 \tag{7.117}$$

outside the cylinder, and the boundary condition

$$U|_C = 0 \tag{7.118}$$

on the cylinder boundary (and does not satisfy the radiation condition).

For the circular cylinder there exists the exact solution (7.7), (7.9) to this problem. At $ka \ll 1$ the diffracted field expressed by the Rayleigh series (7.7), can be substituted by the first term (for $n = 0$) of this series. The coefficient A_0 in the series has the order $1/\ln(ka)$, all the next coefficients A_n, $(n > 0)$ are much smaller, they have the order $(ka)^{2n}$. Dropping the terms of the order $(ka)^2$ and replacing (in the same approximation) $U^0|_C$ by the value $U^0(0)$ of the incident field on the cylinder axis, we obtain the following expression for the diffracted field:

$$U^{\text{diff}}(r, \varphi) = -\frac{H_0^{(2)}(kr)}{H_0^{(2)}(ka)} U^0(0). \tag{7.119}$$

At $kr \ll 1$, this formula becomes

$$U^{\text{diff}}(r, \varphi) = -\frac{\ln(kr)}{\ln(ka)} U^0(0). \tag{7.120}$$

This function satisfies the Laplace equation

$$\Delta U = 0. \tag{7.121}$$

However, in contrast to the similar functions for the three-dimensional small bodies (see, e. g., (7.96)), function (7.120) does not vanish at $r/a \rightarrow \infty$. In the *electrostatic domain*, that is, at $kr \ll 1$, the diffracted field in the two-dimensional problem to which the idealization of "infinitely long cylinder" leads for this polarization, is not a solution to the electrostatic problem.

In electrostatics, the problem about the field into which an infinitely long metallic body directed along the field line, is placed, has no solution. This idealization is not consistent with the electrostatic laws. Introducing such a body into the electrostatic field destroys the field.

The diffracted field (7.120) satisfies the Laplace equation (7.121) and the boundary condition following from (7.118), but at the "static infinity" ($kr \ll 1$, $r/a \gg 1$) it does not decrease, it is proportional to $\ln(kr)$ and increases; moreover, this field depends on the wave number.

We find the current J flowing on the cylinder along the axis. It differs from the integral of the component $H_\varphi = -(i/k)\partial U/\partial r$ over the contour C, only by the factor $c/(4\pi)$. This integral equals

$$\int_0^{2\pi} H_\varphi a\, d\varphi = -\frac{2\pi i}{k\ln(ka)} U^0(0). \tag{7.122}$$

In this integral, the term containing $\partial U^0/\partial r$ is small, proportional to $(ka)^2$, and it is omitted here. At $ka \ll 1$, the induced current weakly depends on the cylinder radius. At $a \rightarrow 0$, the current decreases to zero very slowly.

For the cylinder of arbitrary cross-section, the function U^{diff} should be determined from equation (7.121), the boundary condition

$$U^{\text{diff}}\Big|_C = -U^0\Big|_C, \tag{7.123}$$

equivalent to (7.118), and the condition that U^{diff} is proportional to $\ln r$ far from the cylinder. After U^{diff} is found, the current J induced on the cylinder, can be calculated. In the same approximation as formula (7.122), the current is proportional to

$$-\frac{i}{k}\int_C \frac{\partial U^{\text{diff}}}{\partial N} ds. \tag{7.124}$$

The problem about finding the field U^{diff} and calculating the integral (7.124) can be solved by the conformal mapping method. We outline the backgrounds of this method.

Introduce the Cartesian coordinates x, y instead of the polar ones r, φ. Associate the complex value $Z = x + iy$ with the point (x, y). Then, introduce the auxiliary plane with the Cartesian coordinates u, v and associate the complex value $W = u + iv$ with the point (u, v) (see Fig. 7.7). Denote the main plane by Z and auxiliary plane by W. Connect the complex numbers W and Z by the equation

$$Z = F(W), \tag{7.125}$$

where the function F will be found below. This equation connects any point of the plane W with a point of Z. Separating the real and imaginary parts in (7.125), we obtain the two equivalent equations

$$x = x(u, v), \quad y = y(u, v). \tag{7.126a}$$

When a point of the plane W draws a certain line, then the point of Z, corresponding to it by (7.126a), draws a line as well.

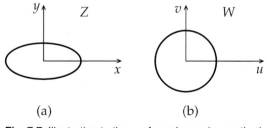

(a)   (b)

Fig. 7.7 Illustration to the conformal mapping method

The function F can be found from the following two demands. Firstly, at $|W| \to \infty$, that is, far from the coordinate origin, the ratio Z/W should tend to unity, at the limit, both planes must coincide. Secondly, when a point of the plane W draws the circle of certain radius a_{ef} (which should be found as well), then the corresponding point of Z draws the given contour C, participating in formulas (7.123), (7.124).

It is known that for any closed contour C there always exist an analytical function F and a number a_{ef} for which the above conditions are fulfilled. From the analyticity of the function F follows that the mapping (7.126a) has the following property: any function satisfying the Laplace equation in the variables u, v, satisfies the same equation in the variables x, y connected with u, v by the substitution

$$u = u(x, y), \quad v = v(x, y), \tag{7.126b}$$

inverse to (7.126a). Obviously, the value of $U(u, v)$ at any point on the circle of radius a_{ef} equals the value $U(x, y)$ at the respective point on the contour C (for simplicity, we use the same notation for these two functions). Behavior of the function $U(x, y)$ at $x^2 + y^2 \to \infty$ coincides with behavior of the function $U(u, v)$ at $u^2 + v^2 \to \infty$. Consequently, if $U(u, v)$ solves the problem consisting of equation (7.121), the boundary condition (7.123) and the condition that $U(u, v)$ is proportional to $\ln(u^2 + v^2)$ at $(u^2 + v^2) \to \infty$, then $U(x, y)$, obtained by substitution (7.126b), satisfies equation (7.121), condition (7.118) on the contour C and the condition that $U(x, y)$ is proportional to $\ln(x^2 + y^2)$ at $(x^2 + y^2) \to \infty$. If we know the function $U^{\mathrm{diff}}(u, v)$ for the circular cylinder, then the function $U^{\mathrm{diff}}(x, y)$ for the cylinder with the cross-section contour C is calculated after determining F and a_{ef} corresponding to this contour.

The conformal mapping of the plane W onto Z is accompanied with an extension (or compression) and hence the surface current density $\partial U^{\mathrm{diff}}/\partial N_{xy}$ on the contour C does not equal the surface current density $\partial U^{\mathrm{diff}}/\partial N_{uv}$ on the circle. However, the extension is the same in any direction (as a mapping by the analytical function), and the element $\partial U^{\mathrm{diff}}/\partial N_{xy} ds_{xy}$ in the integral (7.124) is invariant at this mapping. The total current is invariant as well, and, according to (7.122), it equals

$$\int_C H_s \, ds = -\frac{2\pi i}{k} \frac{1}{\ln(ka_{\mathrm{ef}})} U^0(0) \tag{7.127}$$

for any contour C.

We give an example in which all calculations are elementary. Let C be the ellipse with half-axes a, b. It is easy to check that in this case equation (7.125) has the form

$$Z = W + \frac{A}{W}, \quad A = \frac{a^2 + b^2}{4}. \tag{7.128}$$

The mapping (7.126a) is

$$x = u + \frac{Au}{u^2 + v^2}, \quad y = v - \frac{Av}{u^2 + v^2}. \tag{7.129}$$

When a point moves in the plane W along the circle of radius a_{ef}, then $u = a_{\mathrm{ef}} \cos \alpha$, $v = a_{\mathrm{ef}} \sin \alpha$, where the angle α varies in the plane W from

0 to 2π. Then, according to (7.129), $x = (a_{\rm ef} + A/a_{\rm ef})\cos\alpha$, $y = (a_{\rm ef} - A/a_{\rm ef})\sin\alpha$. Consequently, the corresponding point in the plane Z moves along the ellipse with the half-axes $a = a_{\rm ef} + A/a_{\rm ef}$, $b = a_{\rm ef} - A/a_{\rm ef}$. The coefficient A introduced in (7.128), is calculated from these equations, and the value $a_{\rm ef}$ of the effective radius is

$$a_{\rm ef} = \frac{a+b}{2}. \tag{7.130}$$

For instance, for the strip ($b = 0$) of the width $2a$ we have $a_{\rm ef} = a/2$, that is, the current arising on such a strip is the same as on the circular cylinder of the diameter a (twice smaller than the strip width) (Fig. 7.8).

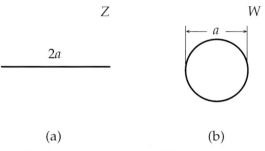

(a)   (b)

Fig. 7.8 Geometry of equivalent diffraction problems

7.4.6
The current on a metallic surface of small curvature

In general case, the current induced on the surface of metallic body can be found from the integral equation obtained from the demand that the electric field created by this current has the tangential component on the surface, equal to the same component of the incident field with the opposite sign. If the surface curvature radius is large in comparison with the wavelength, then there exists an approximate solution to this problem. The solution is local: according to it, at any point of the surface, the current does not depend on the current on other parts of the surface. The solution is exact only for the case of the plane wave falling onto the plane surface. In this case the surface current, that is, the tangential component of the magnetic field equals the twofold value of the same component in the incident wave.

In the approximate solution, the current at any point of the surface is assumed to be equal to the current which would arise on the metallic plane tangent to the surface at this point, when the same wave fell onto this plane. This approach is called the *physical optics method*.

There exists an improved version of the method often called as the *physical diffraction theory*. The improvement relates to surfaces with fractures and to the shaded domains arisen when illuminating surfaces with small curvature. We consider these two case separately.

Let the surface have a fracture: the line on which two planes or surfaces of small curvatures meet. Locally, such a body is a wedge. In the physical optics approximation, the currents on both of these surfaces do not depend on the distance to the wedge edge. They discontinue on the edge. If the geometrical shadow arises, that is, one of the planes is not illuminated, then the current does not overflow onto this surface.

In the physical diffraction theory, the current distribution near the edge is assumed to be the same as for the wedge, for which the exact solution exists. There is a strip near the edge of the width of the wavelength order, where the currents depend on the distance l from the edge. The current component parallel to the edge decreases when approaching it. For instance, for the half-plane, this component tends to infinity as $l^{-1/2}$ (compare with (7.47)). The second component is continuous on the edge. The current exists not only in the illuminated domain but also it overflows into the shaded one.

The second improvement relates to the case, when the geometro-optical shadow arises behind the smooth convex surface (Fig. 7.9). On the one side of the light-shadow interface the surface is illuminated, on the other one it lies in the shadow. In the physical optics approximation, the current on the illuminated side is constant (twice larger than the magnetic field of the incident wave), on the shaded one it equals zero. On the interface the current is discontinuous.

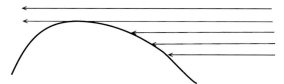

Fig. 7.9 Geometro-optical rays near a convex surface

In the physical diffraction theory, the current is continuous. There exists a domain along the interface, in which a smooth change from light to shadow takes place. The main result of the theory, transferred from the problems of diffraction on the large sphere and large cylinder, is that the field in the semi-shaded domain is described by a universal function. On the one side of the interface it tends to the value 2, on the second one to 0. The argument of the function is the ratio of the distance from the border to a number d defining the order of the transient domain width. The value of d depends only on the curvature radius a and the wavelength, and equals

$$d = \frac{a}{(ka)^{1/3}}. \tag{7.131}$$

At $ka \to \infty$, the transient domain width tends to zero; but not too fast.

These results are also valid for large dielectric bodies. The field on their shadowed surface varies approximately by the same law, as the currents on the metallic bodies. Even at finite $|\varepsilon|$, the transient domain has the order, given by (7.131). For instance, for the Earth ($a = 64,000$ km) and the wavelength $\lambda = 0.1$ m we have $d \approx 40$ km.

Hence, the physical diffraction theory consists, mainly, in the fact that the results obtained when solving the standard problems about the wedge, cylinder, and sphere, are extended onto the problems which have no analytical solution, but are similar to the above ones.

The approximate methods explained in this subsection allow us to find the currents induced on the surface of large bodies. In the next subsection, we give an approximate method for finding the field in the whole domain adjacent to such surfaces; not only on the surfaces themselves. The main interest of this method lies in finding the field in the domain of the optical shadow. Below we describe only this part of the method.

7.4.7
Ray structure of the field in shadow: geometro-optical theory of diffraction

If the field structure is close to that of the plane wave, then the field can be described in terms of the ray optics. The field at any point depends on properties of the medium on a certain line (*ray*) passing through this point. Although, formally, the field at any point depends on the medium properties and on the bodies located in the whole space, this dependence is weak under conditions permitting the ray interpretation. In the homogeneous medium the rays are straight lines, in nonhomogeneous one they are curved (see Subsection 5.1.5). The rays reflect and deflect on the medium interfaces. The field energy slowly varies along the ray, except for the segments near the points and lines (*focuses* and *caustics*) where the ray optics is nonapplicable. The rays are orthogonal to the surfaces of constant phase (equiphase surfaces). They satisfy the Fermat principle. The optical path (integral (5.32)) along the ray between the source and any point is extreme (as a rule, minimal) in comparison with all close paths connecting these two points. In the homogeneous medium, the ray is a line of the smallest length, connecting the source with observation point.

When diffracting on large bodies, the field has the ray structure in almost the whole illuminated domain. The rays do not penetrate into the shaded domain. The rays do not envelop the obstacles and not come into the shadow. The field in the shadow is zero in the geometro-optical description.

The ray concept can be extended in such a way that the field in the shaded domain is also described in terms of the ray optics. In this generalized approach, the rays envelop obstacles. The rules, by which this process proceeds, make up the base of the method called the *geometrical* (more exactly *geometro-*

optical) diffraction theory. The geometro-optical and diffraction methods are combined in this theory. Similarly as the physical diffraction theory, it is based on the solutions of several standard problems. The Fermat principle is completely preserved. We explain the method on the problems about the shaded domains behind a fraction and behind a smooth convex surface.

Let the wedge (fraction) be illuminated by the source located at the point A (Fig. 7.10). The point B is located in the shadow, that is, under the light-shadow line ACD. The ray (in the extended notion) connecting points A and B, consists of the two straight segments. The first one connects the point A with a certain point C on the edge, whereas the second one connects the points C and B. If A and B lie in the plane perpendicular to the edge, then C lies in this plane as well. If there is no plane perpendicular to the edge and containing A, B, then C is located on the edge in such a place that the lines AC and BC make the same angles with edge. It is easy to check that the summarized length of the segments AC and BC is minimal in this case, that is, the ray ACB satisfies the Fermat principle.

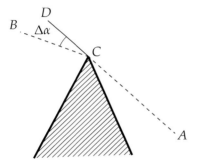

Fig. 7.10 Geometric-diffraction rays near a wedge

The same ray image can be described in other terms. The rays outgo from the source A. The rays coming to the edge create secondary sources there. Each of them generates the rays but not in all directions. The ray fan outgoing from a secondary source consists of the rays making the same angle with edge, which is made by the ray AC. If one of these rays passes through the point B, then the sought point C is the point at which this secondary source is located. For any point B there exists a point C on the edge, corresponding to ("illuminating") it.

The rays generated by a secondary source have different energy. The energy of each ray CB depends on the angle $\Delta\alpha$ made by the ray with the light–shadow interface CD. The energy distribution among the rays can be obtained from the exact solution of the problem about diffraction on the wedge. At $\Delta\alpha \ll 1$, the energy of the ray CD is smaller than that of AC by the factor $1 - N\Delta\alpha$, where N is calculated from the same problem.

If an opaque body has the vertex instead of the edge (as the cone or pyramid), then the vertex is the only secondary source, the rays of which penetrate into the shaded zone.

The second example relates to the ray structure of the field in the shadow behind a convex surface (Fig. 7.11). Let the surface be illuminated by the source A, the line AC_1D tangent to the source, separates the illuminated and shaded domains. The point B is located in the shadow. According to the geometro-optical diffraction theory, the ray consisting of three segments comes into the point B: the tangent segment AC_1, the line C_1C_2, following the curved body surface, and the tangent segment C_2B. The "creeping" ray C_1C_2 "breaks away" the surface at the point C_2 and comes into B along the straight line. The creeping ray goes from C_1 to C_2 along a geodesic arc of the surface. The length of the whole ray AC_1C_2B is minimal among all the lines connecting A and B in presence of the opaque body.

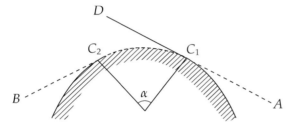

Fig. 7.11 Geometric-diffraction rays near a convex surface

The deeper the point B is located in the shadow, the larger the segment C_1C_2 is, and the smaller the energy of the ray coming to this point. The ray energy decreases exponentially on this segment. This fact conforms with the above result that the energy decreases by the factor $1 - N\Delta\alpha$ on a small fraction. Indeed, the bend can be treated as a limit of many fractions by small angles. After passing many fractions the energy decreases by the product of such factors. Denote the angle between the tangents at the points C_1 and C_2 by α, and each small fraction of the ray by $\Delta\alpha$. Then the number of the fractions is $\alpha/\Delta\alpha$, and the factor describing the total energy decrease is $(1 - N\Delta\alpha)^{\alpha/\Delta\alpha}$. At the limit, this value describes the exponential decrease as $\exp(-\alpha N)$.

7.4.8
The method of auxiliary sources

The *method of auxiliary sources* is based on the concept of a set of point sources, which are located in the "nonphysical domain." We have introduced this term in Subsection 6.3.7 while finding the possible antenna surface, which creates the desired radiation pattern. In the problem about diffraction on the metallic body the nonphysical domain is the one occupied by the body. When using this concept, we compare the diffraction problem with an auxiliary one about the field of sources located in the nonphysical domain. They radiate into the space free from bodies. Outside the domain bounded by the same surface S as in the diffraction problem, the sources create a field equal to the diffracted one. On the surface this field satisfies the same demand: the tangential component of the electric field is equal in magnitude and opposite in sign to corresponding component of the incident field. The field in the nonphysical domain is analytical on the surface, that is, it transforms into the diffracted one continuously together with all its derivatives. The medium filling the nonphysical domain in the auxiliary problem can be chosen arbitrarily; it is not connected in any way with medium filling the actual body. As a rule, it is taken to be the same as the exterior medium.

In contrast to the problem considered in Subsection 6.3.7, in the diffraction problems the filed to be created by the auxiliary sources is a priori unknown, and therefore the direct methods for localization of the singularity points of analytical continuation of the field are nonapplicable.

Denote the location of the nth source and its amplitude by $\vec{r}_n$ and a_n, respectively; $n = 1, 2, \ldots, N$, where N is the number of sources. The field created by the nth source at the point $\vec{r}$ equals $a_n H_0^{(2)}(k|\vec{r} - \vec{r}_n|)$ in the two-dimensional scalar problem or $a_n \exp(-ik|\vec{r} - \vec{r}_n|)/|\vec{r} - \vec{r}_n|$ in the three-dimensional scalar one. In the three-dimensional vector problem, the sums of such terms are the Debye potentials.

The coordinates $\vec{r}_n$ are chosen arbitrarily to some extent and the coefficients a_n are found (as is shown below) from the boundary condition. Thus, as in the spectral methods, the solution to the wave equation (the Maxwell equation system) is expressed as a sum of the partial solutions of the homogeneous equation. However, these partial solutions correspond to the same values of the frequency and the medium parameters as those given in the diffraction problem. The partial solutions have the singularities in their sources and therefore differ from zero. In contrast to the eigenfunctions in the spectral methods, the partial solutions in the method of auxiliary sources are not orthogonal.

The auxiliary sources are located on a certain surface Σ. If their locations are uniform (in some sense) and the number of them tends to infinity, then the sums

$$\sum_{n=0}^{N} a_n H_0^{(2)}(k|\vec{r} - \vec{r}_n|) \tag{7.132}$$

approximate any field outside the surface Σ. Here we confine ourselves to the two-dimensional scalar problem. The coefficients a_n are found from the system of N linear algebraic equations obtained from the demand that the sum (7.132) at the points $\vec{r} = \vec{r}_m$, $m = 1, 2, \ldots, N$ (*collocation points*) located on the body surface S, satisfies the boundary condition, that is, equals the values of the incident field with opposite sign. The points $\vec{r}_n$ and $\vec{r}_m$ should be chosen in such a way that the numbers a_n tend to some limits when the number of equations increases. After finding these numbers we can calculate the diffracted field by formula (7.132) (or similar one for the Maxwell equations).

The efficiency of the method depends on the location of the surface Σ with respect to S (Fig. 7.12). The surface Σ should be located inside S. If Σ coincides with S and $\vec{r}_m \neq \vec{r}_n$ for any n, m, then the system of algebraic equations coincide with that obtained in the collocation method applied to the integral equation for the induced field. If Σ lies fully inside S, then the collocation points and the source location ones are placed in the finite distances and all coefficients of the equation system for a_n are finite.

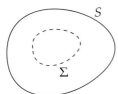

Fig. 7.12 Location of auxiliary sources for closed domains

In general, the computational procedure of the equation system solving becomes unstable at $N \to \infty$. To improve the stability, the surface Σ, the number of source points and their location on it should be carefully accorded. The main demand to the surface Σ is that Σ should surround all singularity points of the analytical extension of the diffracted field. This means that Σ should not be too small. If this demand is not fulfilled and Σ is located too far from S, then the coefficients a_n do not tend to limiting values, because the field of auxiliary sources has no singularities outside Σ. There exist qualitative methods for approximate localization of the singularity points of the analytical extension of diffracted field by the incident field structure and specific properties of the body surface S. The method of auxiliary sources is the most effective if the sources are located in the singularity points (i.e, the surface Σ passes through them) or close to them.

There exist different variants of the auxiliary sources method, applicable to a wide class of the problems. For instance, when applying the method to the diffraction problem on the dielectric body, we should construct the surface Σ consisting of two parts Σ_1 and Σ_2 (Fig. 7.13). One of them, Σ_1 lies inside the body surface S, the second one, Σ_2 surrounds it. The field of the sources located on Σ_1 approximates the diffracted one outside S, the field of sources located on Σ_2 approximates the field inside the dielectric body. In the field of the latter sources, the wave number k must be replaced by $k\sqrt{\varepsilon}$, where ε is the permittivity of the body material. At the collocation points on S, the continuity conditions for the components E_t, H_t must be satisfied by the fields of all auxiliary sources together with incident field. In another variant of the method, the functions $H_0^{(2)}(k|\vec{r} - \vec{r}_n|)$ in the sum (7.132) are replaced by the functions $H_m^{(2)}(k|\vec{r} - \vec{r}_n|) \exp(im\varphi)$, $m > 0$.

The method of auxiliary sources is effective for the diffraction problems on large bodies, as well as for the nonstationary problems, that is, for the problems with nonmonochromatic dependence of time.

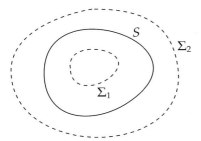

Fig. 7.13 Location of auxiliary sources for dielectric bodies

The question on existence of the so-called *back waves* in the diffracted field is also connected with the auxiliary sources method. This notion was introduced in Subsection 6.3.7 as well. The problem is formulated as the question whether the diffracted field can be expressed in the whole domain outside the body boundary S only by the outgoing waves, or the field near the body also contains incoming waves: in other words, whether anywhere outside S the field is expressible as the series

$$\sum_{n=-\infty}^{\infty} A_n H_n^{(2)}(kr) \exp(in\varphi) \tag{7.133a}$$

by the functions $H_n^{(2)}(kr)$ describing the outgoing (*direct*) waves (*Rayleigh hypothesis*), or we must supply this series by the second one

$$\sum_{n=-\infty}^{\infty} B_n H_n^{(1)}(kr) \exp(in\varphi) \tag{7.133b}$$

by the functions $H_n^{(1)}(kr)$ describing the incoming ("back") waves.

Note that the question is rather mathematical than physical, since it is connected with coordinate system choice. The above functions describe the direct and back waves only in cylindrical coordinates. In other systems the separation of the waves into these two types is different. For instance, the field containing the terms $H_n^{(1)}(kr)$ (back waves in the cylindrical system) may not contain back waves in another coordinate system (e. g., in the elliptic one).

A simple answer exists to the above question, based on using the location of the analytical extension singularities. If the smallest circle centered at the origin, inside which all singularity points are located, lies inside S, then the back waves are not presented in the diffracted field (Fig. 7.14(a)). If this circle partially lies outside S (although all singularity points lie, of course, inside S), then the back wave may be present in the field (Fig. 7.14(b)). For instance, it is known that when the plane wave falls onto the ellipse, the singularity points lie on the segment connecting the focuses. The diameter of the smallest circle containing this segment equals the distance between the focuses, that is, $2\sqrt{a^2 - b^2}$, where a, b are the ellipse half-axes, $a \geq b$. If the diameter is smaller than the axis $2b$ (i. e., the eccentricity $e < 1/\sqrt{2}$), then the minimal circle lies inside the ellipse and back waves do not participate in the diffracted field. For the more prolate ellipse, when $\sqrt{a^2 - b^2} > b$ (i. e., $e > 1/\sqrt{2}$) the back waves are present in the field.

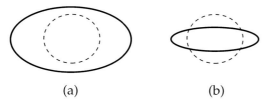

(a) (b)

Fig. 7.14 Illustration to the existence of back waves

7.4.9
The optical theorem

The scattering pattern of the field, arisen at the incidence of the plane wave onto an arbitrary body without losses, is subjected to the condition stated by the so-called *optical theorem*. This theorem immediately follows from the Maxwell equations together with radiation condition and does not depend neither on the body shape nor on its material.

First, we derive this condition for the two-dimensional scalar problem. When the plane wave of the unit amplitude, propagating along the x-axis, falls onto any cylinder, then the field in the far field zone consists of the two terms

$$U(r, \varphi) = \exp(-ikr \cos \varphi) + F(\varphi) \frac{\exp(-ikr)}{\sqrt{kr}}. \tag{7.134}$$

The sought condition on the pattern $F(\varphi)$ can be obtained from the formula

$$\int_0^{2\pi} \operatorname{Im}\left(U \frac{\partial U^*}{\partial r}\right) ds = 0, \quad ds = r d\varphi, \tag{7.135}$$

being the scalar analog of the energy conservation law. This formula is obtained from the wave equation

$$\Delta U + k^2 \varepsilon U = 0. \tag{7.136}$$

At $\operatorname{Im} \varepsilon = 0$, it follows from (7.136) that

$$\operatorname{div} (U \operatorname{grad} U^* - U^* \operatorname{grad} U) = 0. \tag{7.137}$$

Integrating over the domain bounded by the circle of radius r, we obtain (7.135).

If the circle radius is so large that (7.134) holds on it, then the integrand in (7.135) equals

$$\frac{1}{kr} |F(\varphi)|^2 + \frac{2}{\sqrt{kr}} \operatorname{Re} \left[F(\varphi) \exp \left(-ikr (1 - \cos \varphi) \right) \right]. \tag{7.138}$$

The integral in (7.135) is calculated by the stationary phase method at $kr \gg 1$. Only the highest term of the order $1/\sqrt{kr}$ is found. Since the integration with respect to the φ-coordinate is performed only over the small interval around the stationary phase point $\varphi = 0$, the factor $\cos \varphi$ before the exponent is replaced by unity. Terms of the next order of smallness, and those disappearing at the integration over φ are dropped. Since, in the higher order,

$$\int_0^{2\pi} F(\varphi) \exp \left[-ikr (1 - \cos \varphi) \right] d\varphi = \sqrt{\frac{\pi}{kr}} (1 - i) F(0), \tag{7.139}$$

it follows from (7.135) that

$$\int_0^{2\pi} |F(\varphi)|^2 \, d\varphi = -2\sqrt{\pi} \operatorname{Re} \left[(1 - i) F(0) \right]. \tag{7.140}$$

This is the two-dimensional scalar formulation of the optical theorem. It is easy to check that the above solution of the problem about diffraction on the metallic cylinder satisfies condition (7.140). The diffracted field (7.7) with coefficients (7.11), has the pattern $F(\varphi) = \sum_{n=0}^{\infty} \alpha_n \cos(n\varphi)$, where

$$\alpha_n = \frac{-\sqrt{i}}{(1 + \delta_{0n})} \sqrt{\frac{8}{\pi}} \frac{J_n(ka)}{H_n^{(2)}(ka)}. \tag{7.141}$$

Condition (7.140) holds if

$$\pi \sum_{n=0} (1 + \delta_{0n}) |\alpha_n|^2 = 8 \sum_{n=0} \frac{1}{(1 + \delta_{0n})} \operatorname{Re} \frac{J_n(ka)}{H_n^{(2)}(ka)}. \tag{7.142}$$

The coefficients α_n (7.141) satisfy this equation.

We repeat this derivation for the three-dimensional vector field. Let the plane wave having the two components $E_x = \exp(-ikz)$, $H_y = \exp(-ikz)$, propagate along the z-axis. It diffracts on a certain body and creates a divergent spherical wave with pattern defined, according to (1.33), by the functions $F_1(\vartheta, \varphi)$ and $F_2(\vartheta, \varphi)$. In the highest order of $1/(kR)$ the components of the total field in the far field zone are equal to

$$
\begin{aligned}
E_\vartheta &= F_1(\vartheta, \varphi) \frac{\exp(-ikR)}{kR} + \cos\vartheta\cos\varphi\exp(-ikR\cos\vartheta), \\
E_\varphi &= F_2(\vartheta, \varphi) \frac{\exp(-ikR)}{kR} - \sin\varphi\exp(-ikR\cos\vartheta), \\
H_\vartheta &= F_2(\vartheta, \varphi) \frac{\exp(-ikR)}{kR} + \cos\vartheta\sin\varphi\exp(-ikR\cos\vartheta), \\
H_\varphi &= -F_1(\vartheta, \varphi) \frac{\exp(-ikR)}{kR} + \cos\varphi\exp(-ikR\cos\vartheta).
\end{aligned}
\tag{7.143}
$$

By definition, there is no energy absorption inside the body on which the diffraction takes place. The energy flux through a sphere encircling the body is zero. The equality

$$
\int_0^{2\pi}\int_0^\pi \mathrm{Re}\left(-E_\vartheta H_\varphi^* + E_\varphi H_\vartheta^*\right)\sin\vartheta\, d\vartheta d\varphi = 0
\tag{7.144}
$$

holds. We substitute the expressions (7.143) for the fields in far field zone into this condition. Dropping the terms which disappear during the integration over ϑ, and replacing the factor $\cos\vartheta$ before the exponent by unity, we obtain

$$
\int_0^{2\pi}\int_0^\pi \left\{|F_1(\vartheta, \varphi)|^2 + |F_2(\vartheta, \varphi)|^2\right\}\sin\vartheta\, d\vartheta d\varphi
$$

$$
= -2kR \cdot \mathrm{Re}\int_0^{2\pi}\int_0^\pi \{-F_1(\vartheta, \varphi)\cos\varphi
$$

$$
+ F_2(\vartheta, \varphi)\sin\varphi\}\exp\left[-ikR(1-\cos\vartheta\right]\sin\vartheta\, d\vartheta d\varphi. \tag{7.145}
$$

Denote the Cartesian components of the scattered field in the far field zone by $1/(kR)E_x(0)$ and $1/(kR)E_y(0)$. At small ϑ,

$$
\begin{aligned}
F_1(\vartheta, \varphi) &= E_x(0)\cos\varphi + E_y(0)\sin\varphi, \\
F_2(\vartheta, \varphi) &= -E_x(0)\sin\varphi + E_y(0)\cos\varphi.
\end{aligned}
\tag{7.146}
$$

Consequently,

$$\int_0^\pi F_1 (\vartheta, \varphi) \cos \varphi \, d\varphi = \pi E_x (0),$$

$$\int_0^\pi F_2 (\vartheta, \varphi) \sin \varphi \, d\varphi = -\pi E_x (0).$$

(7.147)

In the higher order of $1/(kR)$,

$$\int_0^\pi \exp\left[-ikR\left(1 - \cos \vartheta\right)\right] \sin \vartheta \, d\vartheta = -\frac{i}{kR}.$$

(7.148)

Substituting the above formulas into (7.145) we obtain the optical theorem for the three-dimensional vector case

$$\int_0^{2\pi} \int_0^\pi \left[|F_1 (\vartheta, \varphi)|^2 + |F_2 (\vartheta, \varphi)|^2\right] \sin \vartheta \, d\vartheta d\varphi = -4\pi \operatorname{Re} E_x (0).$$

(7.149)

Note that the right-hand side of (7.149) contains only the component of electric field of the scattered wave, parallel to that of the incident plane one.

In particular, from the optical theorem (7.140), (7.149) it follows that if the plane wave falls on any nonabsorbent body, then the scattering pattern cannot be zero in the direction of the wave propagation. Although the shaded domain appears immediately behind the body, there is no shading further, in the far field zone, where the cylindrical or spherical wave is already formed. The geometro-optical notions (ray, shade) are not applicable in the wave zone.

Formulas (7.140), (7.149) explain the mechanism of swapping the energy from the plane wave into the divergent spherical one. This swapping is carried out by the interference of the fields of both waves along the line $\varphi = 0$ or $\vartheta = 0$. On these lines, both waves propagate in the same direction. The notion "interference" emphasizes that the right-hand sides of these formulas contain the cross-products of the terms corresponding to the spherical (cylindrical) and the plane waves.

Certain fuzziness of the assertion about interference is explained by its application to the plane wave being an idealized object. Such a wave carries an infinite energy. The real ("quasi-plane") wave, field of which is close to (7.10) in a large but finite domain, is distorted by diffraction, and its energy decreases. Formulas (7.134), (7.143) cease to be valid in the distance large enough for the incident wave not to be a plane one. The wider the front of the incident quasi-plane wave, the larger this distance.

7.4.10
Spectral method for the diffraction problems

In the *spectral method*, the field created by certain given source (and hence satisfying the nonhomogeneous equation) is found as a sum of the field of the same source in free space and the series by the eigenfunctions of some auxiliary homogeneous problem. In some variants of the method, the total field is also represented as the series of the eigenfunctions. One of the parameters involved in the nonhomogeneous problem is chosen to be a *spectral* one. The homogeneous problem is solvable only at some values of this parameter (*eigenvalues*). Such a solution is an *eigenfunction* corresponding to the eigenvalue. In different variants of the spectral method, different parameters play a role of the spectral one. In the problem for a closed resonator, the frequency is usually chosen as such a parameter (see Section 4.1). In the problem of diffraction on a certain body, it is more convenient to choose one of its electrodynamic parameters as a spectral one. In the two examples below such a parameter is the surface impedance and the dielectric permittivity of the body material, respectively.

We begin with the problem about the field of the plane wave diffracted on the metallic cylinder, already solved in Section 7.1. Series (7.7) can be considered as a series by the eigenfunctions $u^n(r, \varphi)$ of certain homogeneous problem. In this problem the functions $u^n(r, \varphi), n = 1, 2, \ldots$ satisfy the wave equation (7.2) with $\varepsilon = 1$, radiation condition, and condition of the impedance type

$$u^n(a, \varphi) = -iw_n \left. \frac{\partial u^n(r, \varphi)}{\partial r} \right|_{r=a} \qquad (7.150)$$

on the cylinder surface. The constant (independent of φ) quantities w_n are the eigenvalues in this problem. They have a physical meaning of the impedance (*eigenimpedance*) of the cylinders considered in this auxiliary problem.

In Subsection 4.2.2 we noted that the auxiliary problems generating the system of the eigenfunctions have an independent physical sense, that is, they describe the electrodynamic oscillations in a certain domain. In the problem considered here such a domain is the exterior of the cylinders having the surface impedance w_n (note that each eigenvalue corresponds to different physical object: the cylinder with same geometry but different surface impedances). The oscillations are described by the functions

$$u^n(a, \varphi) = H_n^{(2)}(kr) \cos(n\varphi), \qquad n = 0, 1, \ldots \qquad (7.151)$$

(as above, we confine ourselves only to oscillations, symmetrical with respect to the x-axis).

In this variant of the spectral method the frequency in the homogeneous problem remains the same *real* value, as in the diffraction one. Therefore, the eigenoscillations are not dumping in time. Since the radiation losses are present in the auxiliary (exterior) problem, the surface impedance must posses the property

$$\text{Re}\, w_n < 0, \tag{7.152}$$

opposite to $\text{Re}\, w > 0$ for the impedance of any surface absorbing the energy. In the auxiliary problem, the material of the body surface has "negative losses." Such a material radiates (does not absorb) the energy, proportional to the squared electric field in it.

According to (7.140), in the homogeneous problem generating the eigenfunction system (7.151), the eigenimpedances are

$$w_n = ik \frac{H_n^{(2)}(ka)}{H_n^{(2)\prime}(ka)}. \tag{7.153}$$

The value $\text{Re}\, w_n = -2\pi ka / |H_n^{(2)\prime}(ka)|^2$ is negative, which is in accordance with (7.152).

Existence of the impedance with property (19.56) does not contradict the Maxwell equations. The auxiliary problem describes the eigenoscillations in the infinite domain $r > a$, which exist in presence of objects having the properties which do not violate the physical laws.

The system of functions (7.151) allows us also to solve the problem of diffraction on the cylinder with nonzero surface impedance w ($\text{Re}\, w > 0$). The field U is found in the form $U = U^{\text{diff}} + U^0$, where U^{diff} is series (7.7). The coefficients A_n are calculated from the boundary condition at $r = a$

$$ik \left(U^0 + \sum_{n=0}^{\infty} A_n u^n \right) = w \left(U^{0\prime} + \sum_{n=0}^{\infty} A_n u^{n\prime} \right). \tag{7.154}$$

It is easily seen that the formula for A_n has the same structure as (4.67). It generalizes formula (7.9) and coincides with it at $w = 0$.

Now we apply the above spectral method to the diffraction problem on the dielectric cylinder. Choose the dielectric permittivity of the body on which the diffraction proceeds as the spectral parameter. The eigenfunctions $u^N(r, \varphi)$ solve the homogeneous problem consisting of the equations

$$\Delta u^N + k^2 \varepsilon_N u^N = 0, \tag{7.155a}$$

$$\Delta u^N + k^2 u^N = 0, \tag{7.155b}$$

which hold in the finite domain V^+ occupied by the body, and the infinite one V^- outside the body, respectively, the continuity conditions for u^N and its normal derivative on the body boundary, and the radiation condition

$$u^N \Big|_{r \to \infty} \simeq F^N(\varphi) \frac{\exp(-ikr)}{\sqrt{kr}} \tag{7.156}$$

at infinity. In these formulas the index N is aggregate, $N = (n, p)$, $n = 0, 1, 2, \ldots, p = 0, 1, 2, \ldots$, numbers ε_N are the eigenvalues.

The orthogonality condition for the functions u^N has the form

$$\int_{V^+} u^N(r, \varphi) u^M(r, \varphi)\, dV = 0, \qquad \varepsilon_N \neq \varepsilon_M. \tag{7.157}$$

To prove it, we construct the equality

$$\operatorname{div}(u^M \operatorname{grad} u^N - u^N \operatorname{grad} u^M) = \left\{ \begin{array}{ll} k^2(\varepsilon_M - \varepsilon_N)u_N u_M & \text{in } V^+, \\ 0 & \text{in } V^-, \end{array} \right\} \tag{7.158}$$

which follows from (7.155). Integrate this equality over V^+ and V^- and add the results. The sum of the surface integrals on the left-hand side, to which the volume integrals are reduced, equals zero owing to (7.156) and the boundary conditions; from this condition (7.157) follows. It can be shown that the eigenfunctions make up a complete system for a large class of functions.

The eigenvalues ε_N of this homogeneous problem have the property $\operatorname{Im}\varepsilon_N > 0$. It is obtained from (7.158) after replacing u^M by u^{N*}, ε_M by ε_N^*, integrating over the whole volume and using the asymptotic (7.156). As a result, we obtain

$$\operatorname{Im}\varepsilon_N = \frac{1}{2k^3} \frac{\int_0^{2\pi} |F^N(\varphi)|^2\, d\varphi}{\int_{V^+} |u^N|^2\, dV} > 0. \tag{7.159}$$

Inequality (7.159) is opposite to $\operatorname{Im}\varepsilon_N < 0$, which is valid for any dielectric with losses. It has the same meaning as inequality (7.152) for the eigenimpedances. Note that the material having "negative losses," really exists (see Subsection 1.2.10). It is a medium with inverse population of the quantum levels.

In the diffraction problem, the field is expressed in the form

$$U(r, \varphi) = U^0(r, \varphi) + \sum_N \alpha_N u^N(r, \varphi), \tag{7.160}$$

(recall, that $N = \{n, p\}$ and the sum is double). The series termwise satisfies the boundary condition at $r = a$, the radiation one, and the wave equation in V^-, outside the body. In V^+, the functions U, U^0, and u^N satisfy the wave equation with different values of the dielectric permittivity, namely, ε, 1 and ε_N, respectively. Coefficients α_N are found from the demand that the right-hand side in (7.160) satisfy the wave equation $\Delta U + k^2 \varepsilon U = f$ in V^+, where f is the source (if it exists in V^+). Substituting (7.160) into this equation and using equations (7.155a) and $\Delta U^0 + k^2 U^0 = f$ for the eigenfunctions and incident field, respectively, we obtain the equality

$$\sum_N \alpha_N (\varepsilon_N - \varepsilon) u^N = k^2 (\varepsilon - 1) U^0, \tag{7.161}$$

valid in V^+. The functions u^N are orthogonal and make up the complete system in this domain. Applying (7.157) to (7.161) gives

$$\alpha_N = \frac{\varepsilon - 1}{\varepsilon_N - \varepsilon} \frac{\int_{V^+} U^0 u^N \, dV}{\int_{V^+} (u^N)^2 \, dV}. \tag{7.162}$$

The factor $\varepsilon_N - \varepsilon$ in the denominator differs from zero owing to (7.159).

In the case of the circular cylinder of radius a the eigenfunctions u^N have the explicit form

$$u^N(r, \varphi) = \begin{cases} J_n(k\sqrt{\varepsilon_N}r) \cos(n\varphi), & r < a, \\ H_n^{(2)}(kr) \cos(n\varphi), & r > a. \end{cases} \tag{7.163}$$

The eigenvalues ε_N are obtained from the demand that u^N and their normal derivatives are continuous at $r = a$, which leads to the equation

$$\sqrt{\varepsilon_N} H_n^{(2)}(ka) J_n'(k\sqrt{\varepsilon_N}a) - H_n^{(2)'}(ka) J_n(k\sqrt{\varepsilon_N}a) = 0. \tag{7.164}$$

The technique described above in application to the diffraction problem for the electric polarization can be carried over to the case of the magnetic polarization with small modifications, connected with fact that ε appears also in the boundary condition in this problem. The method may be generalized for the vector problem, as well as for the problem of diffraction on the body with variable $\varepsilon = \varepsilon(\vec{r})$.

This subsection has only a methodical value. It illustrates the possibility to use the spectral methods for solving the diffraction problems in the open domains, which was noted in Section 4.2. In this case one of the variants of the method may be used, in which the frequency k in the auxiliary problem is the same as in the diffraction one, but one of other physical parameters is chosen as the spectral one.

Of course, if the solution can be obtained by the variable separation method (as in the problems of this subsection), then the spectral method does not simplify the solving procedure. However, in more complicated diffraction problems, use of the spectral method (and, in particular, the variational technique for the eigenvalues finding) may turn out to be expedient.

General References [1]

1 S. A. Schelkunoff. *Electromagnetic Waves.* Van Nostrand, New York, 1943

2 S. A. Schelkunoff, H. T. Friis. *Antennas, Theory and Practice.* Wiley, New York, 1952

3 R. F. Harrington. *Time-Harmonic Electromagnetic Fields.* McGraw-Hill, New York, 1961

4 A. F. Harvey. *Microwave Engineering.* Academic Press, London, 1963

5 N. Marcuvitz (Ed.). *Waveguide Handbook.* Dover, New York, 1965

6 R. E. Collin. *Foundations of Microwave Engineering.* McGraw-Hill, New York, 1966

7 E. C. Jordan, K. G. Balmain. *Electromagnetic Waves and Radiating Systems, 2nd ed.* Prentice-Hall, Upper Saddle River, NJ, 1968

8 M. Schwartz. *Principles of Electrodynamics.* Dover, New York, 1972

9 R. S. Elliot. *Antenna Theory and Design.* Prentice-Hall, Upper Saddle River, NJ, 1981

10 H. C. Chen. *Theory of Electromagnetic Waves.* McGraw-Hill, New York, 1983

11 S. Silver (Ed.). *Microwave Antenna Theory and Design.* Peter Peregrinus, London, 1984

12 R. E. Collin. *Antennas and Radiowave Propagation.* McGraw-Hill, New York, 1985

13 C. A. Balanis. *Advanced Engineering Electromagnetics.* Wiley, New York, 1989

14 D. K. Cheng. *Field and Wave Electromagnetics.* Addison-Wesley, Reading, MA, 1990

15 J. A. Kong. *Electromagnetic Wave Theory, 2nd ed.* Wiley, New York, 1990

16 R. E. Collin. *Field Theory of Guided Waves, 2nd ed.* IEEE Press, Piscataway, NJ, 1991

17 R. S. Elliot. *Electromagnetics—History, Theory, and Applications.* IEEE Press, Piscataway, NJ, 1993

18 R.S. Elliott. *An Introduction to Guided Waves and Microwave Circuits.* Prentice-Hall, Upper Saddle River, NJ, 1993

19 L. B. Felsen, N. Marcuvitz. *Radiation and Scattering of Waves.* IEEE Press, New York, 1994

20 S. Ramo, J. R. Whinnery, T. van Duzer. *Fields and Waves in Communication Electronics, 3rd ed.* Wiley, New York, 1994

21 S. Cornbleet. *Microwave and Geometrical Optics.* Academic Press, London, 1994

22 C. A. Balanis. *Antenna Theory, Analysis and Design, 2nd ed.* Wiley, New York, 1996

23 G. S. Smith. *An Introduction to Classical Electromagnetic Radiation.* Cambridge University Press, Cambridge, 1997

24 C. G. Someda. *Electromagnetic Waves.* Chapman and Hall, London, 1998

25 A. F. Peterson, S. L. Ray, R. Mittra. *Computational Methods for Electromagnetics .* IEEE Press, New York, 1998

26 W. L. Stutzmann, G. A. Thiele. *Antenna Theory and Design, 2nd ed.* Wiley, New York, 1998

27 J. D. Jackson. *Classical Electrodynamics, 3rd ed.* Wiley, New York, 1998

28 D. A. de Wolf. *Essentials of Electromagnetics for Engineering.* Cambridge University Press, Cambridge, 2001

29 E. J. Rothwell, M. J. Cloud. *Electromagnetics.* CRC Press, Boca Raton, FL, 2001

30 P. Russer. *Electromagnetics, Microwave Circuit and Antenna Design for Communications Engineering.* Artech House, Boston, 2003

[1]Compiled by M. Thumm

Complementary References

1* L. D. Landau, E. M. Lifshitz. *Electrody-namics of Continuous Media*. Pergamon, Oxford, 1960 [1.2; 6.2.6; 7.2; 7.4.3]

2* A. Priou, A. Sihvola, S. Tretyakov, A. Vinogradov (Eds.). *Advances in Complex Electromagnetic Matherials*. Kluwer, London, 1977 [1.2.5; 2.1.3; 2.1.4]

3* V. G. Veselago. Formulating Fermat's principle for light travelling in negative refraction materials. *Physics-Uspekhi*, **45** (2002), 10, 1097–1099 [2.2.7]

4* V. G. Veselago. Electrodynamics of materials with negative index of refraction. *Physics-Uspekhi*, **46** (2003), 7, 764–768 [2.2.7]

5* P. Markos, C. M. Soukoulis. Transmission properties and effective electromagnetic parameters of double negative metamaterials. *Optic Express*, **11** (2003), 7, 649–661 [2.2.7]

6* Yu. N. Kazantsev, O. A. Kharlashkin. On oversize waveguides of rectangular cross-section with small losses. *Radio Engineering and Electronic Physics*, **16** (1971), 6, 1063-1065 [3.2.3]

7* V. V. Shevchenko. Excitation of waveguides in the presence of associated waves. *Soviet J. Comm. Tech. Electron.*, **31** (1986), 7, 26–35 [3.2.5]

8* B. Katsenelenbaum, L. Marcader del Rio, M. Perreyaslavets, M. Sorolla Auza, M. Thumm. *Theory of Nonuniform Waveguides*. IEE Series, London, 1998 [3.4.1; 3.4.2]

9* A. G. Ramm, N. N. Voitovich, O. F. Zamorska. Numerical implementation of cross-section method for irregular waveguides. *Radiophysics and Radioastronomy*, **5** (2000), 3, 274–283 [3.4.1]

10* L. Weinstein. *The Theory of Diffraction and the Factorization Method*. Golem Press, Boulder, CO, 1969 [3.4.6; 7.2]

11* B. Noble. *Methods Based on the Wiener–Hopf Technique, 2nd ed.* Chelsea Press, New York, 1988.

12* M. S. Agranovich, B. Z. Katsenelenbaum, A. N. Sivov, N. N. Voitovich. *Generalized Method of Eigenoscillations Method in Diffraction Theory*. Wiley-VCH, Berlin, 1999 [4.2.3; 7.4.10]

13* L. Weinstein. *Open Resonators and Open Waveguides*. Golem Press, Boulder, CO, 1969 [5.1; 6.3.3]

14* B. Z. Katsenelenbaum. Excitation of an arbitrary-cross section dielectric waveguide at near-critical frequencies *Radio Eng. Electron. Phys.*, **25** (1980), 2, 12–17 [5.1.2]

15* N. N. Voitovich, B. Z. Katsenelenbaum, A. N. Sivov, A. D. Shatrov. Characteristic propagation modes of dielectric waveguides with a composite cross-section. *Radio Eng. Electron. Phys.*, **24** (1979), 7, 1–16 [5.1.2]

16* V. Shevchenko. *Continuous Transitions in Open Waveguides*. The Golem Press, Boulder, CO, 1971 [5.1.4]

17* V. V. Shevchenko. On the completeness of spectral expression of the electromagnetic field in the set of dielectric circular rod waveguide eigen waves. *Radio Science*, **17** (1982), 1, 229–231 [5.1.4]

18* B. Z. Katsenelenbaum. *Electromagnetic Fields—Restrictions and Approximations*. Wiley-VCH, Weinheim, 2003 [5.3.1–5.3.5; 6.2.4; 6.3.6]

19* O. O. Bulatsyk, N. N. Voitovich. Properties of nonlinear Hammerstein integral equations connected with modified phase problem. *Direct and Inverse Problems of Electromagnetic and Acoustic Wave Theory (DIPED-2003)*, Lviv, Ukraine, 2003, 135–138 [5.3.6]

20* N. N. Voitovich. Antenna Synthesis by Amplitude Radiation Pattern and Modified Phase Problem. Appendix in [18*], 195–233 [5.3.6; 6.2.4]

21* N. N. Voitovich, Yu. P. Topolyuk, O. O. Reshnyak. Approximation of compactly supported functions with free phase by functions with bounded spectrum. *Fields Institute Communications, AMS*, **25** (2000), 531–541 [5.3.6; 6.2.4]

22* B. Z. Katsenelenbaum. The field in the vicinity of a transmitting antenna. *Journal of Communications Technology and Electronics*, **47** (2002), 8, 844–847 [6.3.6]

23* A. N. Tikhonov, V. Y. Arsenin. *Solutions of Ill-Posed Problems*. Winston and Sons, Washington, 1977 [6.3.6]

24* A. G. Kyurkchan. Analytical continuation of wave fields. *Soviet Journal of Communications Technology and Electronics*, **31** (1986), 11, 59–69 [6.3.7]

25* A. A. Izmest'ev. One parameter wave beams in free space. *Radiophysics and Quantum Electronics*, **13** (1970), 9, 1062–1068 [7.3.3]

26* J. Keller. Geometrical theory of diffraction. *Journal of the Optical Society of America*, **52** (1962), 2, 116–130 [7.4.7]

27* V. A. Borovikov, B. Ye. Kinber. *Geometrical Theory of Diffraction*. IEE Series on Electromagnetic Waves, London, 1994 [7.4.7]

28* R. Zaridze, G. Bit-Babik, K. Tavzarashvili, N. K. Uzunoglu, D. Economou. The method of auxiliary sources (MAS)—solution of propagation, diffraction and inverse problems using MAS. In: *Applied Computation Electromagnetics. State of the Art and Future Tends*, NATO Advances Science Institute Series, Series F: *Computer and Systems Sciences*, **171**, Springer, 2000, 33-45 [7.4.8]

29* F. G. Bogdanov, D. D. Karkashadze, R. S. Zaridze. The Method of Auxiliary Sources in Electromagnetic Scattering Problems in: *Mechanics and Mathematical Methods, A Series of Handbooks. First Series: Computational Methods in Mechanics.* **Vol. 4**, *Generalized Multipole Techniques for Electromagnetic and Light Scattering (Ed. T. Wreidt)*, North-Holland, Amsterdam, 1999, 143–172. [7.4.8]

Index